从机器学习到无人驾驶

宋哲贤 编著

清华大学出版社
北京

内容简介

本书以机器学习为出发点，使用简易的代码讲解机器学习的核心算法（深度神经网络和强化学习），在算法学习的基础上使用增量方法开发包含定位、预测、路径规划和业务控制等一系列自动驾驶模块。本书代码实例涉及自动驾驶的普遍业务方法，可使读者理解自动驾驶背后的设计思想和原理，快速入门自动驾驶的算法和开发流程。

本书示例代码丰富，涵盖实际开发中所有的重要知识点，适合无人驾驶从业者、想要学习机器学习和无人驾驶的开发人员阅读，也可用作培训机构和高校相关专业的教学参考书。

图书在版编目（CIP）数据

从机器学习到无人驾驶/宋哲贤编著.— 北京：清华大学出版社，2020.4（2025.1 重印）

ISBN 978-7-302-55215-4

Ⅰ. ①从… Ⅱ. ①宋… Ⅲ. ①机器学习—研究②无人驾驶—汽车—智能技术 Ⅳ. ①TP181②U469.79

中国版本图书馆 CIP 数据核字（2020）第 049030 号

责任编辑：王金柱
封面设计：王 翔
责任校对：闫秀华
责任印制：丛怀宇

出版发行：清华大学出版社
网　　址：https://www.tup.com.cn, https://www.wqxuetang.com
地　　址：北京清华大学学研大厦 A 座　　**邮　　编**：100084
社 总 机：010-83470000　　**邮　　购**：010-62786544
投稿与读者服务：010-62776969，c-service@tup.tsinghua.edu.cn
质量反馈：010-62772015，zhiliang@tup.tsinghua.edu.cn
印 装 者：三河市龙大印装有限公司
经　　销：全国新华书店
开　　本：190mm×260mm　　**印　张**：27.5　　**字　　数**：704 千字
版　　次：2020 年 5 月第 1 版　　**印　　次**：2025 年 1 月第 2 次印刷
定　　价：99.00 元

产品编号：083861-01

致 谢

感谢王金柱老师在本书写作过程中的悉心指导，包括章节设计、写作思路等方面都给了我非常专业的帮助，王老师不但站在图书编辑的专业角度给予指导，而且通过调研给出了读者在内容接受方面的专业意见。

同时感谢慕课网这个平台，本书前一半内容都是我与慕课网平台合作制作的网课的精炼。在网课制作完成并且和学生深入交流后，我才形成了撰写这本书的想法，在写作过程中，我的许多朋友都提供了帮助，直至内容最终呈现，在这里不一一列举大家的名字了，但是请让我真诚地说声谢谢。

感谢我的妻子徐佳女士，感谢她包容我在写作期间由于写作压力产生的情绪阶段性低落，更加感谢她曾经在我抑郁时帮助我走出阴霾，you are all my reasons。

感谢家中的 4 位老人，我们的父母，由于我独自长期在北京工作，写作期间占用了大量休息时间，是你们给了我无私的支持。

我的两个可爱的小天使 Echo 和 Max，老爸也要感谢你们。Echo 日常的鼓励和提醒帮助老爸工作和写作得更加顺利，Max 富有感染力的笑容是我们全家人的宝藏。

宋哲贤

2020 年 1 月

前　言

机动车是灵活性高、业务环境复杂的交通工具，一直和我们的生活、工作息息相关。近年来，结合感知、融合、决策、控制的自动驾驶技术无疑是最有前景的研发领域之一。在《中国制造 2025》中，已经将智能车联网提升到国家战略高度，车作为智能化终端的概念逐渐浮出水面，并在我们的心中逐渐加强。近几年，各项国家地方政策层出不穷，甚至开放了包括北京、上海部分道路在内的一部分路段进行路测。在世界范围内，围绕自动驾驶产生了多家发展迅速的独角兽企业，这些企业有的专注地图的高精化，有的设计智能化边缘计算硬件，有的则着眼于智能驾驶算法系统。智能驾驶和交通安全息息相关，虽然目前测试的结果都是很惊人的，但是大多数结果都是基于特定路段和特定环境的。在直观的理解中，自动驾驶包含车辆与环境的交互过程。这里面涉及众多环节，比如车辆定位、路径规划、状态感知、车辆控制等。如此复杂的流程必须依靠包括深度学习、强化学习等在内的机器学习技术进行支撑，机器通过大量离线和在线数据的采集与特征提取，在一定算法的基础上，模型自主地完成优化和学习，从而最终得出一个具有统计学意义的结果。这个统计学结果的得出其实是值得探讨的。由于数据量不足及模型本身的种种限制，自动驾驶还没有实现 100%的全路段、全时况的无人驾驶。

汽车行业的智能化技术不同于普通的 Web 技术，具有很强的边缘计算结合性，需要从业人员了解车辆本身的包括运动、机械等方面的基础知识，更需要了解不同传感器的数据微观特征，通过结合近年来发展迅猛的机器学习算法（深度神经网络和强化学习）开发包含定位、预测、路径规划和业务控制等一系列自动驾驶模块，通过驾驶测试来进一步完善系统能力。想要进入此领域的初级开发人员往往因为系统繁杂而不能很快掌握开发技术的核心。本书涵盖机器学习的基本概念、神经网络算法和强化学习算法，在学习算法的基础上通过增量方式开发进行无人驾驶各个功能模块的开发学习。

本书致力于通过生动的实例讲解机器学习的核心算法，代码实例涉及自动驾驶的普遍业务方法，通过具体代码实例对自动驾驶背后的设计思想和原理进行详细分析，力图使读者通过实际操作快速入门和理解自动驾驶的算法和开发流程。

宋哲贤

2020 年 1 月

目　录

第一篇　机器学习基础

第三篇 无人驾驶

第一篇　机器学习基础

第 1 章

机器学习与无人驾驶

机器学习是人工智能的一个分支，也是人工智能的一种实现方法。它从样本数据中学习得到知识和规律，然后用于实际的推断和决策。它和普通程序的一个显著区别是需要样本数据，是一种数据驱动的方法。如果只用一句话解释这个概念，那么最简单直观的答案是：机器学习用计算机程序模拟人的学习能力，从实际例子中学习得到知识和经验。近几年，无人驾驶在机器学习的基础上也得到了长足的发展。从原理来说，无人驾驶软件实践最大程度强调软件的自身识别和决策能力；从实际来说，无人驾驶设计团队很大程度上放弃了基于规则的人工智能软件，最终功能使用机器学习的策略来完成他们的核心软件决策部分。

本章主要讲解机器学习的一些宏观概念以及无人驾驶和机器学习之间的定性关系。内容分为两个部分，第一部分着重介绍机器学习中的核心概念，给出深度学习和强化学习两种目前业界热点算法的直观解释；第二部分立足于通过无人驾驶发展历史中的核心实践过程进行定性分析，推断出使用机器学习来完成无人驾驶软件设计的优势。

1.1　机器学习简介

1.1.1　机器学习

人工智能（Artificial Intelligence，AI）一直是科技的前沿，也是一门涉及多个领域的综合科学。由于概念的广泛性，各种权威的人工智能资料定义上都有很大的区别，换句话说，很多定义都有范围和倾向性。在2009年，Russell和Norvig总结前人的研究实践，提出一种相对全面的概念集合论述，这种论述通过对比的方式阐述了两对核心概念。第一，通过推理判断来获得结果的能力与通过行为经验获得结果的能力相对比；第二，通过实践来完善推理模型的过程与从原理建模来得到应用或产品的未来的不同。以这个对比讨论为蓝本，可以得到比较细致的关于人工智能的4点子定义论述。

1. 子定义论述一

人类能够通过人工智能的发展过程得到进化发展所需要的算法概念，从而更新自我体系，通过类似机器学习的训练技巧将人脑智能进一步优化开发，从而完成以智能进化为核心目标的人工智能过程。换句话说，计算机的进步会推动人脑的进步，但是最终的落实点还是人类自身。举一个具体的例子，目前我们在互联网产品中使用了很多智能问答机器人，其核心是计算机通过机器学习的过程获得类似人类回答问题的过程和能力，我们接收外界提出的问题，通过思考得出结论，最终将结论以合适的形式表达出来。仔细分析这个过程，我们可以分别列出具体过程以及具体过程所依赖的算法能力。为了能够流畅地完成问题接收的过程，需要使用当前飞速发展并取得骄人成绩的机器视觉和机器听觉方面的相关技术，针对不同语言的问题，自然语言处理技术（Natural Language Processing，NLP）是不可或缺的，通过建立相关的知识库，机器学习能够进一步模型化相关范围的问题，最终得到相应的答案。不断完善问答机器人的精确度和泛化性能的过程，就会给人类带来技术的提升和思考模式的更新，这个过程有点像反向仿生学，通过更好的机器、更好的回答让我们重新定义自身的问题回答机制。

2. 子定义论述二

这个论述是大家最为关注的方向，也是科幻电影中经常使用的题材，那就是计算机具有和人类一样的智能能力，我们也可以称这个概念为狭义的人工智能概念。在这里，我们回顾一下大名鼎鼎的图灵测试。艾伦·图灵（Alan Turing）提出了一种测试机器是否具备人类智能的方法。假设有一台计算机，其运算速度非常快，记忆容量和逻辑单元的数目也超过了人脑，而且还为这台计算机编写了许多智能化的程序，并提供了合适种类的大量数据，是否就能说这台机器具有思维能力？艾伦·图灵因此对智能问题从行为主义的角度提出了一个著名假想：一个人在不接触对方的情况下，通过一种特殊的方式和对方进行一系列的问答，如果在相当长时间内无法根据这些问题判断对方是人还是计算机，就可以认为这个计算机具有同人相当的智力，即这台计算机是能思维的。这就是著名的"图灵测试"（Turing Testing）。这是一种"模仿游戏"，远处的人类测试者在一段规定的时间内，根据两个实体对他提出的各种问题的反应来判断是人类还是计算机。通过一系列这样的测试，从计算机被误判断为人的概率就可以测出计算机智能的成功程度。

3. 子定义论述三

通过对计算机本源化的研究来抽象模型用来升级解决智能模型获得理性化的过程。这里包含通过计算机模型对人类精神行为的建模，以及通过建模完成对认知、推理和执行过程的模拟。这类研究在20世纪80、90年代一度盛行。

4. 子定义论述四

通过机器学习数学模型的使用来解决具体的产品推断决策类问题。这其实是人工智能目前真正的阶段，比如发展非常好的机器视觉，其实最简单的模型就是通过图像的像素矩阵来进行分类推断，正是因为有了这些包含机器视觉的一系列人工智能算法，我们才可以完成本书中提及的自动驾驶的感知、路径规划和控制等功能。

由于人工智能是一门融合科学，我们通过列举各个学科的具体相关问题来进一步说明上面 4 点子定义的细节。例如，偏哲学方面的问题，能否通过具体的规则得到现实中的实际事物的准确判断？计算机的自我意识是如何实现的？人力的知识是如何在历史中出现的？知识怎样作用于人脑最终产生人的自主判断及行动？例如，偏数学方面的问题，通过不完整或者非确定性的信息，我们是如何通过数学模型进行推断的？如何对物质世界进行合理分类，哪些类别可以通过数学进行建模计算？例如，偏经济学方面的思考，利润最大化模型的研究，比如德州扑克 AI 实际就在这个研究范围之内，在宾夕法尼亚州匹兹堡的 Rivers 赌场，卡耐基梅隆大学（Carnegie Mellon University，CMU）开发的人工智能系统 Libratus 战胜 4 位德州扑克顶级选手，获得最终胜利。当然不止这些领域，其他很多领域也与人工智能的思考紧密相连，包括神经学中对大脑认知思考过程的研究、通过人工智能算法改进计算机软硬件设计以及语言学中想法与表达的关系和流程等。

AI 的发展历史是很悠久的，除去人类对认知的各种早期思考，单单考虑图灵实验提出之后的正式发展阶段，也有接近一个世纪的岁月。在这个过程中，人工智能经历了阶段前期的概念研究到连接主义的兴起，直到 21 世纪，随着互联网，特别是移动互联网的规模增长，应用维度扩展，引发的数据量以及配套运算能力的几何级增长造就了机器学习在人工智能各个领域中一枝独秀的完美表现，首先在机器视觉领域，而后普及至人工智能的大多数问题领域。

简单来说，机器学习就是根据数据推断出模型，或者更新模型参数，进一步使用模型完成预测，得出我们需要的结论。通常机器学习可以分为 3 大类：监督学习、非监督学习和强化学习。在监督学习中，我们使用的数据集合是标注过的，类似 X~Y 的形式；在非监督学习中，数据集合的情况恰恰相反，是没有标注的，类似 X 的形式。在监督学习中，标签数据实际上存在连续（身高）或者非连续（性别）两种形态，因此针对这两种标签数据，我们需要分别使用回归算法分析或者分类算法分析来解决。非监督学习是一个从无标签数据中进行模式分析并提取归纳相应模型的过程，例如，聚类分析和核密度估计。表示学习也是一种典型的非监督学习算法。但是深度学习可以划归为表示学习的范畴，原因是为了使得深度网络结构变得更加容易训练，并且强化深度网络的特征提取和函数逼近能力，需要对深度学习网络采用更高效的网络表达方式。网络的表达方式是指网络采用何种结构上的连接方式来抽象表达输入模式的内部结构，或表示输入样本之间的关系。

深度学习（或称为深度神经网络）是机器学习中自成体系的一类算法，通常在监督学习和非监督学习中使用，并且能够结合强化学习，在强化学习中进行状态表达或者用作函数优化器。监督学习和非监督学习的学习过程往往是单次的，忽视数据集合中不同数据条目之间的关联关系。在这方面，强化学习有很大的不同，强化学习通过连续性的数据算法构造了一种长期积累的激励关系，通过当次学习和积累的学习结果综合分析得到合理的推断结论，因此强化学习往往针对的是连续性的行为对结果影响的场景。不同于监督学习和非监督学习，在强化学习中，我们会使用一些持续的可评估的系统回馈来代替简单的标签结果。这更加增加了强化学习在学习过程中的过程复杂度和算法难度。强化学习和最优控制学紧密相连，和前面提及的人工智能及其衍生学科也有着很紧密的联系，比如心理学、神经学等。

机器学习的基础是数据，通过数据科学中的数据挖掘、数据处理等过程得到原始数据集合，这些数据可能是图像信息、语言模型，也可能是任意相关联的多组数据的高维组合，结合数学和统计学（最大似然估计）的基础算法，我们可以完成图像识别、自然语言处理、机器人控制等应用。当然，机器学习并不是人工智能领域的全部，但是针对近 20 年人工智能领域取得的核心成就，机器学习技术的重要性是其他学派鞭长莫及的。

当我们深入机器学习算法实践过程本身，就会发现有些算法组件是具有通用性的，包括数据集合、代价函数、学习优化器和这些模块围绕的核心模块——算法模型。我们一般会将数据集合分为训练子集合、交叉验证子集合和测试子集合，它们一般是由同一份数据集合按照合适的比例分割而成。代价函数的目的是对模型能力进行有效评估，比如使用均方差作为代价函数来描述线性回归模型的误差率。学习过程中的核心是通过优化器使用训练数据集最小化训练误差。训练误差描述的是模型在训练数据集上的泛化效果，而测试误差（或者称为泛化误差）是模型在当前问题的新输入数据上的泛化效果，也可以理解为模型对应问题真正的预测能力。机器学习算法不但试图最小化训练误差，而且希望能够达成训练误差和测试误差的一致性，综合这两点，才能保证模型在实际生产中有很好的泛化效果。当模型训练误差很大时，在机器学习中被称为拟合不足（Under-Fitting），而更加常见的是过拟合（Over-Fitting），在过拟合的情况下，虽然在训练数据集上得到了足够小的训练误差，但是在测试数据集合上的误差会非常大。拟合能力的强弱会使用模型容量这一指标来表征。人工智能科学家对模型的判断和我们的直观判断是一致的，同样能力的模型中最简单的最好。误差（包括训练误差和测试误差）和模型容量的关系往往是 U 型的。在机器学习的过程中，我们也可以根据这个关系来确定能够达到产品需求的模型的容量，保证最终的训练误差以及测试误差都在合理化范围之内。为了说明模型容量这一模型核心指标，我们需要介绍偏差（Bias）和方差（Variance）。偏差反映的是模型在样本上的输出与真实值之间的误差，即模型本身的精准度。方差反映的是模型每一次输出结果与模型输出期望值之间的误差，即模型的稳定性。当模型容量增加的时候，偏差会相应地有所降低，而方差则会随之增加。在这样的过程中，我们最终会选择合适的模型容量，在泛化误差和模型容量的关系图上表现选择的容量数据点的左侧模型容量大小会导致拟合不足，右侧模型容量大小会产生过拟合问题。

正则化成本函数通过添加惩罚项能够有效减少泛化误差，但是训练误差不会减小。没有免费的午餐定理指出，没有具有普遍的最佳模型或者适用于所有模型的正则项。因此，我们在深度学习获得巨大成功的同时，必须清楚地认识到针对某些问题深度神经网络可能不是最好的模型。正则项的参数矩阵称为超参数（Hyper-Parameters）。为了调优超参数从而使偏差和方差之间取得平衡，因此得到最优模型，我们需要使用交叉验证技术。

最大似然估计（Maximum Likelihood Estimation，MLE）是一种常用的参数估计方法。最大似然估计是一种统计方法，用来求一个样本集的相关概率密度函数的参数。这个方法最早是遗传学家以及统计学家罗纳德·费雪爵士在 1912—1922 年间开始使用的。“似然”是对 Likelihood 的一种较为贴近文言文的翻译，“似然”用现代汉语说即“可能性”，故而，称之为“最大可能性估计”更加通俗易懂。

在实际数据训练环节，特别是针对深度神经网络模型，梯度下降法（Gradient Descent）是求解优化问题的一种常用方法。随机梯度下降法（Stochastic Gradient Descent）扩展了梯度下降法，每次处理一个样本。当然我们也可以扩大学习样本的规模，每次处理一批样本而不是全部，这种方式称为批量梯度下降法。

当然，也有针对机器学习算法设计的数学分析框架。Kolmogorov 复杂度（或称为算法的复杂性）具体可以使用奥卡姆剃刀原理的算法表达，可能近似正确（Probably Approximate Correct，PAC）学习算法的目的是选择一个实现低泛化误差有高概率的泛化函数，VC 维度法能够很好地测量二元分类器的容量。

1.1.2　深度学习

为了能够清晰地理解深度学习这个概念的提出过程，我们需要首先研究一些回归深度学习以外的机器学习算法，可以不太严谨地将它们统一命名为浅度学习，比如在书中将详细介绍的线性回归算法、逻辑回归算法以及没有详细介绍的支持向量机（Support Vector Machine，SVM）算法、决策树算法等。浅度学习算法都设计了输入层和输出层，并且输入层经常需要进行手动特征向量提取，换句话说，机器学习开发者需要在开发进行之前完全了解问题，能够在很大程度上定性地预测问题，而机器学习算法只是量化我们的定性预测经验。而深度学习算法最显著的结构特征就是在输入层和输出层直接添加了多层隐藏层。除了输入层外的每一层的运算逻辑都是类似的，一般来说，每个运算单元接收上一层的所有运算单元的输入，首先计算输入的加权和，然后使用非线性转换函数来处理这个加权和，处理结果作为输入进入下一层运算单元中。非线性转换函数在深度学习中一般被称为激活函数，ReLU 函数目前是使用最为广泛的激活函数，另外深度学习神经元也可以使用 Logistic、Tanh 等函数。这样的模型构造使权重矩阵保持了层与层之间的连续传递性。神经网络向前传播（从输入层开始到第一个隐藏层，再到第二个隐藏层，以此类推，直到最后的输出层），计算各个层的每个节点的输出值，得到所有的输出值，我们就可以从后向前计算误差，通过梯度下降完成反向传播算法，从而达到更新整个神经网络的权重矩阵链的目的。我们可以根据神经元内部算法设计上的区别对深度神经网络进行分类表述，最简单的称为全连接神经网络或者前馈神经网络，当然它还有一个更加形象的名字—多层感知机（Multi-Layer Perceptron，MLP）。在这种最简单的神经网络中，每个神经元都是统一标准的个体，层与层之间只有输入和输出规模不同，计算的流动途径是从输入层到输出层，中间不存在逆向流程。卷积神经网络（Convolutional Neural Networks，CNN）其实是一种特殊的前馈神经网络，包括卷积层、池化层和全连接层。卷积神经网络设计主要是用来处理具有多层数组的数据结构，比如图像、文本、音频以及视频，通过多次卷积和池化操作能够将输入的特征分类提取出来，这种结构在很大程度上参考了视神经的工作方式。残差神经网络（Residual Networks，ResNets）是针对特别深的神经网络设计出来的，因为随着网络越来越深，训练变得越来越难，网络的优化也变得越来越难，残差神经网络设计了直接向后传递输出结果而不经过激活环节的捷径，这种捷径能够保证残差神经网络在训练集上训练的效率不会有明显的减弱。循环神经网络（Recursive Neural Network，RNN）通常用来处理时序数据，比如语音或者语句，在时序数据中的数据是按照先后顺序组织在一起的，因此循环神经网络中的隐藏单元可以存储部分过往数据以维持数据集合的结构关系。换句话说，循环神经网络可以简单理解为在正演运算过程中模型的所有计算层都使用相同的权重配置。标准循环神经网络的问题在于很难存储很长时间的历史数据，如果模型简单储存和利用了更多的历史数据，就会导致模型梯度消失。为了应对这一关键问题，长短期记忆网络（Long Short Term Memory network, LSTM）和门控循环单元（Gated Recurrent Unit，GRU）应运而生。在这两种神经网络中增加了专门从事对过往数据的处理单元。虽然深度神经网络的网络设计有很多变化，但是对于深度神经网络的训练方法却是殊途同归的，这种处理技巧就是梯度下降算法。

深度神经网络训练过程的优化技巧对于学习结果十分重要。最重要的优化方法是 Dropout，Dropout 的原理是在学习过程中随机移除一部分节点，使用部分网络进行数据训练。当然，优化技巧还有很多，它们针对优化的方面也有些许不同，例如为了提升训练效率所使用的批正则化方法和层正则化方法。

最后，我们从直观上分析深度神经网络结构的意义。深度神经网络的层模型能够自然地提取原始数据输入的特征，而且随着网络结构的特异化，深度神经网络能首先提取低价的特征向量，这些特征向量随着运算的进行组合成高阶的特征向量，最好的例子就是深度神经网络最成功的实践图像识别。在图像识别中，离输入层比较近的隐藏层能够提取出颜色、线条等低级特征，而在后面的层中，这些低阶特征进行组合，输出形状、物体的部分图像等高阶特征。这种深度学习的特性称为分布式表示（Distributed Representation），这种特性意味着特性和输入数据集合之前是多对多的逻辑关系，某种抽象特征可能存在于多个输入当中，另外每个输入中也含有海量的抽象特征。因此，最近的深度学习模型会使用或部分使用端对端训练技术，端对端技术会使用完整的原始数据集合，比如图像识别中使用整张图片（AlexNet）、语音识别中使用完整的对话（Seq2Seq）以及强化学习中使用整个游戏界面数据（Deep Q-Learning，DQN）。

1.1.3 强化学习

在机器学习中的一个重要业务场景和核心课题是连续决策。连续决策是指通过经验数据在不确定的环境中实现某些目标而采取的有顺序的任务执行过程。连续决策任务应用非常广泛，涵盖当前众多热门领域，比如机器人、医疗、智能电网和金融，特别是后面将介绍的自动驾驶汽车会有很多实际场景的应用。

针对连续决策问题目前最有进展的算法是强化学习（Reinforcement Learning，RL）。通过借鉴行为心理学的思想，强化学习设计了一个完整的学习框架来解决连续决策问题。在强化学习框架中，首先需要建立一个类似生物体的计算机代理终端，它能够接收相关的环境信息的输入，通过输入数据集的运算能够得到一个决策结果，接下来程序会将结果输出到环境中并收集环境的变化状况，根据业务要求定义变化的程度，这种程度一般在强化学习中称为奖励，通过对奖励变化的跟踪来动态调整终端的运算方式，最终达到解决当前连续决策问题的要求。这种方法原则上适用于依赖于过去的经验的连续决策问题。针对不同的环境，强化学习的策略也可以有细微的不同：一种环境是完全开放的，在这种环境下，代理终端只能观察部分当前状态的信息；另一种是受限环境，在这种环境下，数据特征规模可能较少，但是相对的每次的信息输入能够保证一定的完整性。

在过去的几年中，强化学习已经成为人工智能算法领域非常热门的算法之一。而强化学习与深度神经网络技术的结合成为热点中的热点，这种技术称为深度强化学习。深度学习具有较强的感知能力，但是缺乏一定的决策能力；而强化学习具有决策能力，对感知问题束手无策。因此，将两者结合起来，优势互补，为复杂系统的感知决策问题提供了解决思路。其中的典型算法 DQN 算法融合了神经网络和 Q-Learning 的方法。DQN 不用 Q 表记录 Q 值，而是用神经网络来预测 Q 值，并通过不断更新神经网络从而学习最优的行动路径。DeepMind 公司就是用 DQN 从玩各种电子游戏开始，直到训练出打败人类围棋选手的阿尔法狗。

1.2　无人驾驶与机器学习

1.2.1　无人驾驶的历史

无人驾驶是指通过软件算法代替人类操作过程使交通工具能够自行完成行驶过程的整套流程。本书介绍的无人驾驶软件算法主要针对汽车主体，无人驾驶智能汽车是能够完成通过车载传感系统感知道路环境，自动规划行车路线并控制车辆到达预定目标的汽车。无人驾驶智能汽车利用车载传感器来感知车辆周围的环境，并根据感知所获得的道路、车辆位置和障碍物信息控制车辆的转向和速度，从而使车辆能够安全、可靠地在道路上行驶。

无人驾驶技术集自动控制、体系结构、人工智能、视觉计算等众多技术于一体，是计算机科学、模式识别和智能控制技术高度发展的产物。通常在业界将车辆自动驾驶的水平分为 Level 0~Level5 共 6 个级别，分别说明如下。

- Level 0（无自动化）:由人类驾驶者全权操作汽车，在行驶过程中可以得到警告和保护系统的辅助。
- Level 1（驾驶支援）:通过驾驶环境对方向盘和加减速中的一项操作提供驾驶支援，其他的驾驶动作都由人类驾驶员进行操作。
- Level 2（部分自动化）：通过驾驶环境对方向盘和加减速中的多项操作提供驾驶支援，其他的驾驶动作都由人类驾驶员进行操作。
- Level 3（有条件自动化）:由无人驾驶系统完成所有的驾驶操作。根据系统请求，人类驾驶者提供适当的应答。
- Level 4（高度自动化）:由无人驾驶系统完成所有的驾驶操作。根据系统请求，人类驾驶者不一定需要对所有的系统请求做出应答、限定道路和环境条件等。
- Level 5（完全自动化）：由无人驾驶系统完成所有的驾驶操作。人类驾驶者在可能的情况下接管。在所有的道路和环境条件下驾驶。

自动驾驶的算法和工程实践探索几乎贯穿了 20 世纪的整个历史，在 21 世纪的发展更加突出商业化特点。世界上第一台自动驾驶车辆（非乘用汽车）大约是在 1912 年出现的，发明家约翰·哈蒙德（John Hammond）和本杰明·梅森纳（Benjamin Miessner）使用一个电子回路和一对光感性硒光电管作为自动驾驶核心单元自动引导旅行箱大小的一辆小车。当车上的光感性电管受到光线照射时，会发送前进方向给底层控制系统，控制系统进行转向操作从而使小车朝向光源方向行驶，通过测定比对两侧感光电管的数据，达到不偏离行驶路线的目的。设计师给这一粗糙的自动驾驶车辆起了一个凶悍的名字——“战争狗”，因为这一设备的研发初衷是应用于军事领域。战争狗的设计理念简单明了，在靠近敌人防线的地方投放出去，它能自动执行破坏性任务，无须人类指引。

在乘用汽车上的自动驾驶研究起步也很早。在汽车问世不久之后，发明家们就开始研究自动驾驶汽车了。1925 年，发明家 Francis Houdina 展示了一辆无线电控制的汽车，他的车在没有人控制方向盘的情况下在曼哈顿的街道上行驶。根据《纽约时报》的报道，这种无线电控制的车辆可以发动引擎、转动齿轮，并按响它的喇叭。报道使用了形象的比喻“就好像一只幽灵的手在方向盘上”来说明自动驾驶给公共大众的最初感觉。

1969 年，人工智能的创始人之一约翰麦卡锡在一篇名为“电脑控制汽车”的文章中描述了与现代自动驾驶汽车类似的想法。麦卡锡提出的想法是关于一名“自动司机”可以通过“电视摄像机输入数据，并使用与人类司机相同的视觉输入”来帮助车辆进行道路导航。他在文章中写道，用户应该可以通过使用键盘输入的目的地来驱使汽车立即自动前往目的地。同时，存在额外的命令可以让用户改变目的地，例如在休息室或餐厅停留时可以放慢速度或者在紧急情况下加速。虽然没有这样的车辆存在，但麦卡锡的文章为其他研究人员的任务设计提供了帮助。

20 世纪 90 年代初，卡内基梅隆大学的研究人员 Dean Pomerleau 写了一篇描述神经网络如何让自动驾驶汽车实时从公路获取原始图像来实现和输出方向控制的博士论文。Pomerleau 并不是唯一一个研究自动驾驶汽车的研究人员，但他使用神经网络的方法比其他尝试手动将图像划分为“道路”和“非道路”类别的尝试更有效。

进入 21 世纪，工程化的实践需求逐渐浮出水面，当然开始来自于学术界。2002 年，美国国防高级研究计划局（Defense Advanced Research Projects Agency，DARPA）宣布了一项重大挑战，他们为顶级研究机构的研究人员设立的条件是：如果能够建造一辆在莫哈维沙漠行驶 142 英里的无人驾驶汽车，他们将提供 100 万美元的奖金。当 2004 年挑战开始时，15 个竞争者中没有一个完成任务。“胜利”号在着火之前，几小时内只能跑不到 8 英里的路程。看来第一次比赛时，DARPA 设置的标准似乎太高。2005 年 10 月，他们举行另一场比赛时，23 支参赛队伍中有 5 支完成了 132 英里的路程，剩下的只有一支队伍未能完成一年前的 7.4 英里的记录。斯坦福大学教授塞巴斯蒂安•特伦（Sebastian Thrun）所领导的 Stanley 获得了第一名，卡内基梅隆大学的 Sandstorm 排名第二。Stanley 是一辆使用了现代标准摄像头、雷达和激光扫描仪的大众 Touareg，在很大程度上依靠机器学习来理解其收集到的数据以及决定如何行驶。Sebastian Thrun 在自动驾驶领域自此开始崭露头角，并且逐年水涨船高。从 2009 年开始，谷歌开始秘密开发无人驾驶汽车项目，该项目就是现在大名鼎鼎的 Waymo。该项目最初就是由 Sebastian Thrun 领导的。

随着机器学习开始崭露头角，汽车公司开始基于机器学习实践推出了各种 L2~L3 级自动驾驶服务。丰田公司的日本普锐斯混合动力车从 2003 年开始提供自动停车辅助服务，而雷克萨斯很快就为其雷克萨斯 LS 轿车添加了类似的系统。福特也在 2009 年加入了自动泊车辅助系统。一年后，宝马推出了自己的平行泊车助手。

1.2.2　为什么要在无人驾驶中应用机器学习

斯坦福大学教授 Sebastian Thrun 能够成为业界的著名人物，不只是完成了无人驾驶汽车的挑战，更重要的是他率领团队重新定义了无人驾驶成功路径的构造方式，这种构造方式主要是指最大程度强调软件的自身识别和决策能力，从实际来说，他们很大程度上放弃了当时盛行的基于规则的人工智能软件，而使用机器学习的策略来完成核心软件决策部分。关于这种抉择，特伦本人曾对其产生的原因做了深入的分析。

“许多参赛者非常关注硬件，于是许多车队研发出了独家的机器人汽车来参赛。我们则认为，比赛重点不在于机器人的力量强弱或汽车底盘的设计优劣。从能力来考虑，如果是人类驾驶，那么任何车都能够顺利通过这段普通的沙漠赛道。于是，我们决定仅研究人工智能，把一台计算机安装在车内，为车子安装眼睛和耳朵，打造出一辆智能化汽车。在汽车智能化的研发中，我们发现简单的规则不足以指导软件完成驾驶，需要成百上千条代码指令才足以应付多种意外情况。某天的测试中，公路旁有一群鸟。当汽车靠近时，鸟群飞起。这时我们才发现在机器人的“眼睛”里无法识别鸟类和岩石！于是我们不得不提高汽车的智能化程度，把鸟和石头区分开来。后来，我们依靠所谓的机器学习或大数据来驱动汽车：与其写出所有的程序，不如用人类学习驾驶的方法来教机器人。我们来到沙漠，我开车，机器人“观看”并“模仿”我所有的动作。后来，我们直接让机器人来驾驶，当它犯错的时候，我们返回数据中解释错误原因，给机器人机会改正错误。”

要搞清楚为什么特伦的机器学习在当时被称为一项重大创新，就需要解释当时十分流行的符号型人工智能的工作原理和机器学习工作原理的不同。最大的不同是算法和数据的上下游关系，基于指令的符号型人工智能是自上而下的，数据驱动型人工智能是自下而上的。自上而下的符号型人工智能要求开发人员完全理解人工智能业务的细节，然后基于自身的理解首先搭建出完整的理论模型，最后通过一系列应对各种状况的指令集合与模型产生交流，产生最终的驾驶决策。不同的是，自下而上的机器学习人工智能依赖于数据集合的表现，因此需要采用大量的数据算法并使用相应的技术进行数据处理，最终使得汽车软件无须人类监管也能自主学习并“识别”固定的模式。

在 2005 年 DARPA 挑战赛中，某支参赛队伍花费数月时间，编写了整套逻辑指令来处理传感器输出的数据流。他们希望当车辆前方的路况数据增长时，系统控制软件能够指示车辆转向，绕开障碍物。经过数月的艰苦努力，他们创建了数据量巨大、详细的数据库。遗憾的是，这支车队赢取百万奖金的旅程停止在比赛途中。他们的车辆由于制动不及时，在隧道入口发生碰撞。对此，该车队表示他们在预先编写中缺乏应对隧道的特殊指令，团队在赛前准备时没有预料到这种状况。在缺乏清晰指令的情况下，系统只能根据数据胡乱猜测：隧道顶部很高，且赫然出现在车道中央，于是系统软件将其判断为一座巨大的高墙。在这样的误解下，系统做出自认为对的处理：踩下刹车，拒绝前进直等到工作人员到来，将它带回安全地段。

假如我们能够统揽前面一届比赛的结果，就能在很大程度上理解特伦团队选择的核心动机。自动驾驶不是两三条指令可以做到的事，需要成百上千条指令方可成事，并且如果指令固定化，从某种意义上来说场景本身就是固定的。指令数量多以及多个场景的细微差异需要的决策的可变化性带来的编程挑战很难通过传统软件开发的方式完整写出来。哪怕在一条空旷的赛道上，也可能发生无数人们始料未及的新状况。因此，使用机器学习完成无人驾驶工作具有先天优越特性。对于自动驾驶的典型场景接下来会概括地讨论一下，包括使用卷积神经网络进行驾驶感知和使用循环神经网络进行驾驶预测。

（1）使用卷积神经网络进行驾驶感知

车辆在行驶过程中不可避免地会和障碍物打交道，障碍物一般包括静态和动态两类。静态障碍物包括墙壁、树木、交通障碍等，动态障碍物包括行人、非机动车及其他机动车辆。我们的感知程序需要检测到所有的障碍物，并且按照一定规则对障碍物进行分类。通过不同的障碍物类型最终决定程序的控制策略。比如我们可能会选择绕过前面的车辆，而对于墙壁必须执行停车策略。因此，程序使用一个卷积神经网络来确定物体在摄像头图片中的位置，然后通过另一个神经网络来完成上一个神经网络提取的物体的分类工作。当然，我们也可以使用一个核心神经网络来完成全部工作，只需要定制两种不同的输出层架构就能完成这个工作。卷积网络有很多变体，比如基于其中的一个变体 YOLO 网络结构建立车辆检测模型，判断图片中是否有车辆及车辆在图片中的位置，判断被检测车辆与摄像头的相对方位及运动趋势，判断被检测车辆对自身车辆的相对关系。

（2）使用循环神经网络进行驾驶预测

循环神经网络通常用来处理时序数据，在时序数据中的数据是按照先后顺序组织在一起的，车辆的行驶过程数据就是一个完美的时序数据模型，比如要完成转弯动作，整个过程就包括准备、转弯和方向的回复，这些过程有着明确的先后顺序，过程中的相关障碍物车道等也能够拥有这种时序特征，因此我们可以使用循环神经网络来预测驾驶路径和方向等。举例来说，比如使用循环神经网络预测车辆的目标车道的算法。它为车道序列提供一个循环神经网络模型，为障碍物提供另一个循环神经网络模型，Apollo 连接这两个循环神经网络的输出，并将它们的反馈输入到另一个神经网络，该网络会估算每个车道序列的概率，具有最高概率的序列就是预测目标车辆将遵循的序列。

1.2.3 无人驾驶商业化的优势

无人驾驶系统的应用将会全面改变整个乘用车和商用车领域的人车关系，在乘用车领域强化人的乘坐属性，最大限度地发挥车辆的媒体终端属性，比如在乘坐过程中根据位置的不同推荐饮食、购物场地。对于商用领域，我们能够对开放道路的 L2 和封闭区域的 L4 分别提取核心商用场景或者综合商用场景，提升运输等商业活动的效率，减少相应的人力成本。综合各种应用场景，无人驾驶的商业优势主要表现在增强高速公路安全、缓解交通拥堵、减少空气污染上。研究表明，无人车会使这 3 个领域有大幅改善。

1. 增强高速公路安全

高速公路事故死亡是全世界面临的重大问题。在美国，每年估计有 35 000 人死于车祸，中国这一数字约为 260 000，日本每年高速公路事故死亡人数为 4 000 左右。根据世界卫生组织统计，全世界每年有 124 万人死于高速公路事故。据估计，致命车祸每年造成 2 600 亿美元的损失，而车祸致伤带来 3 650 亿美元的损失。高速公路伤亡每年导致 6 250 亿美元的损失。美国兰德公司研究显示，2011 年车祸死亡中 39%涉及酒驾。几乎可以肯定，在这方面，无人驾驶汽车将带来大幅改善，避免车祸伤亡。在中国，约 60%的交通事故和骑车人、行人或电动自行车与小轿车和卡车相撞有关。在美国的机动车事故中，94%与人为失误有关，因此可以得到有效避免。美国高速公路安全保险研究所的一项研究表明，全部安装自动安全装置能使高速公路事故死亡数量减少 31%，每年将挽救 11 000 条生命。这类装置包括前部碰撞警告体系、碰撞制动、车道偏离警告和盲点探测。

2. 缓解交通拥堵

交通拥堵是几乎每个大都市都面临的问题。以美国为例，每位司机平均遇到 40 个小时的交通堵塞，年均成本为 1 210 亿美元。在莫斯科、伊斯坦布尔、墨西哥城或里约热内卢，浪费的时间更长，每位司机每年将在交通拥堵中度过超过 100 小时。在中国，汽车数量超过 100 万的城市有 35 个，超过 200 万的有 10 个。在最繁忙的市区，约 75%的道路会出现高峰拥堵。中国私家车总数已达 1.26 亿辆，仅北京就有 560 万辆汽车。Donald Shoup 的研究发现，都市区 30%的交通是由于司机为了寻找附近的停车场而在商务区绕圈造成的，这是交通拥挤、空气污染和环境恶化的重要原因。另外，根据估算都市交通拥堵中有 23%~45%发生在道路交叉处。交通灯和停车标志不能发挥作用，因为它们是静止的，无法将交通流量考虑其中。绿灯或红灯是按照固定间隔提前设定好的，无论某个方向的车流量有多大。一旦无人驾驶汽车逐渐投入使用，并占到车流量比较大的比例，车载感应器将能够与智能交通系统联合工作，优化道路交叉口的车流量。红绿灯的间隔也将是动态的，根据道路车流量实时变动。这样可以提高车辆通行效率，缓解拥堵。

3. 减少空气污染

汽车是造成空气质量下降的主要原因之一。兰德公司研究表明，无人驾驶技术能提高燃料效率，通过更顺畅的加速、减速，比手动驾驶提高 4%~10%。由于工业区的烟雾与汽车数量有关，增加无人驾驶汽车的数量能减少空气污染。一项 2016 年的研究估计，等红灯或交通拥堵时汽车造成的污染比车辆行驶时高 40%。无人驾驶汽车共享系统也能带来减排和节能的好处。德克萨斯大学奥斯汀分校的研究人员研究了二氧化硫、一氧化碳、氮氧化物、挥发性有机化合物、温室气体和细小颗粒物。结果发现，使用无人驾驶汽车共享系统不仅节省能源，还能减少各种污染物的排放。约车公司 Uber 发现，该公司在旧金山和洛杉矶的车辆出行中分别有 50%和 30%是多乘客拼车。在全球范围内，这一数字为 20%。无论是传统车，还是自动驾驶车，拼车越多，对环境越好，也越能缓解交通拥堵。改变一车一人的模式将能大大改善空气质量。

1.2.4 无人驾驶商业化的进展

几十年来，现代汽车行业由于核心零部件关系复杂和零部件供应商的排他关系造成了比较大的行业门槛，将许多专注于产品模块功能和体量较小的竞争者挡在门外。随着汽车智能网联化的基础日益增强，通过整车远程升级（Firmware Over-The-Air，FOTA）不断加强汽车终端化特性，用户也在驾驶使用过程中逐渐增加媒体使用特征。正如我们现在选择智能手机，最重要的可能已经不是声音大不大、按键灵不灵，而是操作系统及其版本号，以及手机生产商对这种能力的保证情况。在这种变化的大环境下，软件公司在汽车业逐渐崭露头角，无人驾驶作为一项软件复杂度和算法先进性并存的汽车核心需求，以 Waymo 为首的自动驾驶公司近几年发展迅速。下面列举一些自动驾驶独角兽企业和汽车制造企业的产品定位和商业规划。

1. Waymo

网约车、物流、私家车和公共交通是 Waymo 确立的 4 个无人驾驶技术部署应用的目标领域，因为这样的商业路线，摩根士丹利将 Waymo 的估值上调到 1 750 亿美元。Waymo 无人驾驶车队的大本营是一座面积为 7 万平方英尺（约 6 5031 平方米，1 平方英尺≈0.0929 平方米）的车库，由技术人员、工程师、机械师、客服代表和产品经理组成的团队就是在这里进行车队维护和管理的。数百辆克莱斯勒（Pacifica）全部装备有先进的无人驾驶硬件和软件，无须人类驾驶员的支持即可安全上路行驶。

过去一年里，Waymo 一直在为其“早期乘客项目”（Early Rider）的 400 名参与者提供出行服务，他们使用 Waymo 的网约车 App 即可免费预约乘车前往在菲尼克斯运行区域内的任何地点。不久之后，Waymo 即将面向公众提供这项服务，当然也会开始收费。在全自动驾驶交通服务运营初期，Waymo 仍然准备在车内安排一位工作人员随行。如果顺利的话，这将是全世界最先推出的全自动驾驶网约车服务。

尽管自动驾驶汽车仍然饱受质疑，但还是有很多人期待其能从根本上改变交通运输方式。车队管理是被很多人忽视的一项重要挑战，这项工作将直接影响 Waymo 无人驾驶技术的成败。无人驾驶汽车装备的传感器、计算机芯片和其他系统使其能在没有人类驾驶员的情况下安全抵达目的地，而为了解决这些技术设备的成本，车队中的所有车辆都将几乎不间断地在路上行驶。这意味着它们的行驶里程会达到数十万公里（远远高于人类驾驶汽车），以此来保障其在经济上的可行性。

2. 特斯拉

Autopilot（自动辅助驾驶）的进化史是“硬件先行，软件后更新”——每一台特斯拉都会配置当时最新的硬件，然后通过整车远程升级不断更新固件，获得更完善的驾驶辅助或自动驾驶功能。庞大的用户群可以源源不断地供给真实路况的驾驶数据，帮助 Autopilot 训练和迭代算法。

2014 年 10 月，Autopilot 首发时采用了 1.0 版硬件，传感器使用了一个前置摄像头、一个前向毫米波雷达以及车身一周的 12 个超声波雷达。官方在召开发布会时介绍，这套硬件及其配套的软件算法最终不能实现全自动驾驶，Autopilot 只是提升舒适性和安全性的辅助功能，车辆的控制权仍然在驾驶员。

从 Enhanced Autopilot（增强自动辅助驾驶）开始，特斯拉使用了 2.0 版硬件，这也是大家常常把增强自动辅助驾驶称为 Autopilot 2.0 的原因。马斯克在发布会上宣布，2.0 版硬件（8 个摄像头、1 个毫米波雷达、12 个超声波雷达以及 NVIDIA Drive PX2 计算平台）在固件更新后，可以开启全自动驾驶功能。虽然马斯克一直表示不会使用激光雷达，但特斯拉“毫米波雷达+摄像头”全自动驾驶传感器方案还是在行业内引发了热议。

马斯克最新表示，预计 2020 年会有 Robotaxi（无人驾驶出租车）投入运营，特斯拉汽车车主可以将他们的汽车加入这项服务中。车主可以选择只将自己的车共享给自己的朋友、同事或社交媒体上的好友，同时他们还可以在特定时间限制 Tesla Network 的可用性。

3. 通用汽车

通用汽车公司在自动驾驶领域的步伐并不慢，大规模的车队测试自不用说，他们还在 2018 年 1 月发布了 L4 级别名为 Cruise AV 的全自动驾驶车，这辆车基于雪佛兰 Bolt 纯电动车打造，最大的特点是座舱内完全没有方向盘存在，按照通用汽车公司的说法，这是人类有史以来第一次大规模量产无人驾驶汽车，对于通用汽车公司而言，这将具有里程碑的意义。

另外，具备 L2 级别的驾驶辅助系统 Super Cruise 超级智能驾驶系统，目前北美上市的 2018 款凯迪拉克 CT6 上可以选配。在 2018 CES AISA 展上，凯迪拉克宣布将会在中国市场投放搭载 Super Cruise 超级巡航系统的 CT6 车型，并且在 2020 年会将该技术拓展至凯迪拉克品牌下的所有车型上。

由此可见，通用汽车公司选择两条腿走路，协同进击自动驾驶商业化的想法很实际，Super Cruise 可以有效促进通用传统汽车的销售，还能够起到教育用户的作用，让更多的用户对自动驾驶技术产生客观公正的认知，为今后的自动驾驶汽车落地打好基础。

4. Navya

Navya（纳维亚）的自动驾驶着力点为自动驾驶巴士，目的是为城市和私人场所带来更多的流动性。在城市或私人场所，Navya 设想的航天飞机拥有一个创新、有效、清洁和智能的移动解决方案。由于其温和的导航，自动驾驶穿梭车保证了自动运输的性能，以及在第一英里和最后一英里的舒适旅行。AUTONOM 接驳巴士可以搭载多达 15 人，它结合了许多优势。自动驾驶穿梭车队使运营商有可能提高私人场所的生产率，并缓解城市中心的道路拥堵。乘客在充分利用旅行时间的同时，也可以享受一次愉快的旅行。AUTONOM 接驳巴士既没有方向盘又没有踏板，它采用了有效的制导和探测系统，结合了各种先进技术，激光雷达传感器、摄像机、GPS RTK、IMU、测程等数据融合在一起，通过深度学习程序进行驾驶控制。接驳巴士可以高效地移动，并做出熟练的决策。

5. 图森未来

图森未来研发的 L4 级别自动驾驶技术以摄像头为主要传感器，并融合激光雷达、毫米波雷达等其他传感器，配合自主研发的核心算法，能够实现环境感知、定位导航、决策控制等自动驾驶核心功能。图森未来宣称感知技术上已实现对 1000 米范围内的可视环境进行像素级辨识。图森未来自主研发区域 L4 级别自动驾驶技术能够实现货运卡车在高速公路货运和港内集装箱码头运输以及相似场景下的全无人驾驶。

参考文献

[1] Russell S J, PN Norvig. Artificial Intelligence: A Modern Approach. Prentice Hall[J]. Applied Mechanics & Materials, 2010, 263(5):2829-2833.

[2] Saygin A P, Cicekli I, Akman V. Turing Test: 50 Years Later[J]. Minds and Machines, 2000, 10(4):463-518.

[3] Brown N, Sandholm T. Superhuman AI for heads-up no-limit poker: Libratus beats top professionals[J]. Science, 2017:eaao1733.

[4] Krizhevsky A, Sutskever I, Hinton G. ImageNet Classification with Deep Convolutional Neural Networks[C]// NIPS. Curran Associates Inc. 2012.

[5] Chen J X. The Evolution of Computing: AlphaGo[J]. Computing in Science and Engineering, 2016, 18(4):4-7.

[6] Pomerleau D, Jochem, T. Rapidly adapting machine vision for automated vehicle steering[J]. IEEE Expert, 2002, 11(2):19-27.

[7] Thrun S. Winning the DARPA grand challenge[C]// European Conference on Machine Learning. 2006.

[8] King D, Manville M, Donald ShoupÃ. The political calculus of congestion pricing[J]. Urban Transport of China, 2013, 14(2):111-123.

第 2 章

TensorFlow 基础

我们可以说机器学习算法是完成人工智能问题的“道”，但是真正解决人工智能问题还需要好用的“术”，换句话说，我们需要功能强大并且容易学习的机器学习软件框架。当然，作为目前计算机学科最热门的领域，各种机器学习框架的研发创新正在如火如荼地进行着。这些框架的设计思路不尽相同，有的框架着眼于模型研究，有的框架面向工程化实践，有的和大数据紧密绑定，还有的在用户易用和框架兼容性方面下工夫。TensorFlow 正是其中的集大成者，强大的研发背景，兼容易用的 API 设计，再加上谷歌在工程化实践的背书，使得 TensorFlow 自诞生以来一直高速持续地在机器学习的各种应用层面迅速发展并开花结果。

本章主要讲解机器学习的一些宏观概念以及无人驾驶和机器学习之间的定性关系。内容分为六部分，第一部分着重介绍机器学习中的主流框架；第二部分详细介绍搭建 TensorFlow 开发环境的过程；第三部分通过 HelloTF 直观引入 TensorFlow 框架机器学习的一般流程；第四部分简单介绍 TensorFlow 的架构；第五部分着重介绍 TensorFlow 的高阶和低阶框架 API；第六部分是扩展阅读部分，主要介绍 tensorflow.js 这一衍生框架。

2.1 机器学习主流框架简介

机器学习属于二进制创新的范畴，在各个阶段都有许多优秀的框架，它们虽然目标都是通过代码发现并反映数据本身的规律，通过训练的模型对于发生的事件进行有效的预测，但是不同框架的侧重点并不相同。有的框架来源于大学实验室，而且创立较早，因此主要针对模型构造的过程，对效率和并行计算等的支持十分有限；有的框架发端于同平台的云服务，针对自己的平台有着不错的优化；有的框架来源于同开源基金会的大数据分析需求，虽然天然对平行计算等有很好的支持，但是使用时却需要机器部署很多依赖框架和服务。

下面就来介绍目前流行的 15 个机器学习框架。

1. Apache Singa

Apache Singa 是一个用于在大型数据集上训练深度学习的通用分布式深度学习平台，它是基于分层抽象的简单开发模型设计的。它还支持各种当前流行的深度学习模型，有前馈模型（卷积神经网络）和能量模型（受限玻尔兹曼机和循环神经网络），还为用户提供了许多内嵌层。

Apache Singa 的软件核心架构包含 3 个主要的组成部分，分别是计算核心（Core）、输入输出（IO）和模型（Model）。图 2.1 中给出了 Apache Singa 的整体软件架构的组成关系以及 3 大组件的主要功能。计算核心提供张量的计算操作以及为了能够提升机器学习模型代码在生产环境下的各种硬件优化（CPU、GPU、内存等）。输入输出主要处理的是从本地或者服务器中读取和写回数据的过程。模型组件致力于提供机器学习特别是复杂机器学习模型构建的基础组件。比如要构建卷积神经网络，我们需要层（Layer）来部署网络，训练中使用 Optimizers/Initializer/Metric/Loss 等组件。

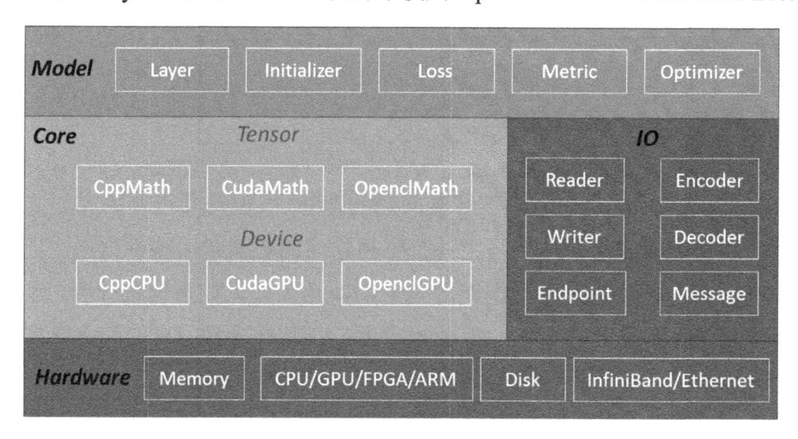

图 2.1 Apache Singa 的整体软件架构

2. Amazon Machine Learning

亚马逊机器学习框架（Amazon Machine Learning，AML）是一种让各种级别使用机器学习技术的开发人员都能够轻松掌握的服务，提供了视觉工具和向导，可以指导开发人员在不必学习复杂的机器学习算法和技术的情况下建立机器学习。其中重要的组成部分是 Amazon SageMaker。Amazon SageMaker 可以帮助开发人员和数据科学家快速轻松地构建、训练和部署任何规模的机器学习模型。从运行实时欺诈检测模型、虚拟分析潜在药物的生物影响到预测棒球比赛中的盗垒成功率，它消除了成功实现机器学习的复杂性。

3. Azure ML Studio

Azure ML Studio 允许微软 Azure 的用户创建和训练模型，随后将这些模型转化为能被其他服务使用的 API。尽管你可以将自己的 Azure 存储链接到更大模型的服务，但是每个账户模型数据的存储容量最多不超过 10GB。在 Azure 中有大量的算法可供使用，这要感谢微软和一些第三方公司。甚至不需要注册账号就可以匿名登录，使用 Azure ML Studio 服务长达 8 小时。

4. Caffe

Caffe 是由伯克利视觉学习中心（Brian Leonard Visual Communications，BLVC）和社区贡献者基于 BSD-2-协议开发的一个深度学习框架，它秉承“表示、效率和模块化”的开发理念。模型和组合优化通过配置而不是硬编码实现，并且用户可根据需要在 CPU 处理和 GPU 处理之间进行切换。Caffe 的高效性使其在实验研究和产业部署中的表现很完美，使用单个 NVIDIA K40 GPU 处理器每天即可处理超过 6000 万张图像。但是由于 Caffe 创立时间较早，数据挖掘工程师一直对它有一些设计上的抵触情绪。因此，Caffe 的作者又重新构建了新的机器学习框架——Caffe 2。

Caffe 2 中的基本单元被称作操作子（Operators）。每一个操作子包含完成机器学习基本的参数输入（w 和 b）、数据输入和计算完成的数据输出。在新一代 Caffe 中，计算参数作为输入值而不是配置项，使用起来编程体验更好。比如我们设计一个全连接的神经元，权重 w、偏置 b 和输入数据 X 是必须全部提供的，在这些输入完成后，操作子会计算出输出值 Y。

图 2.2 所示为 Caffe 与 Caffe 2 架构的对比。

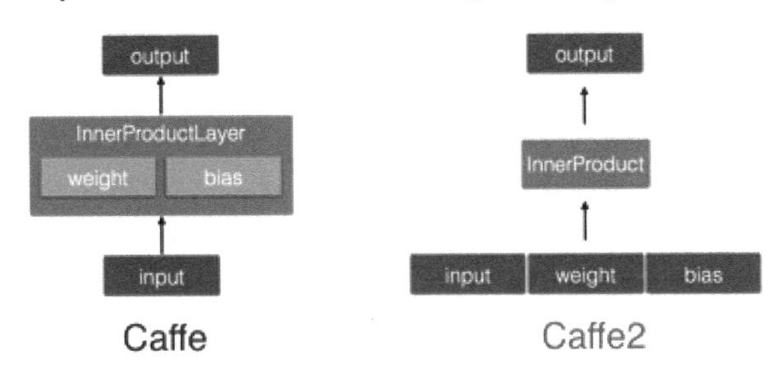

图 2.2　Caffe 与 Caffe 2 架构的对比

5. H2O

H2O 可以轻松地应用数学和预测分析来解决当今极具挑战性的商业问题，它巧妙地结合了目前在其他机器学习平台还未被使用的独有特点——最佳开源技术，易于使用的 WebUI 和熟悉的界面，支持常见的数据库和不同文件类型。使用 H2O 时，你可以使用现有的语言和工具，此外，还可以无缝扩展到 Hadoop 环境中。H2O 与 Hadoop 系统整合架构如图 2.3 所示。

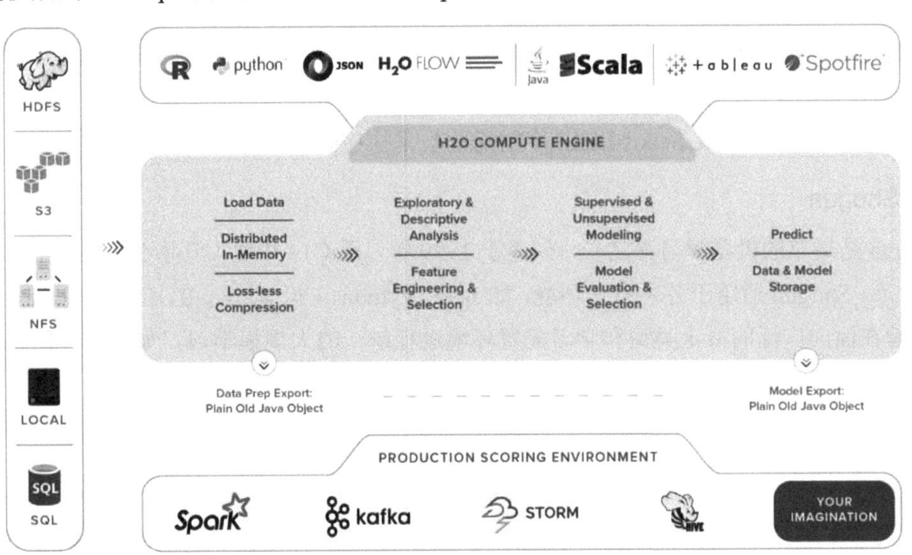

图 2.3　H2O 与 Hadoop 系统整合架构

6. Massive Online Analysis

大规模在线分析框架（Massive Online Analysis，MOA）是目前最受欢迎的数据流挖掘开源框架，拥有一个非常活跃的社区。它包含一系列的机器学习算法（分类、回归、聚类、离群检测、概念漂移检测和推荐系统）和评价工具。和 WEKA 项目一样，MOA 也是用 Java 编写的，但扩展性更好。

7. MLlib (Spark)

MLlib（Spark）是 Apache Spark 的机器学习库，目的是让机器学习实现可伸缩性和易操作性，它由常见的学习算法和实用程序组成，包括分类、回归、聚类，协同过滤、降维，同时包括底层优化原生语言和高层管道 API。

8. Mlpack

Mlpack 是一个基于 C++的基础学习库，最早于 2011 年推出，数据库的开发者声称，它是秉承“可扩展性、高效性和易用性”的理念来设计的。执行 Mlpack 有两种方法：通过快速处理简易的“黑盒”操作命令行执行缓存，或者借助 C++ API 处理较为复杂的工作。Mlpack 可提供简单的能被整合到大型的机器学习解决方案中的命令行程序和 C++的类。

9. Pattern

Pattern 是 Python 编程语言的 Web 挖掘组件，有数据挖掘工具（Google、Twitter、Wikipedia API、网络爬虫、HTML DOM 解析器）、自然语言处理（词性标注、N-Gram 搜索、情感分析、WordNet 接口）、机器学习（向量空间模型、聚类、支持向量机）、网络分析和可视化。

10. Scikit-Learn

Scikit-Learn 用于数学和科学工作，基于现有的几个 Python 包（NumPy、SciPy 和 Matplotlib）拓展了 Python 的使用范围。最终生成的库既可用于交互式工作台应用程序，又可嵌入其他软件中进行复用。该工具包基于 BSD 协议，是完全免费开源的，可重复利用。Scikit-Learn 中含有多种用于机器学习任务的工具，如聚类、分类、回归等。Scikit-Learn 是由拥有众多开发者和机器学习专家的大型社区开发的，因此 Scikit-Learn 中前沿的技术往往会在很短时间内被开发出来。

11. Shogun

Shogun 是最早的机器学习库之一，创建于 1999 年，用 C++开发，但并不局限于 C++环境。借助 SWIG 库，Shogun 适用于各种语言环境，如 Java、Python、C#、Ruby、R、Lua、Octave 和 Matlab。Shogun 旨在面向广泛的特定类型和学习配置环境进行统一的大规模学习，如分类、回归或探索性数据分析。

12. TensorFlow

TensorFlow 是 Google 第二代深度学习系统，是完全开源的框架产品。TensorFlow 是一种编写机器学习算法的界面，也可以编译执行机器学习算法的代码。使用 TensorFlow 编写的运算几乎不用更改就能运行在多种异质系统上，从移动设备（例如手机和平板电脑）到拥有几百台机器和几千个 GPU 之类的运算设备的大规模分布式系统。TensorFlow 降低了深度学习的使用门槛，让从业人员能够更简单和方便地开发新产品。作为Google 发布的“平台级产品”，很多人认为它将改变人工智能产业。

13. Theano

Theano 是一个基于 BSD 协议发布的可定义、可优化和可数值计算的 Python 库。使用 Theano 可以与使用 C 实现大数据处理的速度相媲美，是支持高效机器学习的算法。

14. Torch

Torch 是一种广泛支持把 GPU 放在首位的机器学习算法的科学计算框架。由于使用了简单快速的脚本语言 LuaJIT 和底层的 C/CUDA 来实现，使得该框架易于使用且高效。Torch 的目标是让用户通过极其简单的过程、最大的灵活性和速度建立自己的科学算法。Torch 是基于 Lua 开发的，拥有一个庞大的生态社区，能够找到许多功能独特的驱动库，包括设计机器学习、计算机视觉、信号处理、并行处理、图像、视频、音频和网络等。

15. Veles

Veles 是一套用 C++开发的面向深层学习应用程序的分布式平台，不过它利用 Python 在节点间自动操作与协作任务。在相关数据集中到该集群之前，可对数据进行分析与自动标准化调整，且 REST API 允许将各种训练模型立即添加到生产环境中，侧重于性能和灵活性。Veles 几乎没有硬编码，可对所有广泛认可的网络拓扑结构进行训练，如全卷积神经网络、卷积神经网络、循环神经网络等。

2.2 TensorFlow 开发环境搭建

本书将使用TensorFlow的Python版本进行代码演示，因此我们首先需要安装Python开发环境。为了让 Python 开发环境能够最大程度地保持独立稳定，在演示或进行实验的时候能够最大程度地不受 Python 多版本引发的引用包错误、语法不支持等环境问题的困扰，我们将在环境搭建中引入环境虚拟化方案。本节将介绍两种环境虚拟化方案：一种是基于 Python 语言框架的 Virtualenv；另一种是基于应用容器化的 Docker 方案。

2.2.1 基于 Python 语言框架的 Virtualenv 方案

Virtualenv 是一种基于 Python 语言框架的库工具，目的是能够简单地创建隔离的 Python 开发环境。究其本质原因，笔者认为 Python 开发方面 2.x 版和 3.x 版长期拥有其固定的开发者群体和系统预装的情况难辞其咎，因此，本书使用 Virtualenv 进行开发环境隔离，而且使用 Python 3.7 作为开发语言框架版本，旨在希望能最大程度使读者朋友聚焦于算法实现和算法应用本身，而尽可能少地被开发环境本身干扰。本书的开发思想源于笔者多年从事 IT 开发和技术分享的思考，在分享过程中和学生、朋友以及编程爱好者交流，阻止他们登堂入室的显著问题之一就是无关细节消耗了巨大的信心和时间，因此本书中会有很多预置型的编程约定，目的是加强传递知识本身的效率，提升沟通效果。

1. 安装 Python

安装 Python 需要访问 Python 官网网站（见图 2.4），地址是 https://www.python.org/。

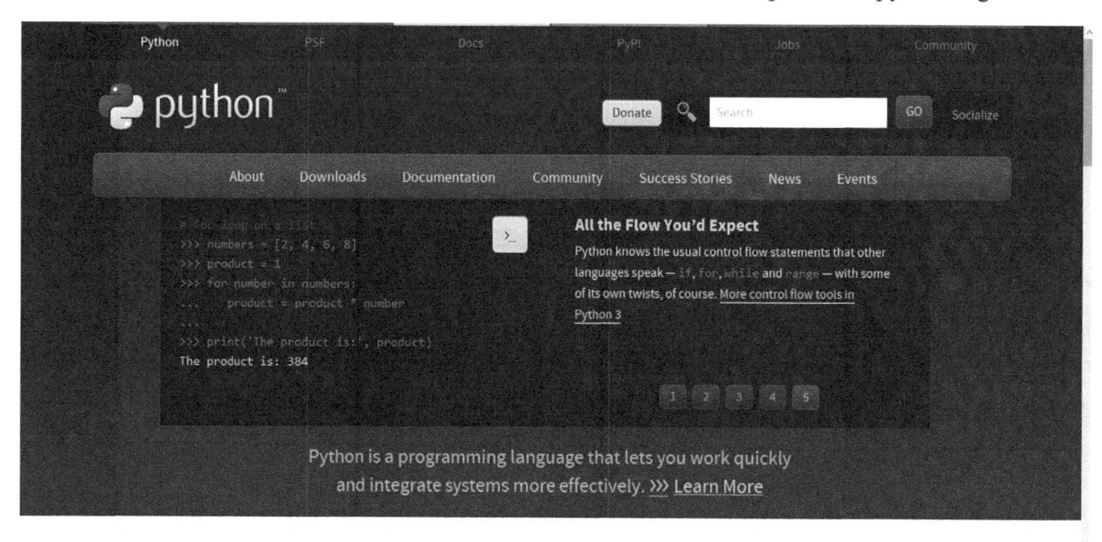

图 2.4 Python 官方网站

然后进入下载页面，选择最新版本的 Python 3.x 安装程序，如图 2.5 所示。特别需要注意的是，请选择 64 位的安装程序，虽然 32 位的安装程序大多可以成功安装到系统中，但是由于目前 TensorFlow 并不支持 32 位操作系统，因此在后续安装 TensorFlow 的步骤中会因为找不到适合的安装包文件而导致安装失败。

Version	Operating System	Description	MD5 Sum	File Size	GPG
Gzipped source tarball	Source release		02a75015f7cd845e27b85192bb0ca4cb	22897802	SIG
XZ compressed source tarball	Source release		df6ec36011808205beda239c72f947cb	17042320	SIG
macOS 64-bit/32-bit installer	Mac OS X	for Mac OS X 10.6 and later	d8ff07973bc9c009de80c269fd7efcca	34405674	SIG
macOS 64-bit installer	Mac OS X	for OS X 10.9 and later	0fc95e9f6d6b4881f3b499da338a9a80	27766090	SIG
Windows help file	Windows		941b7d6279c0d4060a927a65dcab88c4	8092167	SIG
Windows x86-64 embeddable zip file	Windows	for AMD64/EM64T/x64	f81568590bef56e5997e63b434664d58	7025085	SIG
Windows x86-64 executable installer	Windows	for AMD64/EM64T/x64	ff258093f0b3953c886192dec9f52763	26140976	SIG
Windows x86-64 web-based installer	Windows	for AMD64/EM64T/x64	8de2335249d84fe1eeb61ec25858bd82	1362888	SIG
Windows x86 embeddable zip file	Windows		26881045297dc1883a1d61baffeecaf0	6533256	SIG
Windows x86 executable installer	Windows		38156b62c0cbcb03bfddeb86e66c3a0f	25365744	SIG
Windows x86 web-based installer	Windows		1e6c626514b72e21008f8cd53f945f10	1324648	SIG

图 2.5 Python 下载版本选择

很多系统中已经装有 Python 2.x（一般为 Python 2.7），可以利用重新指定路径的方式使系统默认 Python 命令调用 Python 3。当然，不修改也没有问题。下面直接通过 Python 3 创建虚拟环境后，环境中将有且只有一个 Python，而且 Python 的版本与我们的系统 Python 3 保持一致。

在 Python 3.x 的最新安装版本中已经包含 pip 依赖包管理工具，不需要再进行 pip 的安装。完成后，打开命令行操作界面，输入如下命令检测安装是否成功：

```
$ python3 --version
$ pip3 -version
```

当然，熟悉命令行的读者也可以使用Mac下的使用 Homebrew 软件包管理器进行安装。

```
/usr/bin/ruby -e "$(curl -fsSL
https://raw.githubusercontent.com/Homebrew/install/master/install)"
export PATH="/usr/local/bin:/usr/local/sbin:$PATH"
brew update
brew install python  # Python 3
如果是 Ubuntu 那么 apt 将会是一个不错的选择。
sudo apt update
sudo apt install python3-dev python3-pip
```

2. 安装Virtualenv

Python 虚拟环境Virtualenv用于将软件包安装与系统隔离开来，安装主要使用pip工具，通过初始化和激活过程完成虚拟环境的创建。

首先，我们需要完成安装流程，使用pip安装命令，在Ubuntu和Mac OS中还需要使用sudo取得系统权限。

```
(sudo) pip3 install -U virtualenv # system-wide install
```

Ubuntu和Mac OS是类UNIX系统，它们的创建方式很相似。

（1）创建一个新的虚拟环境，方法是选择 Python 解释器并创建一个 ./venv 目录来存放它：

```
virtualenv --system-site-packages -p python3 ./venv
```

（2）使用特定于Shell的命令激活该虚拟环境：

```
source ./venv/bin/activate # sh, bash, ksh, or zsh
```

（3）当Virtualenv 处于有效状态时，Shell提示符带有 (venv) 前缀。

（4）在不影响主机系统设置的情况下，在虚拟环境中安装软件包。首先升级 pip：

```
pip install --upgrade pip
pip list  # show packages installed within the virtual environment
```

（5）之后退出Virtualenv，使用以下命令：

```
deactivate  # don't exit until you're done using TensorFlow
```

在Windows中主要是激活方式有所不同。

（1）创建一个新的虚拟环境，方法是选择 Python 解释器并创建一个 ./venv 目录来存放它：

```
virtualenv --system-site-packages -p python3 ./venv
```

（2）激活虚拟环境：

```
.\venv\Scripts\activate
```

（3）在不影响主机系统设置的情况下，在虚拟环境中安装软件包。首先升级 pip：

```
pip install --upgrade pip
pip list  # show packages installed within the virtual environment
```

（4）之后退出 Virtualenv，使用以下命令：

```
deactivate  # don't exit until you're done using TensorFlow
```

3. 安装 TensorFlow

首先我们要确认安装 TensorFlow 的软硬件需求，为了后续的开发更加顺利，软件系统尽量遵循谷歌官方的要求，而且要注意括号里面的限制内容。笔者由于教学的需要，曾经在以下系统上进行TensorFlow代码运行，综合考虑，在有一定的Linux系统使用经验的条件下，推荐大家使用Ubuntu 16.04 或 18.04。

- Ubuntu 16.04 或更高版本（64 位）
- Mac OS 10.12.6（Sierra）或更高版本（64 位）（不支持 GPU）
- Windows 7 或更高版本（64 位）（仅支持 Python 3）

硬件方面的要求主要体现在对旧版本 CPU 的支持不友好方面。综合考虑时间等成本，如果你的机器很老，或者配置很低，那么推荐更换计算机或者更换硬件，原因是从 TensorFlow 1.6 开始，二进制文件使用 AVX 指令，这些指令可能无法在旧版 CPU 上运行，“工欲善其事，必先利其器”的古语放在这里可以说是很合适的。当然，如果准备从软件上解决，主要有两个途径：一个途径是使用 1.6 以下的版本，但是这种方案的副作用是需要适当修改后面的功能代码（不是太难）；另一个途径是使用适合老式 CPU 的 TensorFlow 版本，推荐下载地址 https://github.com/fo40225/tensorflow-windows-wheel。

接下来，我们会安装 TensorFlow 到启动成功的 Virtualenv 虚拟环境中。当然，你也可以安装到你的系统中。

如果要安装到虚拟环境中，就需要输入下面的命令：

```
pip install --upgrade tensorflow
```

安装到系统中的命令与虚拟环境安装命令类似：

```
pip3 install --user --upgrade tensorflow  # install in $HOME
```

安装完成后，使用 Python 命令打开 Python 命令行，然后输入下面的命令并按回车键。如果 TensorFlow 安装失败，系统就会提示没有安装 TensorFlow 的错误信息；如果安装成功，就不会显示任何错误信息。

```
import tensorflow as tf
```

2.2.2 基于应用容器化的 Docker 方案

Docker 是应用容器化目前流行的解决方案，使用简单，系统支持广泛。如果你使用的是 Ubuntu 系统，那么可以从软件仓库中安装 Docker，Mac OS 和 Windows 10 系统可以选择 Docker 对应的桌面端程序。如果使用的是 Windows 其他低版本的系统，那么建议使用第一种解决方案，因为实际测试结果表明，低版本的 Windows Docker 客户端可能存在很多兼容性问题。

Docker 安装完成后，你可以下载 TensorFlow 的 Docker 镜像并进行运行。

下载 Docker 镜像的命令是：

```
docker pull tensorflow/tensorflow
```

我们使用下载的镜像验证 TensorFlow 的安装。Docker 会在首次运行时下载新的 TensorFlow 映像：

```
docker run -it --rm tensorflow/tensorflow \
python -c "import tensorflow as tf"
```

在配置 TensorFlow 的容器中启动 Bash Shell 会话能够满足我们的一些调试类或测试类的需求。以下命令将会创建一个容器，在此容器中启动 Python 会话并导入 TensorFlow。

```
docker run -it tensorflow/tensorflow bash
```

要在容器内运行在主机上开发的 TensorFlow 程序，请装载主机目录并更改容器的工作目录（-v hostDir:containerDir -w workDir）：

```
docker run -it --rm -v $PWD:/tmp -w /tmp tensorflow/tensorflow python ./script.py
```

向主机公开在容器中创建的文件时可能会出现权限问题，通常情况下，最好修改主机系统上的文件。

2.3 Hello TensorFlow——一个简单的例子

我们已经安装了 TensorFlow 的开发环境，接下来通过简单的例子来说明如何利用 TensorFlow 来进行机器学习应用开发。首先使用 TensorFlow 中的高级 API——Keras 开发一个简单的例子，在这个例子中将训练一个简单的单层神经网络来识别手写数字，使用的训练数据库是大名鼎鼎的 MNIST。闲话少叙，我们先来整体看一下实例的实现代码。

```
import tensorflow as tf
mnist = tf.keras.datasets.mnist

(x_train, y_train),(x_test, y_test) = mnist.load_data()
x_train, x_test = x_train / 255.0, x_test / 255.0

model = tf.keras.models.Sequential([
  tf.keras.layers.Flatten(input_shape=(28, 28)),
```

```
  tf.keras.layers.Dense(512, activation=tf.nn.relu),
  tf.keras.layers.Dropout(0.2),
  tf.keras.layers.Dense(10, activation=tf.nn.softmax)
])
model.compile(optimizer='adam',
          loss='sparse_categorical_crossentropy',
          metrics=['accuracy'])

model.fit(x_train, y_train, epochs=5)
model.evaluate(x_test, y_test)
```

代码虽然非常简洁，但是完全包含了整个机器学习的数据准备、模型构建、模型训练和模型应用（评估）的过程。首先 import tensorflow as tf 引入了 TensorFlow 机器学习框架，如果要使用 TensorFlow 进行机器学习，那么这显而易见是必不可少的。接下来就是数据准备的过程，为了更好地说明数据准备过程，我们先来简单介绍数据集合 MNIST。MNIST 是深度学习的经典入门 Demo，它是由 60 000 张训练图片和 10 000 张测试图片构成的，每张图片都是 28×28 像素大小（见图 2.6），而且都是由黑白色构成的（这里的黑色是一个 0~1 的浮点数，黑色越深，表示数值越靠近 1），这些图片是采集的不同的人手写的 0~9 的数字。TensorFlow 将这个数据集和相关操作封装到了库中。

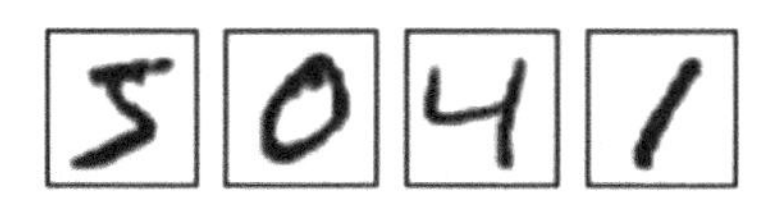

图 2.6 MNIST 图片

图 2.6 是 4 张 MNIST 图片。这些图片并不是传统意义上的 PNG 或者 JPG 格式的，因为 PNG 或者 JPG 格式的图片会带有很多干扰信息（如数据块、图片头、图片尾、长度等），这些图片会被处理成很简易的二维数组，如图 2.7 所示。

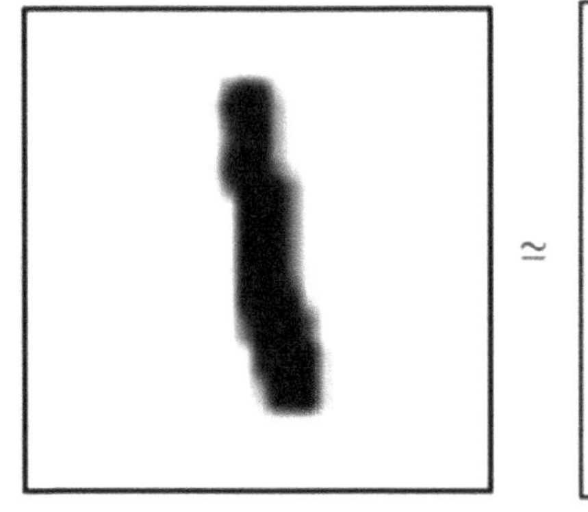

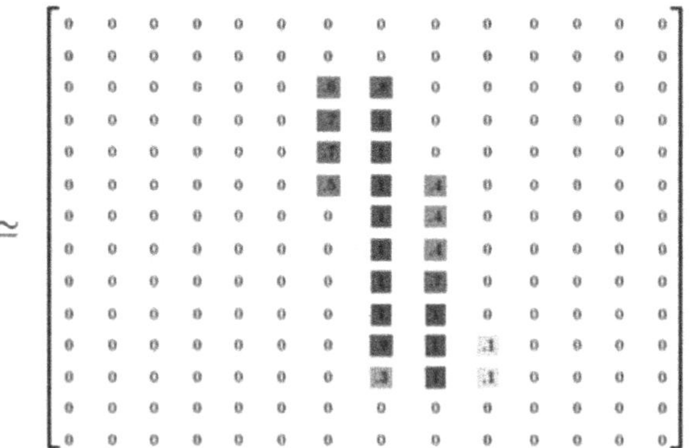

图 2.7 MNIST 图片与二维数组的关系

```
mnist = tf.keras.datasets.mnist
(x_train, y_train),(x_test, y_test) = mnist.load_data()
```

上述代码是数据读取的过程。tf.Keras 将机器学习常用的测试标定数据集合的下载和简单数据处理过程内联设置到 Datasets 的 API 中，包括 MNIST、Cifar10 等。利用数据集合的 load_data()方法可以直接下载对应的数据，并且框架为了能够让人工智能的初学者更快上手，在返回值中直接将训练数据和测试数据区分开，通过 x_train、y_train、x_test、y_test 四个变量一次性保存对应的训练和测试数据的输入和输出值。由于输入值是 28×28 的灰度像素值矩阵，因此利用

```
x_train, x_test = x_train/255.0, x_test/255.0
```

将这些值缩小到 0 到 1 之间，这种手法是在图像识别中经常使用的。

接下来就是模型构建的过程。在进行模型构建之前，我们需要简单介绍 Keras 和 TensorFlow 的关系。Keras 是一个用 Python 编写的高级神经网络 API，它能够以 TensorFlow、CNTK 或者 Theano 作为后端运行。Keras 的开发重点是支持快速的实验，能够以最小的时延把你的想法转换为实验结果。最初的 TensorFlow 是没有直接内嵌 Keras 的 API 的，但是由于使用 Keras 后可以在很大程度上提高开发速度，增加代码的可读性，在应用算法定义阶段受到了很多数据科学家和开发者的欢迎，因此 TensorFlow 在 1.4 版本正式增加了 tf.keras 模块，是 TensorFlow 对 Keras API 规范的实现。这是一个用于构建和训练模型的高阶 API，包含对 TensorFlow 特定功能（例如 Eager Execution、tf.data 管道和 Estimator）的顶级支持。tf.keras 使 TensorFlow 更易于使用，并且不会牺牲灵活性和性能。

```
model = tf.keras.models.Sequential([
  tf.keras.layers.Flatten(input_shape=(28, 28)),
  tf.keras.layers.Dense(512, activation=tf.nn.relu),
  tf.keras.layers.Dropout(0.2),
  tf.keras.layers.Dense(10, activation=tf.nn.softmax)
])
```

Sequential 是 Keras 的线性模型框架，使用这个框架无论是在函数构造的时候填充函数层（Layers）还是使用 add 方法进行填充，结果都是函数层保持线性排列，如图 2.8 所示，而且顺序保持与填充顺序一致。

图 2.8　机器学习线性网络示意图

在 Sequential 传入数组中包含四层，第一层 tf.keras.layers.Flatten(input_shape=(28, 28))的功能是将输入的数据拍平，换句话说，就是将二维数据矩阵转换为一维数据矩阵；第二层 tf.keras.layers.Dense(512, activation=tf.nn.relu) 通过 ReLU 函数进行数据激活；第三层 tf.keras.layers.Dropout(0.2)通过 Dropout 减少网络活性，防止过拟合；第四层 tf.keras.layers.Dense(10, activation=tf.nn.softmax)利用 Softmax 函数输出最终判定。因为本章主要展示 TensorFlow 进行机器学习的基础用法，给大家一个比较直观的认知过程，所以具体函数的实现和推导过程会在后续章节中详细分析。

接下来，使用梯度下降的方法进行训练：model.compile(optimizer='adam', loss='sparse_categorical_crossentropy', metrics=['accuracy'])，compile 函数的 3 个输入值分别设定了梯

度下降使用的优化算法、梯度下降的损失函数以及训练过程的监控指标。在本节展示的例子中，我们使用标准 ADAM 优化算法，使用交叉熵作为损失函数，在例子中会看到网络的预测精确度随着训练而逐步提升，如图 2.9 所示。

```
Epoch 1/5
60000/60000 [==============================] - 27s 450us/sample - loss: 0.2153 - acc: 0.9358
Epoch 2/5
60000/60000 [==============================] - 27s 447us/sample - loss: 0.0953 - acc: 0.9711
Epoch 3/5
60000/60000 [==============================] - 27s 446us/sample - loss: 0.0671 - acc: 0.9788
Epoch 4/5
60000/60000 [==============================] - 26s 440us/sample - loss: 0.0531 - acc: 0.9829
Epoch 5/5
60000/60000 [==============================] - 27s 449us/sample - loss: 0.0425 - acc: 0.9858
10000/10000 [==============================] - 1s 95us/sample - loss: 0.0697 - acc: 0.9798
```

图 2.9　HelloTF 实例训练结果

model.fit(x_train, y_train, epochs=5)是启动训练的过程，通过这样一个简单的函数调用，利用准备好的数据进行模型训练，找到模型的近似最优解。

model.evaluate(x_test, y_test)最后利用测试数据评估模型的预测精确度，通过测试，模型的预测精确度达到了 98%。在 Hello TensorFlow 代码说明的最后需要补充两点：第一，MNIST 模型相对比较简单，其他许多机器学习算法（例如支持向量机）在这个模型上都可以取得不错的预测结果；第二，如果需要将手写数字识别投入实际业务场景中，由于实际业务场景往往和金融、政府服务相关，因此 98%的预测精确度可能还不够，但是不用担心，这只是开端，在后续章节中会介绍能力更加强劲的算法，读者可利用后续知识技能改造 Hello TensorFlow。

2.4　TensorFlow 架构

2.4.1　TensorFlow 架构概述

TensorFlow 运行时是一个跨平台库，图 2.10 展示了它的一般架构，其中的 C API 可将不同语言的用户级代码与核心运行时分开。TensorFlow 在设计时借鉴了 Android 的成功架构，即其生态构建使得层次和背景各异的小微开发者能够简单设计应用，并且易于迁移到类似生态。

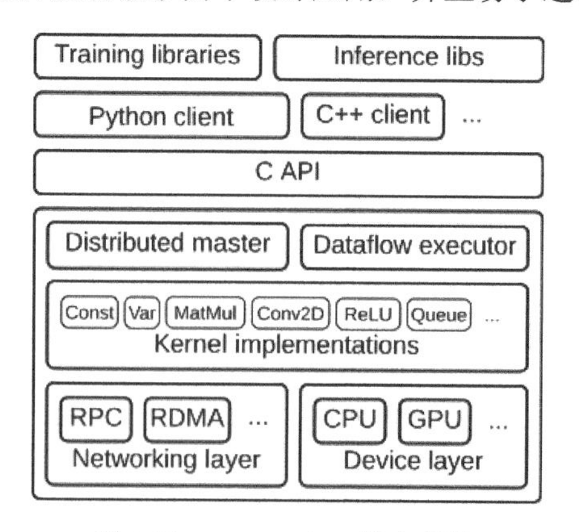

图 2.10　TensorFlow 核心架构

TensorFlow 架构包括以下核心组件。

- 客户端：将计算定义为数据流图和使用会话启动图执行。
- 分布式主服务：根据 Session.run() 的参数定义，从图中裁剪出特定子图；将子图分成在不同进程和设备中运行的多个块；将各个图块分发给工作器服务；通过工作器服务开始执行图块。
- 工作器服务（每项任务一个）：使用适合可用硬件（CPU、GPU 等）的内核实现来安排图操作的执行，向其他工作器服务发送操作结果以及从它们那里接收操作结果。
- 内核实现：对单个图操作执行计算。

图 2.11 说明了这些组件的交互。“/job:worker/task:0”和“/job:ps/task:0”都是工作器服务的任务。PS 代表参数服务器（负责存储和更新模型参数的任务）。其他任务会在优化这些参数时向这些参数发送更新。任务之间的这种特殊分工不是必需的，但对于分布式训练来说很常见。

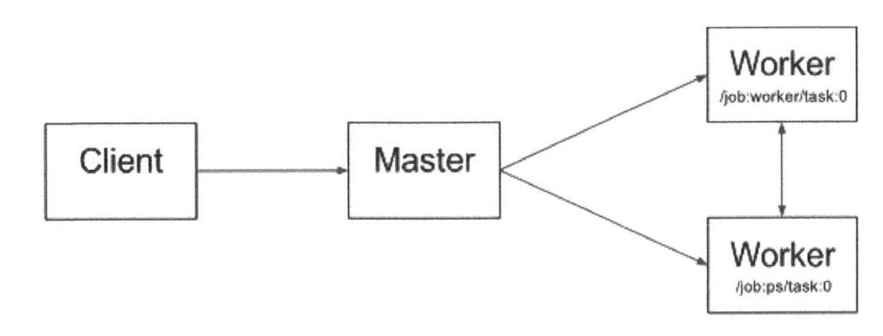

图 2.11 计算操作交互

注意，分布式主服务和工作器服务仅存在于分布式 TensorFlow 中。TensorFlow 的单进程版本包含一个特殊的会话实现，它可以执行分布式主服务执行的所有操作，但只与本地进程中的设备进行通信。

以下各部分将更详细地介绍 TensorFlow 核心层，并逐步说明如何处理示例图。

2.4.2 TensorFlow 客户端架构

用户负责编写可构建计算图的客户端 TensorFlow 程序。该程序可以直接组建各项操作，也可以使用 Keras API 之类的便利库组建神经网络层和其他更高级别的抽象物。TensorFlow 支持多种客户端语言，Python 和 C++是首先支持的，C++是 TensorFlow 框架的实现语言，而 Python 在数据处理工程师群体中有着广泛的使用基础，目前针对 Java 和 JavaScript 的 API 库也日益成熟。随着功能的日益成熟，通常会将它们移植到 C++中，以便用户可以通过所有客户端语言访问经过优化的实现。大多数训练库层面仍然只支持 Python，主要目的是让科研人员更加专注于机器学习算法的构建。

客户端会创建一个会话，以便将图定义作为 tf.GraphDef 协议缓冲区发送到分布式主服务。当客户端评估图中的一个或多个节点时，会调用分布式主服务来启动计算。

在图 2.12 中，客户端构建了一个图，将权重（w）应用于特征向量（x），添加偏差项（b）并将结果保存在变量（s）中。

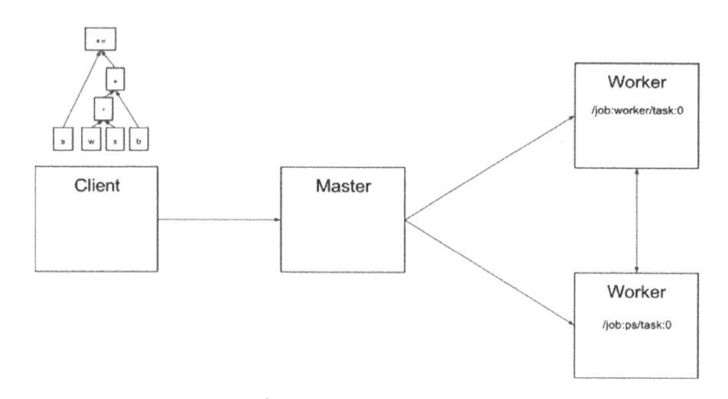

图 2.12　客户端构建图的过程

2.4.3 TensorFlow 分布式主服务架构

分布式主服务包含图的低阶操作，包括：对图进行裁剪以获得评估客户端请求的节点所需的子图，划分图以获取每台参与设备的图块，以及缓存这些块以便可以在后续步骤中重复使用。

由于主服务可以看到某一步的总体计算，因此它会应用常见的子表达式消除或常量折叠等标准优化。然后，它会在一组任务中协调经过优化的子图的执行，如图 2.13 所示。

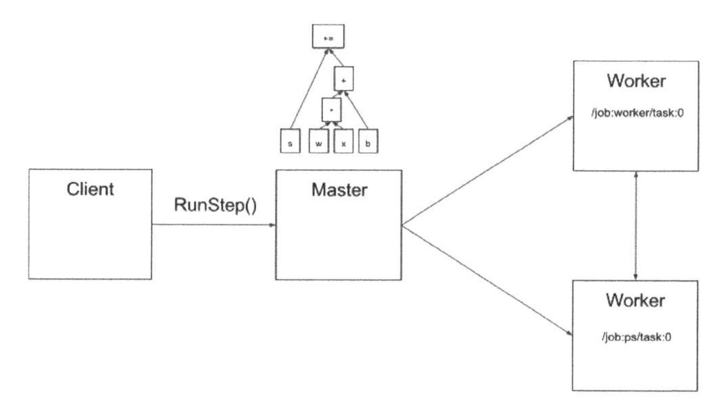

图 2.13　主服务执行计算图开始状态

图 2.14 显示了示例图的可能划分。分布式主服务已对模型参数进行分组，以便将它们一起放在参数服务器上。

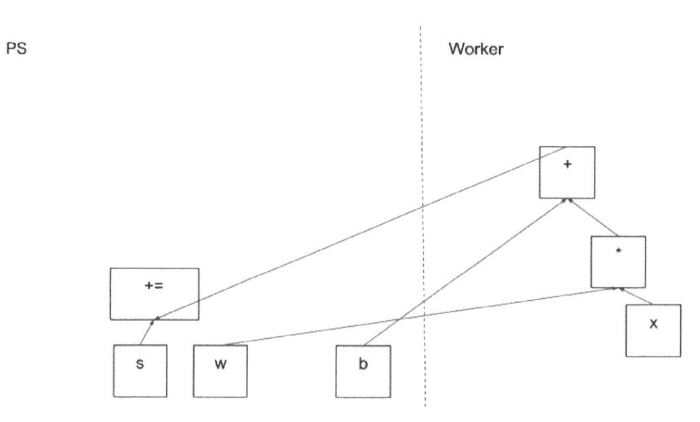

图 2.14　模型参数分组的过程

如果图边缘在划分时被切割掉，分布式主服务就会插入发送和接收节点，以便在分布式任务之间传递信息（见图 2.15）。

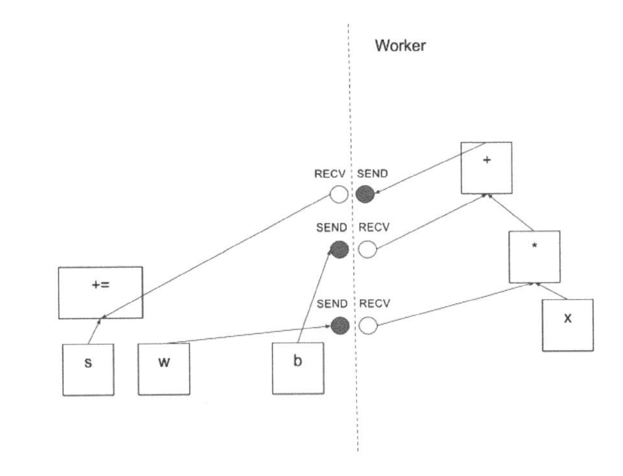

图 2.15　任务之间传递信息的过程

然后，分布式主服务会将各个图块传送到分布式任务，如图 2.16 所示。

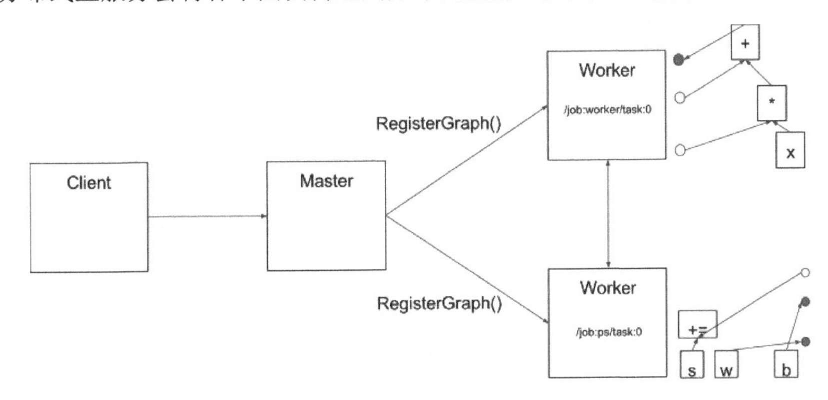

图 2.16　计算图到达分布式任务工作中心

2.4.4　TensorFlow 工作器服务架构

在工作器服务架构中，每项任务中的工作器服务会处理来自主服务的请求，为包含本地子图的操作安排内核的执行时间，以及调解任务之间的直接通信。

我们会优化工作器服务，以便以较低的开销运行大型图。当前的实现每秒可以执行数万个子图，这可让大量副本生成快速、精细的训练步。工作器服务会将各个内核分派给本地设备并尽可能并行运行这些内核，例如通过使用多个 CPU 核或 GPU 流。

本地 CPU 和 GPU 设备之间的传输使用 cudaMemcpyAsync() API 来重叠计算和数据传输，如图 2.17 所示。两个本地 GPU 之间的传输使用对等 DMA，以避免通过主机 CPU 进行代价高昂的复制。对于任务之间的传输，TensorFlow 会使用多种协议，包括：基于 TCP 的 gRPC 和基于聚合以太网的 RDMA。此外，还提供对 NVIDIA 的 NCCL 库的初步支持，以进行多 GPU 通信。

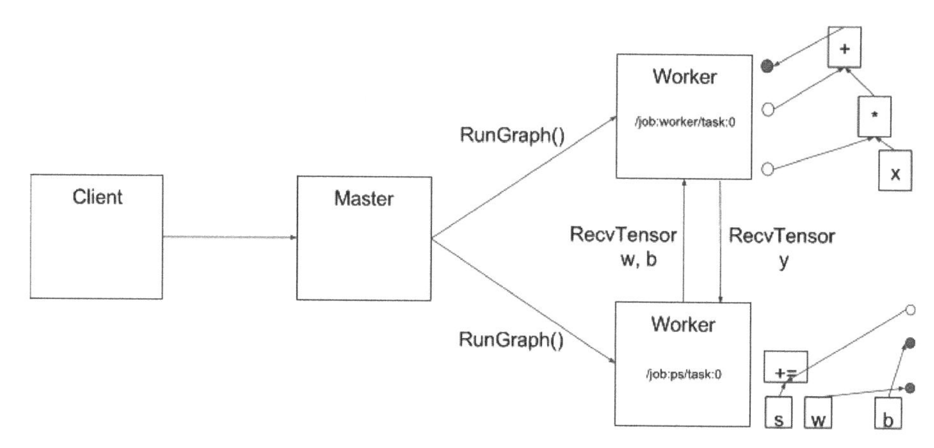

图 2.17　计算图在分布式任务中进行运算和同步

2.4.5　TensorFlow 内核架构

内核是 TensorFlow 真正的能力所在，运行时包含 200 多个标准操作，包括数学、数组、控制流和状态管理操作。其中，每项操作都具有针对各种设备优化的内核实现。许多操作内核都是使用 Eigen::Tensor 实现的，后者使用 C++模板为多核 CPU 和 GPU 生成高效的并行代码。不过，我们可以随意使用 cuDNN 实现更高效内核的库。另外，还实现了量化，能够在移动设备和高吞吐量数据中心应用等环境中实现更快的推断，并使用 GEMMLOWP 低精度矩阵库加快量化计算。

如果将子计算表示为操作组合很困难或效率低下，那么用户可以注册提供高效实现（用 C++编写）的其他内核。例如，建议为一些对性能要求苛刻的操作注册自己的混合内核，例如 ReLU 和 S 型激活函数及其对应的梯度。另外，XLA 编译器具有自动内核混合的实验性实现。

2.5　TensorFlow 核心 API

TensorFlow 中的 API 主要是围绕利用 TensorFlow 框架进行机器学习这一核心目标构建的，包括模型构建、学习过程控制、数据输入输出、数据预处理等功能，本节的主要目的是进一步展示以 Keras 为主的 TensorFlow 高阶 API 的主要功能和能力，另一方面也会介绍 TensorFlow 低阶 API 中的张量等核心概念。讲解低阶 API 主要基于两点考虑：一方面高阶 API 将一些参数和过程封装于框架本身，利用标准或者平均最优进行了固定，这样做虽然能够满足基本的业务需要，但是在实际业务场景中有可能多有掣肘；另一方面，前面介绍过 Keras 是自 1.4 版本才正式内置于 TensorFlow 中的，很多早期资料并没有使用 Keras API，而是使用了 TensorFlow 的低阶 API，因此掌握低阶 API 也是非常重要的。同时，本书会在后续例子中使用低阶 API，甚至使用 NumPy API 完成一些例子，以使读者能够全面掌握 Python 机器学习开发技术。

2.5.1　TensorFlow 低级 API

1. 张量

张量（Tensor）是 TensorFlow 中的核心数据表示方案，任何模型的构建起点都会构建模型的输入输出张量。一个张量由一组形成阵列（任意维数）的原始值组成。张量的阶是它的维数，而它的形式是一个整数元组，指定了阵列每个维度的长度。比如[[[1., 2., 3.]], [[7., 8., 9.]]]是一个 3 阶张量，张量的形式可以用[2, 1, 3]数组来表示。

2. 图和会话

利用 TensorFlow 低阶 API 进行机器学习实践可以分为两部分：第一部分是机器学习模型的设计，这一部分使用的技术就是图（Graph），又称计算图；第二部分是运行设计模型，得到模型的最优解，这一部分使用的技术是会话（Session）。打个比方，比如我们要确定一条直线，第一步是写出直线方程：Y=kX+b，第二步是利用(X,Y)数据集求出 k、b 的值。TensorFlow 框架中的这种设计保持了和机器学习数学过程的一致性，使开发者更加容易理解并实践。

仔细分析计算图，我们可以认为计算图是排列成一个图的一系列 TensorFlow 指令。计算图中含有两种类型的对象：一种是操作（Operator），表示图的节点，描述了消耗和生成张量的计算；第二种是张量（Tensor），表示图的边缘节点，代表将流经图的值。简单来说，操作是过程中的节点，张量是输入输出节点。

要评估张量，需要实例化一个 tf.Session 对象（非正式名称为会话）。会话会封装 TensorFlow 运行时的状态，并运行 TensorFlow 操作。如果说 tf.Graph 像一个 .py 文件，那么 tf.Session 就像一个 Python 可执行对象。

3. 层

构建模型的时候要决定算法的运行方法，在 TensorFlow 模型构建的过程中，你会有一种直观感受，这个过程非常像我们小时候堆叠积木，堆叠的原材料在 TensorFlow 中称为层（Layer），不同层的串联结构是当前机器学习中最常用的，同时也是最有效的架构方式，我们将在高阶 API 中使用适当的方式来描述这个过程，也就是在 Hello TensorFlow 示例中使用的 Sequential。层将变量和作用于它们的操作打包在一起，例如，密集连接层会对每个输出对应的所有输入执行加权和，并应用激活函数（可选）。连接权重和偏差由层对象管理。在一个完整的机器学习层模型的设计使用全流程中，一般会包含层创建、层初始化和层执行 3 个核心阶段。

（1）创建层

下面的代码会创建一个 Dense 层，该层会接收一批输入矢量，并为每个矢量生成一个输出值。要将层应用于输入值，请将该层当作函数来调用，例如：

```
x = tf.placeholder(tf.float32, shape=[None, 3])
linear_model = tf.layers.Dense(units=1)
y = linear_model(x)
```

层会检查输入数据，以确定其内部变量的大小。因此，我们必须在这里设置 x 占位符的形式，以便层构建正确大小的权重矩阵。

我们现在已经定义了输出值 y 的计算，在运行计算之前，还需要处理一个细节，即初始化层。

（2）初始化层

层包含的变量必须先初始化，然后才能使用。尽管可以单独初始化各个变量，但也可以轻松地初始化一个 TensorFlow 图中的所有变量：

```
init = tf.global_variables_initializer()
sess.run(init)
```

调用 tf.global_variables_initializer 仅会创建并返回 TensorFlow 操作的句柄。当我们使用 tf.Session.run 运行该操作时，该操作将初始化所有全局变量。需要特别注意的一点是，global_variables_initializer 仅会初始化创建初始化程序时图中就存在的变量。因此，应该在构建图表的最后一步添加初始化程序。

（3）执行层

我们现在已经完成了层的初始化，可以像处理任何其他张量一样评估 linear_model 的输出张量。例如下面的代码：

```
print(sess.run(y, {x: [[1, 2, 3],[4, 5, 6]]}))
```

会生成一个两个元素的输出向量：

```
[[-3.41378999]
 [-9.14999008]]
```

4. 可视化学习过程

虽然机器学习的 TensorFlow 的 Python 代码看上去相当不错，但是我们依然有让过程和数据更加直观的冲动，因为大脑对于形象理解本身就是过分溺爱的，更别说有了可视化过程，我们可以在很短的时间内把自己的想法解释清楚。当然，谷歌也是深谙此道，因此很早版本的 TensorFlow 就内置了一个学习过程可视化的框架 TensorBoard。

首先将计算图保存为 TensorBoard 摘要文件，具体操作如下：

```
writer = tf.summary.FileWriter('.')
writer.add_graph(tf.get_default_graph())
```

这将在当前目录中生成一个 events 文件，名称格式如下：

```
events.out.tfevents.{timestamp}.{hostname}
```

现在，在新的终端中使用以下 Shell 命令启动 TensorBoard：

```
tensorboard --logdir .
```

接下来，在浏览器中打开 TensorBoard 的图页面，应该会看到与图 2.18 类似的图。

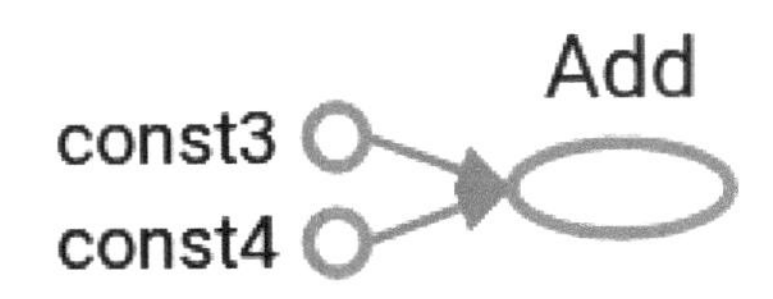

图 2.18　TensorBoard 张量相加的计算图

此外，我们还能够通过 TensorBoard 将记录的运算中间数据可视化，通过图表的形式更加容易地对数据趋势做出准确判断。具体事例会在下面的综合实例中进行展示。

5. 低级 API 综合实例

```
import tensorflow as tf

a = tf.constant(3.1, name='a')
x = tf.placeholder(tf.float32, name='x')
w = tf.Variable(2.0, name='w')
result = tf.add(a, w*x)
tf.summary.scalar('result', result)

sess = tf.Session()
merged = tf.summary.merge_all()
init = tf.global_variables_initializer()
graph = sess.graph
writer = tf.summary.FileWriter('./log', graph)

sess.run(init)
for i in range(100):
    summary_, result_ = sess.run([merged, result], feed_dict={x:float(i)})
    writer.add_summary(summary_, i)

sess.close()
```

这个例子中首先构建了一个 ax+b 的表达式，然后通过 TensorFlow 会话进行了计算，实例中通过 TensorFlow 的 summary 模块进行可视化跟踪。程序运行完成后，会在当前目录建立一个名称为 log 的文件夹，在该文件夹下会存储本次运算的日志文件，日志文件名称包含运行时间戳和运行机器名称。通过 tensorboard –logdir 命令在命令行中启动 TensorBoard，启动完成之后打开浏览器（最好是 Chrome，其他浏览器特别是 IE 浏览器会有很多显示问题），在地址栏输入 http://localhost:6006 就能够看到代码中存储的计算图结构和 result 张量变化的情况，具体如图 2.19 和图 2.20 所示。

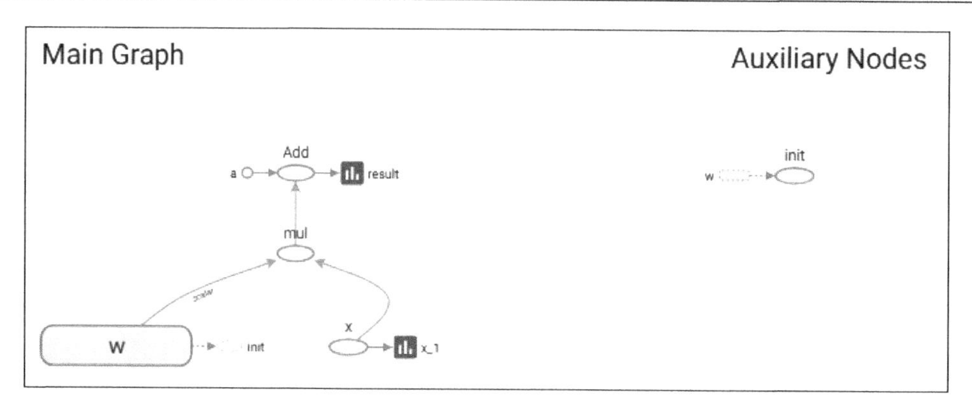

图 2.19　利用 TensorBoard 展示的计算图模型

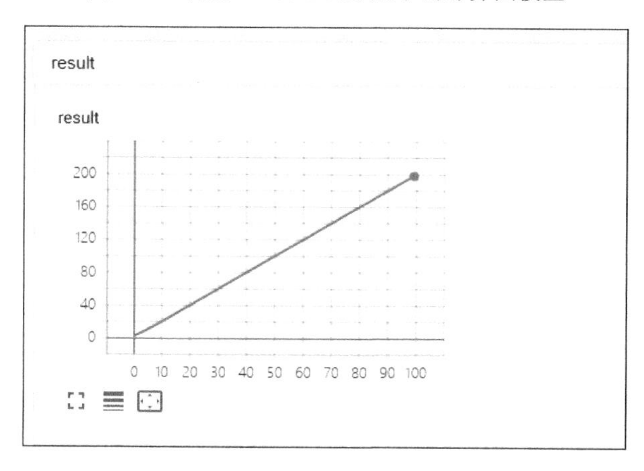

图 2.20　利用 TensorBoard 展示的 result 张量的变化情况

下面我们来分析一下低阶 API 实例中的关键代码。按照我们对于机器学习的理解，代码实际上分为模型构建和模型运行两个阶段，在模型构建阶段应该还可以分为运行准备和实际运行两个阶段。模型构建阶段的代码如下：

```
a = tf.constant(3.1, name='a')
x = tf.placeholder(tf.float32, name='x')
w = tf.Variable(2.0, name='w')
result = tf.add(a, w*x)
```

在这里使用 TensorFlow 的常量、变量和占位符，通过 add 函数和乘法运算符完成了模型的最终构建。接下来是运行准备阶段，这个阶段的主要工作就是实例化会话对象，并且完成张量的初始化。其代码如下：

```
sess = tf.Session()
init = tf.global_variables_initializer()
```

最后，Session（会话）的 run 方法调用就是模型的实际运行，在这里要特别注意的是，我们是通过 feed_dict 这种类字典的简单数据集调用方式完成训练数据的输入的：

```
sess.run(init)
```

```
for i in range(100):
summary_, result_ = sess.run([merged, result], feed_dict={x:float(i)})
sess.close()
```

如果仔细对照代码，就会发现有一些代码我们跳过了，这些代码就是使用 Summary 进行可视化 log 数据记录的过程，包括可视化 log 数据记录的定义：

```
tf.summary.scalar('result', result)
```

可视化 log 数据记录的整合初始化：

```
merged = tf.summary.merge_all()
```

计算图的存储：

```
graph = sess.graph
writer = tf.summary.FileWriter('./log', graph)
```

可视化 log 数据记录的整合初始化的存储：

```
writer.add_summary(summary_, i)
```

2.5.2 TensorFlow 高级 API

Keras 的核心数据结构是 Model，是一种组织网络层的方式。最简单的模型是 Sequential 顺序模型，它由多个网络层线性堆叠。对于更复杂的结构，应该使用 Keras 函数式 API，它允许构建任意的神经网络图。

Keras 有以下三个优点：

第一，用户友好。 Keras 是为人类而不是为机器设计的 API。它把用户体验放在首要和中心位置。Keras 遵循减少认知困难的最佳实践：它提供一致且简单的 API，将常见用例所需的用户操作数量降至最低，并且在用户错误时提供清晰和可操作的反馈。

第二，模块化能力强。模型被理解为由独立的、完全可配置的模块构成的序列或图。这些模块可以尽可能少地限制组装在一起。特别是神经网络层、损失函数、优化器、初始化方法、激活函数、正则化方法，它们都是可以结合起来构建新模型的模块。

第三，易扩展性好。新的模块是很容易添加的（作为新的类和函数），现有的模块已经提供了充足的示例。由于能够轻松地创建可以提高表现力的新模块，因此 Keras 更加适合高级研究。

下面将展示使用 tf.keras 模块的一些配置和技巧。

1. 导入 tf.keras

tf.keras 是 TensorFlow 对 Keras API 规范的实现。这是一个用于构建和训练模型的高阶 API，包含对 TensorFlow 特定功能（例如 Eager Execution、tf.data 管道和 Estimator）的顶级支持。tf.keras 使 TensorFlow 更易于使用，并且不会牺牲灵活性和性能。

导入 tf.keras 以设置 TensorFlow 程序：

```
import tensorflow as tf
from tensorflow.keras import layers

print(tf.VERSION)
print(tf.keras.__version__)
```

在终端打印结果：

```
1.11.0
2.1.6-tf
```

tf.keras 可以运行任何与 Keras 兼容的代码，但请注意：

- 最新版 TensorFlow 中的 tf.keras 版本可能与 PyPI 中的最新 Keras 版本不同。请查看 tf.keras.version。
- 保存模型的权重时，tf.keras 默认采用检查点格式。请传递 save_format='h5'以使用 HDF5。

2. 构建简单的模型

在 Keras 中，你可以通过组合层来构建模型。模型（通常）是由层构成的图，最常见的模型类型是层的堆叠：tf.keras.Sequential 模型。

例如，构建一个简单的全连接网络（多层感知器），可运行以下代码：

```
model = tf.keras.Sequential()
# Adds a densely-connected layer with 64 units to the model:
model.add(layers.Dense(64, activation='relu'))
# Add another:
model.add(layers.Dense(64, activation='relu'))
# Add a softmax layer with 10 output units:
model.add(layers.Dense(10, activation='softmax'))
```

3. 配置层

我们可以使用很多 tf.keras.layers，它们具有一些相同的构造函数参数。

- activation：设置层的激活函数。此参数由内置函数的名称指定，或指定为可调用对象。默认情况下，系统不会应用任何激活函数。
- kernel_initializer 和 bias_initializer：创建层权重（核和偏差）的初始化方案。此参数是一个名称或可调用对象，默认为 Glorot uniform 初始化器。
- kernel_regularizer 和 bias_regularizer：应用层权重（核和偏差）的正则化方案，例如 L1 或 L2 正则化。默认情况下，系统不会应用正则化函数。

以下代码使用构造函数参数实例化 tf.keras.layers. Dense 层：

```
# Create a sigmoid layer:
layers.Dense(64, activation='sigmoid')
# Or:
```

```
layers.Dense(64, activation=tf.sigmoid)

# A linear layer with L1 regularization of factor 0.01 applied to the kernel
matrix:
layers.Dense(64, kernel_regularizer=tf.keras.regularizers.l1(0.01))

# A linear layer with L2 regularization of factor 0.01 applied to the bias vector:
layers.Dense(64, bias_regularizer=tf.keras.regularizers.l2(0.01))

# A linear layer with a kernel initialized to a random orthogonal matrix:
layers.Dense(64, kernel_initializer='orthogonal')

# A linear layer with a bias vector initialized to 2.0s:
layers.Dense(64, bias_initializer=tf.keras.initializers.constant(2.0))
```

4. 训练和评估

构建好模型后，可通过调用 compile 方法配置该模型的学习流程，代码如下：

```
model = tf.keras.Sequential([
# Adds a densely-connected layer with 64 units to the model:
layers.Dense(64, activation='relu'),
# Add another:
layers.Dense(64, activation='relu'),
# Add a softmax layer with 10 output units:
layers.Dense(10, activation='softmax')])

model.compile(optimizer=tf.train.AdamOptimizer(0.001),
              loss='categorical_crossentropy',
              metrics=['accuracy'])
```

tf.keras.Model.compile 采用了以下 3 个重要参数。

- optimizer：此对象会指定训练过程。从 tf.train 模块向其传递优化器实例，例如 tf.train.AdamOptimizer、tf.train.RMSPropOptimizer 或 tf.train.GradientDescentOptimizer。
- loss：要在优化期间最小化的函数。常见选择包括均方误差（Mean Squared Error，MSE）、categorical_crossentropy 和 binary_crossentropy。损失函数由名称或通过从 tf.keras.losses 模块传递可调用对象来指定。
- metrics：用于监控训练。它是 tf.keras.metrics 模块中的字符串名称或可调用对象。

以下代码展示了配置模型以进行训练的两个示例：

```
# Configure a model for mean-squared error regression.
model.compile(optimizer=tf.train.AdamOptimizer(0.01),
              loss='mse',       # mean squared error
              metrics=['mae'])  # mean absolute error
```

```
# Configure a model for categorical classification.
model.compile(optimizer=tf.train.RMSPropOptimizer(0.01),
          loss=tf.keras.losses.categorical_crossentropy,
          metrics=[tf.keras.metrics.categorical_accuracy])
```

2.6 扩展：使用 tensorflow.js 进行机器学习

tensorflow.js 是一个机器学习的前端框架，Google 也在 GitHub 开源了相关代码。GitHub 地址：https://github.com/tensorflow/tfjs。在实现方面，TensorFlow 团队使用了 WebGL 库对运算过程进行优化，使得 tensorflow.js 在学习尤其是网络扩大的时候能够有更好的性能表现。在 API 设计方面，框架更多地考量到了开发人员的易用性，在较为底层的 API 方面使用了 TensorFlow Python 的许多概念，而在高级抽象 API 方面则更多地与 Keras 框架保持一致。

接下来我们通过一个简单的例子了解一下 tensorflow.js 的魅力。

1. 类库引入

（1）script 标签引入

标签的引入是最为直接的方式，引入的地址为 https://cdn.jsdelivr.net/npm/@tensorflow/tfjs@0.9.0。下面给大家提供一个简单的开发模板：

```
<html>
  <head>
    <!-- 引入 tensorflow.js 类库 -->
    <script src="https://cdn.jsdelivr.net/npm/@tensorflow/tfjs@0.9.0">
</script>

    <!-- 在下面的 script 标签里面写机器学习代码-->
    <script>
机器学习代码
    </script>
  </head>

  <body>
  </body>
</html>
```

（2）npm 引入

如果你使用了 Node 进行前端架构的开发，就需要包管理工具 npm 来引入。

```
npm install @tensorflow/tfj
```

下面给大家提供一个简单的开发模板（ES 6）：

```
import * as tf from '@tensorflow/tfjs';
// 在下面写机器学习业务代码
```

2. hello tfjs —— 一个简单的示例

（1）代码编写

```
// 定义模型：线性回归模型
const model = tf.sequential();
model.add(tf.layers.dense({units: 1, inputShape: [1]}));

// 定义模型损失函数和梯度下降算法
model.compile({loss: 'meanSquaredError', optimizer: 'sgd'});

// 准备学习数据
const xs = tf.tensor2d([1, 2, 3, 4], [4, 1]);
const ys = tf.tensor2d([1, 3, 5, 7], [4, 1]);

//模型学习
model.fit(xs, ys).then(() => {
  // 使用训练完成的模型进行预测
  model.predict(tf.tensor2d([5], [1, 1])).print();
});
```

（2）代码分析

代码中具体做的事情是线性回归分析，步骤总结为：模型定义→模型学习→模型使用。如果大家想要深入了解线性回归分析的内容，可以参考笔者的免费课程，课程链接：https://www.imooc.com/learn/972。

（3）运行结果

在浏览器上运行，在命令行中就能看到想要的输出，如图 2.21 所示。

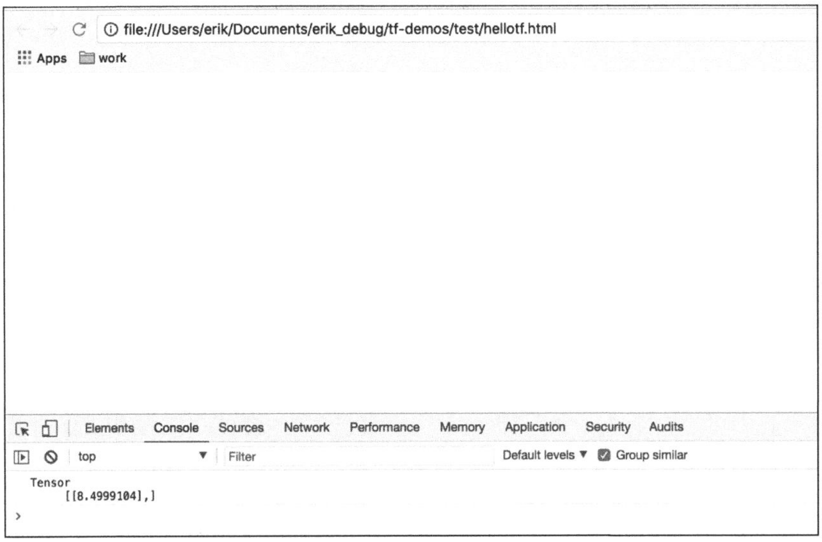

图 2.21 代码运行结果

近年来，随着前端框架（React、Vue、Angular）的崛起和微信小程序的发力，前端从业人员的开发能力得到了长足的进步，人工智能时代不但给予后台通关前后台的能力，而且也给了前端业务更多的想象力，tensorflow.js 就是在这样的环境下应运而生的产物。我们通过上述的入门例子对 tensorflow.js 有了直观的感受，如果你学习了 TensorFlow 的核心知识，那么上手 tensorflow.js 将会非常容易。

参考文献

[1]. Park S J. Evaluation of Deep Learning Frameworks Over Different HPC Architectures[C]// IEEE International Conference on Distributed Computing Systems. 2017.

[2]. TensorFlow: Large-Scale Machine Learning on Heterogeneous Distributed Systems[J]. 2016.

[3]. Yuan T. TF.Learn: TensorFlow's High-level Module for Distributed Machine Learning[J]. 2016.

[4]. Bahrampour S, Ramakrishnan N, Schott L, et al. Comparative Study of Deep Learning Software Frameworks[J]. Computer Science, 2015.

[5]. Vishnu A, Siegel C, Daily J. Distributed TensorFlow with MPI[J]. 2016.

[6]. Andor D, Alberti C, Weiss D, et al. Globally Normalized Transition-Based Neural Networks[J]. 2016.

[7]. Abadi, Martín, Barham P, Chen J, et al. TensorFlow: A system for large-scale machine learning[J]. 2016.

[8]. Jia Y, Shelhamer E, Donahue J, et al. Caffe: Convolutional Architecture for Fast Feature Embedding[J]. 2014.

第 3 章

线性回归

线性回归是最简单的机器学习模型，也是机器学习算法学习的最好起点。在机器学习中，线性回归（Linear Regression）是利用线性回归方程的最小平方函数对一个或多个自变量和因变量之间的关系进行建模的一种回归分析。这种函数是一个或多个回归系数的模型参数的线性组合，只有一个自变量的情况称为简单回归，大于一个自变量情况的叫作多元回归。

本章主要介绍线性回归模型以及通过机器学习进行模型求解的过程。内容分为五部分：第一部分着重介绍线性回归的模型定义；第二部分分析线性回归中使用最小二乘法模型求解模型的数学过程；第三部分是线性回归中使用最小二乘法模型求解模型的代码实例；第四部分分析线性回归中使用梯度下降模型求解模型的数学过程；第五部分是线性回归中使用梯度下降模型求解模型的代码实例。

3.1 什么是线性回归

本节主要介绍线性回归的概念、历史与模型。

3.1.1 线性回归的概念

线性回归从词法构成来说包括“线性”和“回归”两部分内容。“线性”是指线性关系，对于最简化的场景而言，两个变量之间存在一次方函数关系，就称它们之间存在线性关系。通俗一点讲，如果把这两个变量分别作为点的横坐标与纵坐标，其图像是平面上的一条直线，则这两个变量之间的关系就是线性关系。即如果可以用一个二元一次方程来表达两个变量之间的关系，这两个变量之间的关系就称为线性关系，因而，二元一次方程也称为线性方程。推而广之，含有 n 个变量的一次方程也称为 n 元线性方程，不过这已经与直线没有什么关系了。因此，我们需要使用向量来表述这种一般性的线性关系。给定向量组 A：α1,α2,...,αn，以及向量 b，若存在一组数 $k_1,k_2,\ldots,k_n$，使得 b= k_1α1+ k_2α2+...+ knαn，则称向量 b 可由向量组 A 线性表示，也称向量 b 是向量组 A 的一个线性组合，$k_1,k_2,\ldots,k_n$ 称为这个线性组合的系数。

3.1.2 线性回归的历史

线性回归中的“回归”实际是一个颇具争议的名称，这个名称的提出者是高尔顿（Frramcia Galton，1882－1911 年）。高尔顿早年在剑桥大学学习医学，但医生的职业对他并无吸引力，后来他接受了一笔遗产，这使他可以放弃医生的生涯，并于 1850－1852 年期间去非洲考察，他所取得的成就使其在 1853 年获得了英国皇家地理学会的金质奖章。此后，他研究过多种学科（气象学、心理学、社会学、教育学和指纹学等），在 1865 年后他的主要兴趣转向遗传学，这也许是受他表兄达尔文的影响。高尔顿开始思考父代和子代相似，如身高、性格及其他种种特质的相似性问题。于是他选择了父母平均身高 X 与其子身高 Y 的关系作为研究对象。他观察了 1 074 对父母及每对父母的一个儿子，将结果描成散点图，发现趋势近乎一条直线。总的来说，父母平均身高 X 增加时，其子的身高 Y 也倾向于增加，这是意料中的结果。但有意思的是，高尔顿发现这 1 074 对父母平均身高的平均值为 68 英寸（英国计量单位，1 英寸=2.54cm）时，1 074 个儿子的平均身高为 69 英寸，比父母平均身高高 1 英寸。于是他推想：当父母平均身高为 64 英寸时，1 074 个儿子的平均身高应为 64+1=65 英寸；当父母的身高为 72 英寸时，他们儿子的平均身高应为 72+1=73 英寸，但观察结果与此不符。高尔顿发现前一种情况是儿子的平均身高为 67 英寸，高于父母平均值达 3 英寸，后者儿子的平均身高为 71 英寸，比父母的平均身高低 1 英寸。高尔顿研究后得出的解释是自然界有一种约束力，使人类身高在一定时期是相对稳定的。现代遗传学研究表明：基因遗传是决定身高的主要因素，表现为多基因遗传。若父母身高比较高（或矮），其子女比他们更高（矮），则人类身材将向高、矮两个极端分化。自然界不这样做，它让身高有一种回归到中心的作用。例如，父母平均身高 72 英寸，这超过了平均值 68 英寸，表明这些父母属于高的一类，其儿子也倾向于高的一类(其平均身高 71 英寸，大于子代的平均身高 69 英寸)，但不像父母离子代那么远((71-69) <（72-68））。反之，父母平均身高 64 英寸，属于矮的一类，其儿子也倾向于矮的一类（其平均身高为 67 英寸，小于子代的平均身高 69 英寸），但不像父母离中心那么远（(69 -67)<(68-64)）。因此，身高有回归于中心的趋势，由于这个性质，高尔顿创立了“回归”并应用到问题的讨论中，这就是“回归”名称的由来。回归分析研究的是多个变量之间的关系。它是一种预测性的建模技术，研究的是因变量（目标）和自变量（预测器）之间的关系。这种技术通常用于预测分析、时间序列模型以及发现变量之间的因果关系。

3.1.3 线性回归模型

线性回归比较严格的定义是数据集 D，样本有 n 个属性进行描述，在数据集内输入数据（X）和标签（Y），对应的关系可以表示为（X1,X2,X3,...,Xn）~Y，我们试图找到或求得一种关系，这种关系是线性的，能够使输入 X 得到 Y。换一种表述方法，就是我们会找到一组输入变量的系数，能够完成输入变量的线性方程。因此，线性（多元）回归可以表述为如下表达式：

$$Y = \sum_{n}^{1} \theta_n X_n + \theta_0 = \sum_{n}^{0} \theta_n X_n$$

在上面的算术表达式中，为了完成最终加和公式，我们进行了合理假设，输入变量的 $X_0 = 1$。在这里面临一个问题，X 有多个要素，而数据集中理所当然含有多条数据，为了能够准确地把这样的二维数据结构描述清楚，我们需要引入向量运算。向量运算如下：

$$\theta = [\theta^0, \cdots, \theta^n]$$

$$X = [X^0, \cdots, X^n]$$

$$Y = [\theta^T X]$$

线性回归中的线性模型虽然比较简洁，但是是机器学习过程中一个非常好的起点，特别是线性模型非常直观地体现了模型本身的可解释性。例如，我们要来判断一个主播是不是女装大佬，可从声音、肤质和体型方面进行判断，如果最后得到的线性模型是：$f_{\text{女装大佬}} = 0.2 \cdot x_{\text{肤质}} + 0.6 \cdot x_{\text{声音}} + 0.2 \cdot x_{\text{体型}} + 1$，说明判断依据中的声音是决定性的。当然这种情况下，如果某个男士会假嗓，也会很容易骗过群众的耳朵。

3.2　线性回归中的最小二乘法模型

在 3.1 节线性回归的概念学习中，我们明确了线性回归的模型定义和数学表达式，接下来的任务是利用数据集求解出模型参数。首先我们希望利用公式推导来解出表达式的最优解。这种解决方案就是著名的最小二乘法。所谓最小二乘，其实也可以叫作最小平方和，其目的就是最小化预测值或拟合值与实际观测值之间的偏差平方和，使得拟合对象无限接近目标对象。换句话说，最小二乘法可以用于对函数的拟合。在线性回归中的最小二乘法应该准确地称为线性最小二乘。为了能够更加清晰地反映最小二乘法的实践过程，我们从简单的线性回归模型、一元线性回归模型（示意图见图 3.1）出发展开线性分析求解过程。

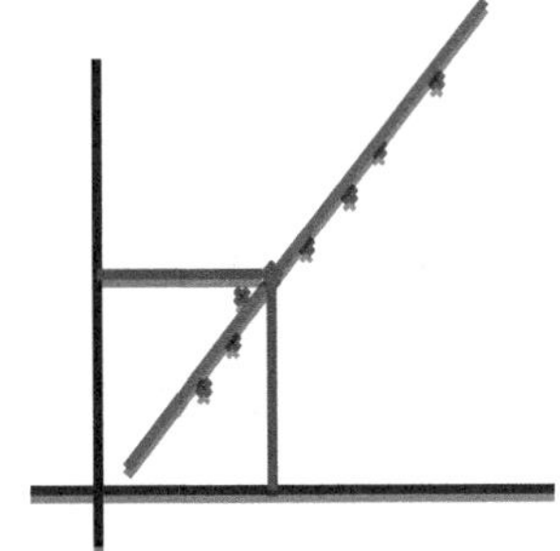

图 3.1　一元线性回归模型示意图

一元线性回归只输入一个要素 X，回归方程 $Y=\Theta X$ 可转化成 $Y=aX+b$，这个形式是标准直线方程，a（Θ_1）为直线的斜率，b（Θ_0）是直线的截距。如图 3.1 所示，我们的目标就是在二维坐标系中确认一条直线，使得各个数据点到直线的投影距离和最小。对于一元线性回归来说，可以看成 Y 的值是随着 X 的值变化的，每一个实际的 X 都会有一个实际的 Y 值，叫作 Y 实际，我们要求出一条直线，每一个实际的 X 都会有一个直线预测的 Y 值，叫作 Y 预测。所以（Y 实际-Y 预测）就可以写成（Y 实际-（aX 实际+b）），于是平方和可以写成 a 和 b 的函数。只需要求出让 Q 最小的 a 和 b 的值，回归线就求出来了。

首先，一元函数最小值点的导数为零，比如 Y=X^2，X^2 的导数是 2X，令 2X=0，求得 X=0 的时候，Y 取最小值。因此，对于函数 Q，分别对 a 和 b 求偏导数，然后令偏导数等于 0，就可以得到一个关于 a 和 b 的二元方程组，进而求出 a 和 b，这个方法被称为最小二乘法。下面是具体的数学演算过程。

先把公式展开一下：

$$
\begin{aligned}
&\mathrm{Q}（a,b）=\sum_{i=1}^{n}(Y_i-(aX_i+b))^2 \\
&=(Y_1-(aX_1+b))^2+(Y_2-(aX_2+b))^2+\ldots+(Y_n-(aX_n+b))^2 \\
&=Y_1^{\ 2}-2Y_1aX_1-2Y_1b+a^2X_1^{\ 2}+2aX_1b+b^2 \\
&+Y_2^{\ 2}-2Y_2aX_2-2Y_2b+a^2X_2^{\ 2}+2aX_2b+b^2 \\
&\ldots \\
&+Y_n^{\ 2}-2Y_naX_n-2Y_nb+a^2X_n^{\ 2}+2aX_nb+b^2 \\
&=(Y_1^{\ 2}+\ldots+Y_n^{\ 2})-2a(Y_1X_1+\ldots+Y_nX_n)-2b(Y_1+\ldots+Y_n) \\
&+a^2(X_1^{\ 2}+\ldots+X_n^{\ 2})+2ab(X_1+\ldots+X_n)+nb^2
\end{aligned}
$$

Q 函数表达式展开后，利用平均数把上面式子中每个括号里的内容进一步化简，例如：

$$(Y_1^{\ 2}+\ldots+Y_{\mathrm{n}}^{\ 2})/n=\overline{Y^2}$$

通过定义 Y2 的平均能够得到：

$$(Y_1^{\ 2}+\ldots+Y_{\mathrm{n}}^{\ 2})=n\overline{Y^2}$$

上式两边乘以 n，于是：

$$\mathrm{Q}(a,b)=n\overline{Y^2}-2an\overline{XY}-abn\overline{Y}+a^2n\overline{X^2}+2abn\overline{X}+nb^2$$

得到 Q 的最终化简结果后，分别对 Q 求 a 的偏导数和 b 的偏导数，令偏导数等于 0。

$$
\begin{aligned}
&\frac{\partial \mathrm{Q}}{\partial a}=-2n\overline{XY}+2an\overline{X^2}+2bn\overline{X}=0 \\
&\frac{\partial \mathrm{Q}}{\partial b}=-2n\overline{Y}+2an\overline{X}+2nb=0
\end{aligned}
$$

Q 分别对 a 和 b 求偏导数，令偏导数为 0。

进一步化简，可以消掉 2n，最后得到关于 a 和 b 的二元方程组：

$$-\overline{XY}+a\overline{X^2}+b\overline{X}=0$$

$$-\overline{Y}+a\overline{X}+b=0$$

通过上述关于 a 和 b 的二元方程组，最后得出 a 和 b 的求解公式：

$$a=\frac{\overline{X}\overline{Y}-\overline{XY}}{(\overline{X})^2-\overline{X^2}}$$

$$b=\overline{Y}-a\overline{X}$$

这样就使用最小二乘法求出了直线的斜率 a 和截距 b。

在一元线性回归的参数求解成功的基础上，我们继续推导多元线性回归的参数集合 θ 的求解方法。首先用矩阵来描述数据集合和对应的标签集合，数据集合可以表示成一个 $m\times$（n+1）大小的矩阵，m 表示数据集合的大小，n 表示数据维度。由于在 3.1 节中将偏置参数（Θ_0）统一化于多元线性回归的向量模型中了，因此数据集中最后一列需要补充恒为 1 的一列数据。X 和 Y 的矩阵如下：

$$X=\begin{bmatrix} x_{11} & \cdots & x_{1n} & 1 \\ x_{21} & \cdots & x_{2n} & 1 \\ \cdots & \cdots & \cdots & \cdots \\ x_{m1} & \cdots & x_{mn} & 1 \end{bmatrix} \quad Y=\begin{bmatrix} y_1 \\ y_2 \\ \cdots \\ y_m \end{bmatrix}$$

接下来，需要求得使预测值或拟合值与实际观测值之间的偏差平方和最小的最优参数解，用数学表达式表述如下：

$$\theta_{best}=\arg\min_{\theta}(Y-X\theta)^T(Y-X\theta)$$

令 $E_\theta=(Y-X\theta)^T(Y-X\theta)$，对 E 进行求导能够得到：

$$\frac{\partial E_\theta}{\partial\theta}=2X^T(X\theta-Y)$$

令上式等于零就可以解出 θ 的最优解：

$$\theta_{best}=(X^TX)^{-1}X^TY$$

其中，$(X^TX)^{-1}$ 为 (X^TX) 矩阵的逆矩阵，因此这里要求 (X^TX) 是满秩矩阵或正定矩阵。

3.3　最小二乘法模型实例

本节展示使用 Python 进行最小二乘法求得线性回归模型的过程，为了使运算简洁，我们使用 NumPy 这个著名的 Python 矩阵和数学运算的利器。NumPy（Numerical Python）是 Python 语言的一个扩展程序库，支持大量的维度数组与矩阵运算，此外也针对数组运算提供大量的数学函数库。

NumPy 的前身 Numeric 最早是由 Jim Hugunin 与其他协作者共同开发的，2005 年，Travis Oliphant 在 Numeric 中结合了另一个同性质的程序库 Numarray 的特色，并加入了其他扩展而开发了 NumPy。NumPy 是开放源代码的，并且由许多协作者共同维护开发。

NumPy 是一个运行速度非常快的数学库，主要用于数组计算，而且包含线性代数、傅里叶变换、随机数生成等功能。NumPy 的程序接口 API 设计十分优秀，兼顾了功能性和易用性两方面，TensorFlow 在 API 设计中也大量借鉴了 NumPy 的 API 设计思想，我们在 TensorFlow 的实例中可以逐渐体会。

```
import numpy as np
from numpy.linalg import inv
from numpy import dot
from numpy import mat

#y=2x
X = mat([1,2,3]).reshape(3, 1)
Y = 2*X


theta = dot(dot(inv(dot(X.T, X)), X.T), Y)
print(theta)
```

在代码中首先引入 NumPy 工具库，大多数语言的工具库引入都支持两种方式：整包引入或者部分引入。整包引入会包含整个 API 功能集合，引入的时候一般会指定别名，例如 import numpy as np，整包引入了 NumPy 并指定别名为 np。调用相应功能模块使用“别名+.+函数名称”，例如 np.sum。模块引入时使用 from 包 import 函数名称，例如实例中的这两个语句：

```
from numpy.linalg import inv
from numpy import dot
```

引用完成后，在代码中直接使用方法名称来调用相应的方法，首先构建数据集合：

X = mat([1,2,3]).reshape(3, 1)

Y = 2*X

为了能够清楚地反映整个最小二乘法的过程，使用了一个简单的一元线性回归数据集合，X 是正整数 1,2,3，Y 符合方程 Y=2X，因此 Y 的值就是 2,4,6。这里使用了 NumPy 的 reshape 方法，reshape 方法是 NumPy 矩阵操作最大的利器。numpy.reshape 函数可以在不改变数据的条件下修改数组矩阵形状，格式为：numpy.reshape(arr, newshape, order='C')。

- arr：要修改形状的数组。
- newshape：整数或者整数数组，新的形状应当兼容原有形状。
- order：'C'为按行，'F'为按列，'A'为原顺序，'k'为元素在内存中出现的顺序。而且 TensorFlow 中的 reshape 操作和 NumPy 保持操作的一致性，比如：

```
import tensorflow as tf
t = tf.range(start=1,limit=13,delta=1)
```

```
    t = tf.reshape(t, [3, 4])
    t = tf.reshape(t, [-1])
    t = tf.reshape(t, [-1, 6])
    t = tf.reshape(t, [-1])
    t = tf.reshape(t, [3, -1, 2])
```

回到实例中，接下来的一行代码使用 NumPy 进行矩阵运算求解参数，也就是求解直线的斜率的过程。

theta = dot(dot(inv(dot(X.T, X)), X.T), Y)

在代码中，dot 方法为矩阵的点乘，inv 方法是矩阵的求逆运算，.T 为矩阵的转置。通过矩阵运算可以迅速得出 theta 的值是 2.0。

3.4　线性回归的梯度下降模型

利用最小二乘法求解多元线性回归参数集合的方法虽然在数学表达上十分简洁，但是完美中往往存在瑕疵，而定理大多存在边界，就像牛顿力学，在宏观低速场景中威力无穷，遇山开路，遇水搭桥，但是到了光速的世界，原子甚至更小的维度就完全失效了。最小二乘法求解多元方程存在两个主要的应用障碍：第一个障碍来源于数学表达式本身，由于表达式中存在求矩阵的逆运算操作，结合线性代数中逆的定义，我们要求(X^TX)是满秩矩阵或正定矩阵；第二个障碍来自于数据集大小的挑战，进入大数据时代，特征数据集的数量与日俱增，超巨型的数据集合产生的数据矩阵在进行 3.3 节的计算机运算时很容易达到瓶颈，运算速度超出我们接受的范围。

在说明梯度最小的最初，首先要改变我们的最终目标，不再追求完美解（E 的全局最优解，局部极小），而是希望得到一个不错的近似解（E 的局部最优解，全局最小）。比如，我们考试的目的是要得到 100 分，而考 99 分这个分数已经足够上一所满意的大学，也许不完美，可是接受不完美并不断向前探索正是人类的美好属性。从直观的向量空间可以这样解释，这是一个沿着当前降低的趋势向着空间局部最低不断逼近的过程，如图 3.2 所示。

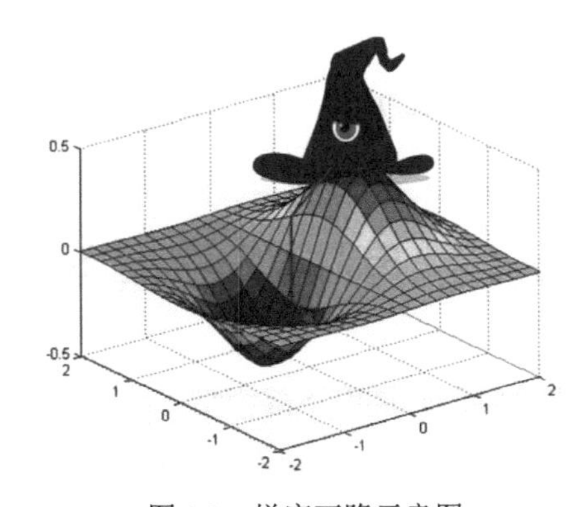

图 3.2　梯度下降示意图

局部极小值 E 对应的参数解集合是参数向量空间中的一个点，这个点相邻的点的预测值和标签值的偏差平方和均不小于该点的偏差平方和；全局最小值 E 对应的参数解集合则是我们放眼整个参数向量空间，空间中每一个点的预测值和标签值的偏差平方和均不小于该点的偏差平方和。参数向量空间中会有多个局部极小值，而只有一个全局最小值。求解最小值的方法就是在这个空间里面打个“滑梯”，向着梯度下降的方向滑动，最终能够到达某个空间的洼地，如图 3.3 所示。

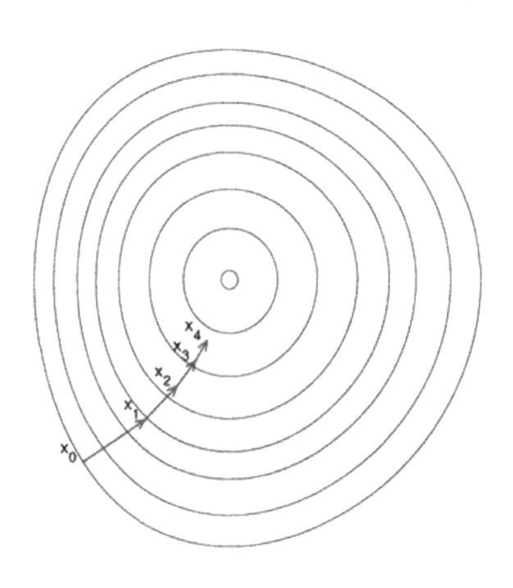

图 3.3　梯度下降结果逼近过程

用数学语言较为准确地描述，从函数的某个初始解出发，通过迭代循环的方法进行搜索求解。如果每次都向着函数值下降（梯度为负指示）的方向移动，当误差函数在当前梯度为零时，就停止循环，这时参数矩阵的值就是近似最优解（如果函数只有一个局部最小值，这个近似解就等于最优解）。

首先对损失函数 $J(\theta)$求偏导，得到：

$$\frac{\partial}{\partial\theta}J(\theta)=\frac{\partial}{\partial\theta}\frac{1}{2}\sum_{i=1}^{m}(h_\theta(x)-y)^2=(h_\theta(x)-y)\cdot x_i$$

然后 θ_i 进行参数更新，更新过程可以理解成 θ_i 沿着梯度下降最快的方向进行递减的过程。等式左边的 θ_i 表示更新之前的值，等式右边表示沿着梯度方向减少的量，α 表示步长，也称作学习速度，这个值需要人工手动设置。

$$\theta_i := \theta_i-\alpha\frac{\partial}{\partial\theta}J(\theta)=\theta_i-\alpha(h_\theta(x)-y)\cdot x_i$$

3.5　梯度下降模型实例

实例 1：使用 Python 实现线性回归梯度下降模型

在 3.3 节中我们使用最小二乘法完成了一元线性回归的参数求解过程，本节的第一个例子就使用 Python 完成梯度下降算法对 3.3 节数据集合对应的一元线性回归模型进行求解。代码如下：

```
import numpy as np
```

```
from numpy import dot
from numpy import mat

#y=2x
X = mat([1,2,3]).reshape(3, 1)
Y = 2*X
theta = 1.
alpha = 0.1
for i in range(100):
    theta = theta + np.sum(alpha * (Y- dot(X, theta))*X.reshape(1,3))/3.
print(theta)
```

数据准备部分与3.3节完全相同，在这里就不赘述了。接下来的代码是模型参数的初始化，我们人为设定起始参数为1.0，也可以使用随机算法生成随机参数。另外，我们还需要指定学习速率，在本例中使用0.1。

theta = 1

alpha = 0.1

然后指定梯度下降的学习速率，学习速率在这里指定为0.1。一般来说，梯度下降学习速率太快（α过大）容易导致算法无法收敛，梯度下降学习速率太慢（α过小）容易造成计算资源的浪费和模型运行时间过长。

```
for i in range(100):
    theta = theta + np.sum(alpha * (Y- dot(X, theta))*X.reshape(1,3))/3.
```

接下来的代码进行参数更新，即梯度下降的实际运行过程通过NumPy矩阵运算，我们很容易就可以完成更新量的计算，通过循环更新参数集合100次。整个过程比较清晰，但是有一点需要特别注意，如果是多元线性回归模型，就需要在一次计算中同步更新参数矩阵的所有要素。

例如下面的三元线性回归模型中，每次循环中需要对参数矩阵的4个要素同时进行更新。

```
for i in range(10000):
    temp[0] = theta[0] + alpha*np.sum((Y-dot(X, theta))*X0)/150.
    temp[1] = theta[1] + alpha*np.sum((Y-dot(X, theta))*X1)/150.
    temp[2] = theta[2] + alpha*np.sum((Y-dot(X, theta))*X2)/150.
    temp[3] = theta[3] + alpha*np.sum((Y-dot(X, theta))*X3)/150.
    theta = temp
```

实例2：使用TensorFlow框架实现线性回归梯度下降模型

本例介绍使用TensoFlow进行线性回归的过程。

```
import tensorflow as tf

x = tf.constant([[1], [2], [3], [4]], dtype=tf.float32)
y_true = tf.constant([[0], [-1], [-2], [-3]], dtype=tf.float32)
```

```
linear_model = tf.layers.Dense(units=1)

y_pred = linear_model(x)
loss = tf.losses.mean_squared_error(labels=y_true, predictions=y_pred)

optimizer = tf.train.GradientDescentOptimizer(0.01)
train = optimizer.minimize(loss)

init = tf.global_variables_initializer()

sess = tf.Session()
sess.run(init)
for i in range(100):
  _, loss_value = sess.run((train, loss))
  print(loss_value)

print(sess.run(y_pred))
```

虽然使用 TensorFlow 看起来代码量比使用 Python 的 NumPy 框架进行数学运算时多不少，但是本实例代码完美地体现了 TensorFlow 的设计思想，具有平滑的学习曲线这样一个内在优势。因此，线性回归过程完成机器学习代数化向语义的进化，这样一套代码完全遵守 TensorFlow 机器学习的标准流程（模型构建、模型初始化、模型训练），而且将复杂的数学表达式封装成了 mean_squared_error 等方法，也使初级开发者更加容易掌握，而高级开发者可以聚焦模型构建本身。

首先是框架的引入和数据的准备，为了能够聚焦到线性回归模型本身，我们依然使用非常简单的数据集合，使用 TensorFlow 中的常量来直接存储。

```
x = tf.constant([[1], [2], [3], [4]], dtype=tf.float32)
y_true = tf.constant([[0], [-1], [-2], [-3]], dtype=tf.float32)
```

接下来的部分是模型的构建过程，也是我们要重点讲解的内容。

```
linear_model = tf.layers.Dense(units=1)
```

此处建立一个一元线性模型，units=1 表示输入一个参数。这里我们使用了 TensorFlow 自己的层模型，也可以用 Keras 模块中的层模型进行替换。

```
y_pred = linear_model(x)
loss = tf.losses.mean_squared_error(labels=y_true, predictions=y_pred)
optimizer = tf.train.GradientDescentOptimizer(0.01)
train = optimizer.minimize(loss)
```

通过 mean_squared_error 告知了 TensorFlow 在框架运行时计算预测值和标签值的偏差平方和，并以最小化这个值作为训练目的（optimizer.minimize(loss)）。

```
init = tf.global_variables_initializer()
```

```
sess = tf.Session()
sess.run(init)
for i in range(100):
  _, loss_value = sess.run((train, loss))
  print(loss_value)

print(sess.run(y_pred))
```

上面的代码是训练过程的详细实现，我们惊喜地发现，训练过程（包含参数初始化）和第 2 章例子的流程完全一致，因此在学习过程中只要聚焦于模型构建本身就可以了。

参考文献

[1] Ober P B. Introduction to linear regression analysis[J]. Journal of the Royal Statistical Society, 2010, 40(12):2775-2776.

[2] Suykens J A K, J. Vandewalle. Least Squares Support Vector Machine Classifiers[J]. Neural Processing Letters, 1999, 9(3):293-300.

[3] Ramsey J B. Tests for Specification Errors in Classical Linear Least-Squares Regression Analysis[J]. Journal of the Royal Statistical Society, 1969, 31(2):350-371.

[4] Bradtke S J, Barto, Andrew G. Linear Least-Squares algorithms for temporal difference learning[J]. Machine Learning, 1996, 22(1-3):33-57.

[5] Jyrki Kivinen M K W. Exponentiated Gradient Versus Gradient Descent for Linear Predictors[M]// Exponentiated Gradient versus Gradient Descent for Linear Predictors. 1997.

[6] Wilson D R, Martinez T R. The general inefficiency of batch training for gradient descent learning.[J]. Neural Networks, 2003, 16(10):1429-1451.

第4章 逻辑回归

有了线性回归工具，我们就可以在数据集合关系比较简单的情况下进行趋势的判断了。但是，当数据标签呈现离散特性时，我们需要将模型替换为适合离散特性的逻辑回归或泛逻辑回归模型。逻辑回归模型其实仅在线性回归的基础上套用了一个逻辑函数，逻辑回归模型的最大优点是对分类可能性进行建模，这样可以很容易地克服由于数据量变大而产生的数据分布评估的困难。换句话说，有了标签化的数据之后，就可以立刻使用逻辑回归来训练模型并且预测分类结果。

本章主要介绍逻辑回归模型以及通过机器学习进行模型求解的过程。内容分为四部分：第一部分着重介绍逻辑回归的概念以及和线性回归的关系；第二部分分析逻辑回归的数学模型；第三部分着重分析泛逻辑回归的模型定义；第四部分是使用逻辑回归结合金融开源数据接口完成股价涨跌幅度预测的代码实例。

4.1 逻辑回归简介

我们在第3章研究了线性回归的模型建立和参数解集的求解策略，但是，当数据关系呈现非线性化特征时，换句话说，我们需要处理分类任务时线性回归模型就遇到了瓶颈，解决方案是将线性回归模型进行升级。因为多元线性回归的结果是一个连续数值，因此需要找到一个函数，这个函数能够将线性回归结果（线性回归的预测值）转换为数据集合中的标签值。比如有一个二分类模型，目的是通过不同的人体数据区分男人和女人，标签值中有两种类型，比如男人是0，女人是1。我们通过人体数据的线性组合可以求得一个实数（比如4.5），因此需要通过一个单调函数将这些实数映射成0或者1。

完成这个工作有很多选择，最简单的方式是建立一个分段函数：当输入值是大于0的实数的时候，输出值是1；当输入值是小于0的实数的时候，输出值为0；0点的输出值是0.5。

$$y=\begin{cases}0 & z<0\\0.5 & z=0\\1 & z>0\end{cases}$$

但是我们的完美选择有一个小小的问题，就是上述函数在实数范围内不可微，这种不可微的特性导致梯度下降算法难以起效。因此，需要找到这个函数的替代函数，这种函数要单调可微，而且要保持分段函数的数据特性。我们可以比较容易地想象出这样的函数在原点附近斜率会急剧变化，因此对数概率函数（或者称作逻辑函数（Logistics Function））成为我们的可靠选择。

$$y=\frac{1}{1+e^{-z}}$$

从图 4.1 可以清楚地看到对数概率函数和分段函数在性质上的统一性。

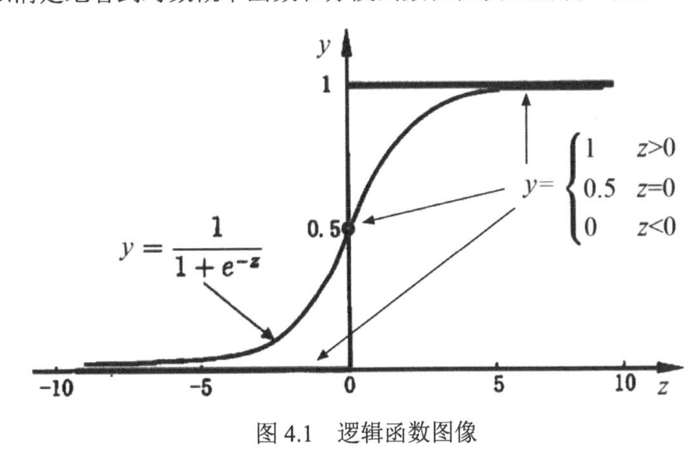

图 4.1 逻辑函数图像

我们将函数表达式中的 z 替换成线性回归表达式，可以得到如下表达式：

$$y=\frac{1}{1+e^{-(w^T x+b)}}$$

整理后可以得到：

$$\ln\frac{y}{1-y}=w^T x+b$$

因此，表达式可以指导我们使用线性回归的模型结果来与当前的标签值的对数概率进行比较，目的是使模型结果能够和数据集的标签的对数概率保持一致。逻辑回归（Logistic Regression，LR）模型其实仅在线性回归的基础上套用了一个逻辑函数，但就是由于这个逻辑函数，使得逻辑回归模型成为机器学习领域一个重要的模型。

4.2 逻辑回归模型

4.1 节介绍了逻辑回归，本节利用最大似然估计来进行模型的参数矩阵求解。首先介绍最大似然估计的核心内容，最大似然估计的意思就是最大可能性估计，其内容为：如果两件事 A、B 相互独立，那么 A 和 B 同时发生的概率满足公式：

$$P(A \mid B) = P(A) * P(B)$$

P(x)表示事件 x 发生的概率。

如何来理解相互独立呢？比如两件事情独立是说这两件事情不能够同时发生，如选择一个硬币的正反两个面，在得到正面向上的情况下，反面向上的情况是不可能发生的，硬币正面和硬币反面这两件事就是相互独立的。

我们使用多元线性回归的目的是总结一些不相关元素的规律，比如以前提到的散列点的表达式，这些点是随机的，所以我们认为这些点没有相关性，也就是独立的。总结不相关事件发生的规律也可以认为是总结所有事件同时发生的概率，所有事件同时发生的概率越大，我们预测到的规律就越准确。

这里重复一下以前提到的观点，回归的意思是用一条直线来概括所有点的分布规律，并不是来描述所有点的函数，因为不可能存在一条直线连接所有的散列点。所以我们计算出的值是有误差的，或者说我们回归出的这条直线是有误差的。我们回归出这条直线的目的是用来预测下一个点的位置。

考虑一下，一件事情我们规律总结得不准，原因是什么？是不是因为我们观察得不够细或者说观察的维度不够多呢？当我们掷一个骰子时，会清楚地知道掷出的高度、落地的角度、反弹的力度等信息，那么上帝视角的我们一定可以知道每次得到的点数。但由于观测不到所有的信息，所以我们认为每次投骰子得到的点数是不确定的，这是符合一定概率的，而未观测到的信息称为误差。

一个事件已经观察到的维度发生的概率越大，对应的未观测到的维度发生的概率就越小，可以说，我们总结的规律就越准确。根据最大似然估计 P(y) = P(x1,x2,...,xn)= P(x1) * P(x2)*...*P(xn)，当所有事情发生的概率为最大时，我们认为总结出的函数最符合这些事件的实际规律。所以可以把总结这些点的分布规律问题转变为求得 P(x1,x2,...,xn)= P(x1) * P(x2)*...*P(xn)发生的概率最大。

具体求解过程是，首先根据最大似然估计确定损失函数，如果想让预测出的结果全部正确的概率最大，就要让所有样本预测正确的概率相乘得到的 P（总体正确）最大化，数学表达形式如下：

$$L(\theta) = \prod_{i=1}^{m} (h_\theta(x_i))^{y_i} (1 - h_\theta(x_i))^{1-y_i}$$

上述公式中的 L 最大时，公式中 θ 的值就是我们需要的最好的 θ。下面对公式进行求解，公式的等号两边同时进行对数运算，这样就把乘法运算转换成了加法运算。

$$l(\theta) = \log L(\theta) = \sum_{i=1}^{m} y_i \log(h_\theta(x_i)) + (1 - y_i)\log(1 - h_\theta(x_i))$$

得到的这个函数越大，证明得到的 θ 就越好。因为在函数最优化的时候习惯让一个函数越小越好，所以我们在前面加一个负号，得到如下公式：

$$\mathrm{J_{log}}(\theta)=\sum_{\mathrm{i=1}}^{\mathrm{m}}-y_i\log(h_\theta(x_i))-(1-y_i)\log(1-h_\theta(x_i))$$

这个函数就是逻辑回归的损失函数，我们把它叫作交叉熵损失函数。

定义完损失函数后，我们利用损失函数进行梯度下降，来求得参数矩阵的解集合。

求解步骤如下：

（1）随机选择一组 θ。

（2）将 θ 代入交叉熵损失函数，让得到的点沿着负梯度的方向移动。换句话说，就是需要对定义的交叉熵损失函数求偏导：

$$\frac{\partial J(\theta)}{\partial\theta_j}=\frac{1}{m}\sum_{i=1}^{m}(h_\theta x_i-y_i)x_i^j$$

（3）重复循环第（2）步。

4.3 泛逻辑回归

我们在 4.2 节学习了标准的逻辑回归，标准的逻辑回归处理是二元分类问题，具体来说就是二元互斥事件的判断预测，例如我们可以判断扔硬币的结果，输入参数有硬币的飞行角度、硬币的制空时间等，通过机器学习训练（包括但不限于梯度下降算法）得到最优化的参数矩阵，通过逻辑回归函数可以判断硬币落地正反面的结果。但是如果我们扔的不是硬币呢？比如我们准备扔的是标准的 6 面骰子怎么办？大体来说有两种方式得到预测结果，第一种方式是进行多次逻辑回归，我们分析一下骰子的情况，首先构建一个逻辑回归模型，预测出结果是 1、2、3 或 4、5、6，然后构建 1 或 2、3 的逻辑回归模型，4 或 5、6 的逻辑回归模型，最后还要构建模型来区分 2 或 3 以及 5 或 6。这个方法太过于烦琐，而且由于训练后续模型的训练数据集合都是之前模型的训练数据集合的子集，因此数据准备会造成大量的资源浪费。我们需要一种方法能够一次性地完成多个互斥事件的分类工作，这就是本节讲解的 Softmax 函数。Softmax 函数（或称归一化指数函数）是逻辑函数（4.2 节）的一种推广，如函数名称所体现的那样，Softmax 函数把所有的输入向量转换为一个小于 1 的正小数，而这些正小数的和为 100%。函数表达式为：

$$S_i=\frac{e^{S_{yi}}}{\sum_j e^{S_j}}$$

其中，S_{yi} 是正确类别对应的线性得分函数，S_i 是正确类别对应的 Softmax 输出。接下来还要推导 Softmax 的损失函数，由于 log 运算符不会影响函数的单调性，因此对 S_i 进行 log 操作：

$$S_i = \log \frac{e^{S_{yi}}}{\sum_j e^{S_j}}$$

我们希望 S_i 越大越好，即正确类别对应的相对概率越大越好，那么可以在 S_i 前面加个负号来表示损失函数：

$$L_i = -S_i = -\log \frac{e^{S_{yi}}}{\sum_j e^{S_j}}$$

对上式进一步处理，把指数约去：

$$L_i = -\log \frac{e^{S_{yi}}}{\sum_j e^{S_j}} = -(e^{S_{yi}} - \log \sum_j e^{S_j}) = -e^{S_{yi}} + \log \sum_j e^{S_j}$$

这样，Softmax 的损失函数就转换成了简单的形式。

最后，使用 Python 进行 Softmax 函数计算，实例代码如下：

```
import math
z = [1.0, 2.0, 3.0, 4.0, 1.0, 2.0, 3.0]
z_exp = [math.exp(i) for i in z]
print(z_exp)  # Result: [2.72, 7.39, 20.09, 54.6, 2.72, 7.39, 20.09]
sum_z_exp = sum(z_exp)
print(sum_z_exp)  # Result: 114.98
softmax = [round(i / sum_z_exp, 3) for i in z_exp]
print(softmax)  # Result: [0.024, 0.064, 0.175, 0.475, 0.024, 0.064, 0.175]
```

在代码中，我们使用 Softmax 函数表达式将输入向量[1,2,3,4,1,2,3]转化成对应的 Softmax 函数的值[0.024,0.064,0.175,0.475,0.024,0.064,0.175]。输出向量中拥有最大权重的项对应输入向量中的最大值 4，显示了该函数通常的意义：对向量进行归一化，凸显其中最大的值并抑制远低于最大值的其他分量。

4.4 实例：股价预测

本节的实例用来展示使用 TensorFlow 进行 Softmax 模型多元逻辑回归分析的过程，实例使用交易量变化趋势来预测股票价格。首先介绍测试数据的获取方法，这里我们需要介绍一下 Tushare Python 库。Tushare 是一个免费、开源的 Python 财经数据接口包，主要实现对股票等金融数据从数据采集、清洗加工到数据存储的过程，能够为金融分析人员提供快速、整洁和多样的便于分析的数据，在数据获取方面极大地减轻工作量，从而更加专注于策略和模型的研究与实现。考虑到 Python Pandas 包在金融量化分析中体现出的优势，Tushare 返回的绝大部分数据格式都是 Pandas DataFrame 类型，非常便于用 Pandas、NumPy 和 Matplotlib 进行数据分析和可视化。当然，如果你习惯用 Excel

或者关系型数据库进行分析，那么可以通过 Tushare 的数据存储功能将数据全部保存到本地后进行分析。从 0.2.5 版本开始，Tushare 同时兼容 Python 2.x 和 Python 3.x，对部分代码进行了重构，并优化了一些算法，确保数据获取的高效和稳定。目前 Tushare 主推的版本是 Tushare Pro，Pro 版本的数据质量有了很大的提升，但是由于平台化策略，在 Pro 版本中会有用户级别和对应的权限，使用时会产生一些级别或访问次数产生的接口异常。

使用之前需要先确定 Tushare 的平台依赖库，在已安装 Python 的前提下，我们要有版本正确的 Pandas 和 LXML。Pandas 库一般会在第一次安装 Tushare（pip install tushare）时级联安装至系统，而 LXML 一般需要手动执行 pip 命令（pip install lxml）。

使用之前还需要进行注册，注册完成后系统会给予相应的访问令牌（Token）。

注册 Tushare 社区用户的流程如下：

（1）访问 Tushare 社区门户（https://tushare.pro），单击右上角的“注册”按钮。

（2）填入相应的注册信息，输入正确的手机号和图片验证码之后，单击“发送验证码”按钮，可以收到验证码，填入后单击“注册”按钮，如图 4.2 所示。

（3）完成后跳转到用户登录页面，即可检测是否注册成功，如图 4.3 所示。

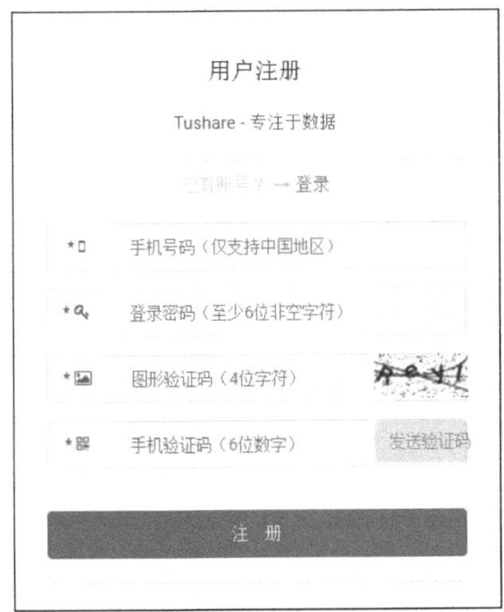

图 4.2　Tushare 注册

图 4.3　Tushare 登录

获取 TOKEN 凭证的步骤如下：

（1）登录成功后，单击右上角的“个人主页”选项，如图 4.4 所示。

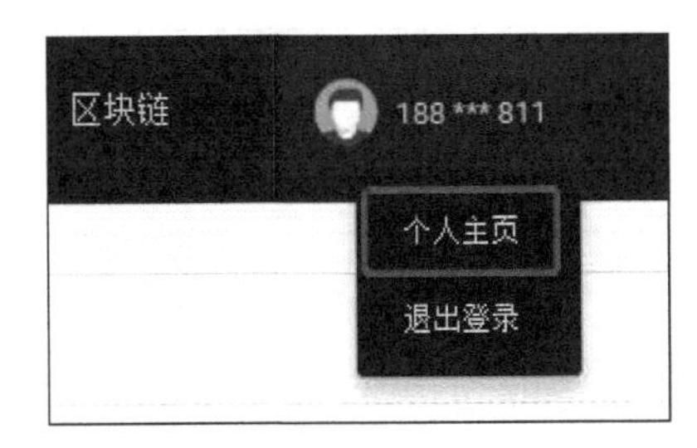

图 4.4　Tushare 个人主页

（2）在“用户中心”单击“接口 TOKEN”选项，如图 4.5 所示。

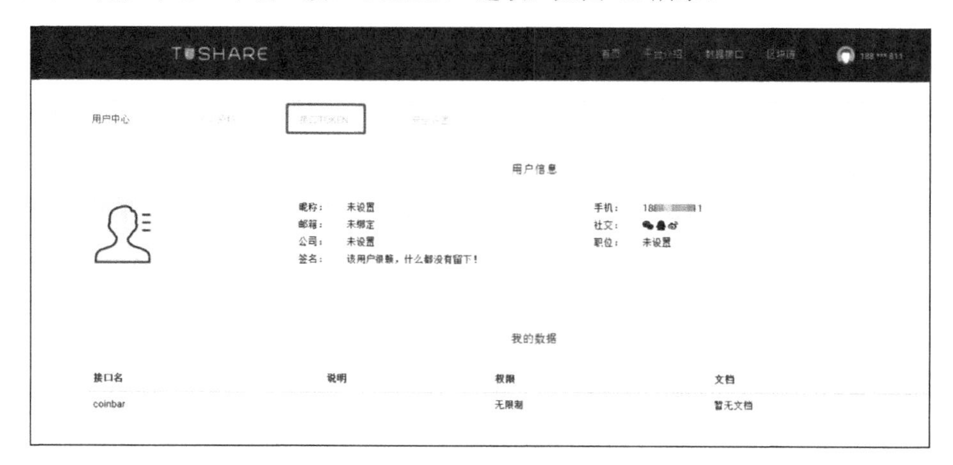

图 4.5　接口 TOKEN

（3）可以单击右侧的复制按钮复制 TOKEN，如图 4.6 所示。

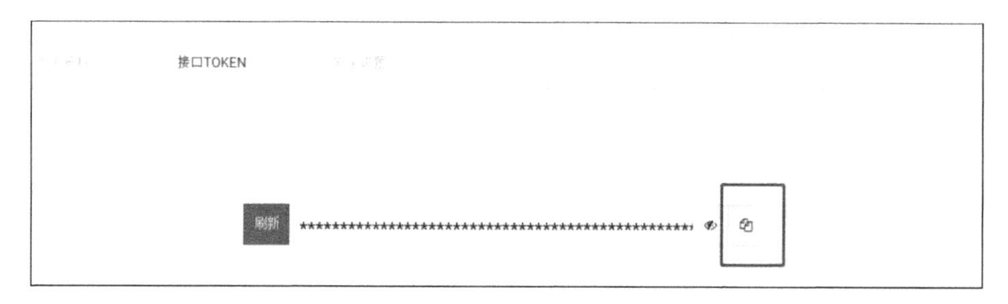

图 4.6　复制 TOKEN

（4）或者单击右侧的睁开眼睛按钮来获取 TOKEN 明文，复制并保存，如图 4.7 所示。

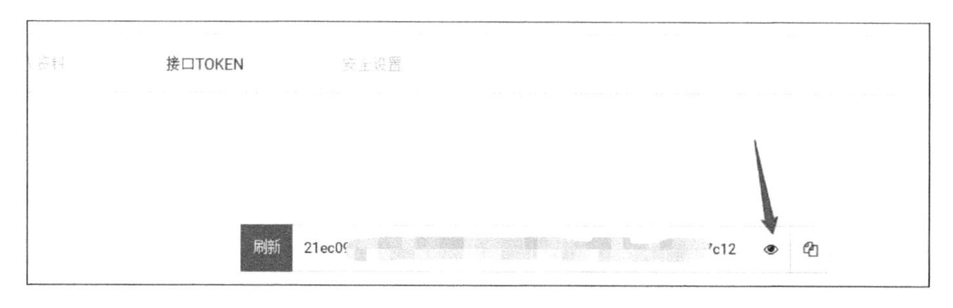

图 4.7　复制 TOKEN 明文

注　意

TOKEN 是调取数据的唯一凭证，请妥善保管，如发现别人盗用，可在本页面单击“刷新”按钮，之前的 TOKEN 将失效。

导入 Tushare：

```
import tushare as ts
```

这里注意，Tushare 的版本需大于 1.2.10。

设置 TOKEN：

```
ts.set_token('your token here')
```

以上方法只需要在第一次或者 TOKEN 失效后调用，完成调取 Tushare 数据凭证的设置，正常情况下不需要重复设置。也可以忽略此步骤，直接用 pro_api('your token')完成初始化。

初始化 Pro 接口：

```
pro = ts.pro_api()
```

如果上一步 ts.set_token('your token')无效或不想保存 TOKEN 到本地，那么可以在初始化接口里直接设置 TOKEN：

```
pro = ts.pro_api('your token')
```

数据调取，以获取交易日历信息为例：

```
df = pro.trade_cal(exchange='', start_date='20180901', end_date='20181001',
fields='exchange,cal_date,is_open,pretrade_date', is_open='0')
```

或者

```
df = pro.query('trade_cal', exchange='', start_date='20180901',
end_date='20181001', fields='exchange,cal_date,is_open,pretrade_date',
is_open='0')
```

调取结果：

```
   exchange  cal_date   is_open pretrade_date
0       SSE     20180901       0      20180831
1       SSE     20180902       0      20180831
2       SSE     20180908       0      20180907
3       SSE     20180909       0      20180907
4       SSE     20180915       0      20180914
5       SSE     20180916       0      20180914
6       SSE     20180922       0      20180921
7       SSE     20180923       0      20180921
8       SSE     20180924       0      20180921
9       SSE     20180929       0      20180928
10      SSE     20180930       0      20180928
11      SSE     20181001       0      20180928
```

有了前面的知识介绍，下面进入实例代码的实际分析过程。代码分为两部分，第一部分是模型学习数据的准备过程，代码如下：

```
# get training data
import tushare as ts
import pandas as pd
import numpy as np

import time
```

```
    ts.set_token(' your token here ')
    pro = ts.pro_api()

    data = pro.stock_basic(exchange='', list_status='L',
fields='ts_code,symbol,name,area,industry,list_date')
    print(data)
    # print(len(data))

    # df = ts.get_realtime_quotes('000581')
    df_new =
pd.DataFrame(columns=['day1','day2','day3','day4','day5','day6','day7','day8',
'day9','day10','Y'])
    for index, row in data.iterrows():
        time.sleep(0.5)
        df = pro.daily(ts_code=row['ts_code'], start_date='20190212',
end_date='20190225')
        dflist = df['amount'].tolist()
        df1 = pro.daily(ts_code=row['ts_code'], start_date='20190226',
end_date='20190226')
        print(df1)
        dflist1 = df1['pct_chg'].tolist()
        # print(dflist1)

        if len(dflist1) > 0 and len(dflist) == 10:
            # print(round((dflist1[0]+10)/2))
            array_new = np.concatenate((np.array(dflist),
np.array([round((dflist1[0]+10)/2)])), axis=0)
            # print(array_new)
            df_new= pd.concat([df_new, pd.DataFrame(array_new.reshape(1,11),
columns=['day1','day2','day3','day4','day5','day6','day7','day8','day9','day10
','Y'])])
            # print(df_new)
    # df_new.to_csv('train01.csv')
    df_new.to_csv('data01.csv')
```

程序开始需要引入 Tushare 等相关 Python 依赖库，然后初始化 Tushare 框架，初始化完成后，使用 stock_basic 方法获取基础信息数据，包括股票代码、名称、上市日期、退市日期等。接口详细说明如表 4.1（输入参数）和表 4.2（输出参数）所示，我们使用 fields 传参指定输出值的维度。

表 4.1 输入参数

名称	类型	必选	描述
is_hs	str	N	是否沪深港通标的，N：否，H：沪股通，S：深股通
list_status	str	N	上市状态：L：上市，D：退市，P：暂停上市
exchange	str	N	交易所：SSE，上交所：SZSE，深交所：HKEX，港交所（未上线）

表 4.2　输出参数

名称	类型	描述
ts_code	str	TS 代码
symbol	str	股票代码
name	str	股票名称
area	str	所在地域
industry	str	所属行业
fullname	str	股票全称
enname	str	英文全称
market	str	市场类型 （主板/中小板/创业板）
exchange	str	交易所代码
curr_type	str	交易货币
list_status	str	上市状态：L：上市，D：退市，P：暂停上市
list_date	str	上市日期
delist_date	str	退市日期
is_hs	str	是否沪深港通标的，N：否，H：沪股通，S：深股通

得到输出数据如下（数据样例）：

```
     ts_code    symbol   name     area   industry  list_date
0    000001.SZ  000001   平安银行  深圳    银行       19910403
1    000002.SZ  000002   万科A    深圳    全国地产    19910129
2    000004.SZ  000004   国农科技  深圳    生物制药    19910114
3    000005.SZ  000005   世纪星源  深圳    房产服务    19901210
4    000006.SZ  000006   深振业A   深圳    区域地产    19920427
5    000007.SZ  000007   全新好    深圳    酒店餐饮    19920413
6    000008.SZ  000008   神州高铁  北京    运输设备    19920507
7    000009.SZ  000009   中国宝安  深圳    综合类     19910625
8    000010.SZ  000010   美丽生态  深圳    建筑施工    19951027
9    000011.SZ  000011   深物业A   深圳    区域地产    19920330
10   000012.SZ  000012   南玻A    深圳    玻璃       19920228
11   000014.SZ  000014   沙河股份  深圳    全国地产    19920602
12   000016.SZ  000016   深康佳A   深圳    家用电器    19920327
13   000017.SZ  000017   深中华A   深圳    文教休闲    19920331
14   000018.SZ  000018   神州长城  深圳    装修装饰    19920616
15   000019.SZ  000019   深深宝A   深圳    软饮料     19921012
16   000020.SZ  000020   深华发A   深圳    元器件     19920428
```

```
17    000021.SZ  000021  深科技     深圳    电脑设备    19940202
18    000022.SZ  000022  深赤湾 A   深圳    港口        19930505
19    000023.SZ  000023  深天地 A   深圳    其他建材    19930429
20    000025.SZ  000025  特力 A     深圳    汽车服务    19930621
```

然后利用 Pandas 迭代器遍历股票的基本数据：

```
for index, row in data.iterrows():
```

接下来，我们使用 time 库的 sleep 方法进行一些系统等时操作，因为 Tushare 对接口的每秒访问时间有要求，如果超出时间，访问接口就会返回访问过快的要求，因此这里增加了等时操作，目的是防止访问频率过高。

```
    time.sleep(0.5)
```

接下来的代码取得连续 10 个交易日的交易量数据，判断一下第 11 个交易日的涨跌幅度，按照-10~10 共 20%的幅度，人为把结果分为 0~9，每一个数字代表涨跌幅在对应 2%的范围内，例如 1 代表-8%~-6%。获取交易数据的代码如下：

```
  df = pro.daily(ts_code=row['ts_code'], start_date='20190212', 
end_date='20190225')
    dflist = df['amount'].tolist()
    df1 = pro.daily(ts_code=row['ts_code'], start_date='20190226', 
end_date='20190226')s
    dflist1 = df1['pct_chg'].tolist()
```

标签化涨跌幅度如下：

```
  if len(dflist1) > 0 and len(dflist) == 10:
        # print(round((dflist1[0]+10)/2))
        array_new = np.concatenate((np.array(dflist), 
np.array([round((dflist1[0]+10)/2)])), axis=0)
```

最后使用 Pandas 将数据集合整理完成，通过 to_csv 方法保存到文件中。

数据整合的核心代码如下：

```
  df_new= pd.concat([df_new, pd.DataFrame(array_new.reshape(1,11), 
columns=['day1','day2','day3','day4','day5','day6','day7','day8','day9','day10
','Y'])])
```

利用 DataFrame 写入文件，保存成 CSV 格式的代码：

```
  df_new.to_csv('data01.csv')
```

第二部分是重点代码，主要是利用 TensorFlow 框架进行 Softmax 泛逻辑回归。

```
  # first tensorflow model training: logistic regression
  import pandas as pd
  import tensorflow as tf
```

```
import numpy as np

data_frame =  pd.read_csv("data01.csv", header=None)
dataset = data_frame.values
x_train = dataset[:, 0:10]
y_train = dataset[:, 10].astype('float64') -1

# data_frame =  pd.read_csv("test01.csv", header=None)
# dataset = data_frame.values
# x_test= dataset[:, 0:10]
# y_test = dataset[:, 10].astype('float64')-1

model = tf.keras.models.Sequential([
  tf.keras.layers.Dense(10, activation=tf.nn.softmax)
])
model.compile(optimizer='adam',
             loss='sparse_categorical_crossentropy',
             metrics=['accuracy'])

model.fit(x_train, y_train, epochs=10, batch_size=64)
# test_loss, test_acc = model.evaluate(x_test, y_test)
# print('Test accuracy:', test_acc)
```

上述代码首先引入对应的库文件，之后的代码主要包含三个步骤，第一步准备数据，即利用Pandas进行文件读取。对应代码如下：

```
data_frame =  pd.read_csv("data01.csv", header=None)
dataset = data_frame.values
x_train = dataset[:, 0:10]
y_train = dataset[:, 10].astype('float64') -1
```

在这里简单介绍一下Pandas的功能，简单来说，Pandas就是一个数据处理、数据挖掘的Python库，Pandas与TensorFlow最大的不同是Pandas对深度学习的支持比较简单，这是由于Pandas设计时深度学习正处于业态低谷的原因。代码中从Data-Frame中提取所有数据值，利用数组取值运算得到输入值x_train和标签值y_train。

第二步是进行模型构建。对应代码如下：

```
model = tf.keras.models.Sequential([
  tf.keras.layers.Dense(10, activation=tf.nn.softmax)
])
model.compile(optimizer='adam',
             loss='sparse_categorical_crossentropy',
             metrics=['accuracy'])
```

模型构建是实例中的核心部分，利用 TensorFlow 中的 Keras 模块进行模型构建是很简单的，而且语音化能力很强。我们构建了一个有 10 个输入和 10 种标签的 Softmax 的计算层，然后把这个层放入一个数组中。当然，我们需要把这个数组作为序列化模型的创建参数，也可以使用 add 方法给一个空的序列化模型添加计算层。添加完计算层后指定学习过程中的优化方法和损失函数，在前面的高级 API 部分已经介绍过，大多学习细节 Keras 模型会有默认值，因此接下来直接使用模型的 fit 方法进行训练就可以了，这就是最后一步的训练和测试。对应代码如下：

```
model.fit(x_train, y_train, epochs=10, batch_size=64)
```

在训练中使用 epochs 指定数据使用的轮数，使用 batch_size 指定每次训练数据集合的大小。

当然，机器学习模型的优劣程度可以使用测试数据来进行测试，测试数据的准备方式和训练数据是相同的。Keras 模块对于测试有专门的方法 test 来支持，参数和训练过程是类似的。具体实现请读者朋友再仔细研究一下实例代码。

补充说明，实例只是利用 Tushare 提供的股票数据演示 Softmax 多元逻辑回归分析的建模和训练过程，和真正的量化交易的概念是不同的。一般的量化交易包括经验和策略通过代码方式的实现、机器学习的评估方法、通过模拟交易的方式来回测整个策略以及评价策略整体在市场中的表现效果等步骤。感兴趣的读者可以自行寻找相应资源深入研究。

参考文献

[1]. Andrew Cucchiara. Applied Logistic Regression[J]. Technometrics, 1992, 34(3):2.

[2] Lemeshow S. A review of goodness of fit statistics for use in the development of logistic regression models[J]. Am J Epidemiol, 1980, 115(4):732-2.

[3] Bewick V, Cheek L, Ball J. Statistics review 14: Logistic regression[J]. Critical Care, 2005, 9(1):112.

[4] Maalouf, Maher. Logistic regression in data analysis: an overview[J]. International Journal of Data Analysis Techniques & Strategies, 2011, 3(3):281-299.

[5] Yuan Z, Li J, Li Z, et al. Softmax Regression Design for Stochastic Computing Based Deep Convolutional Neural Networks[C]// the. ACM, 2017.

[6] Peter Tiňo. Bifurcation structure of equilibria of iterated softmax[J]. Chaos, Solitons and Fractals, 2009, 41(4):1804-1816.

第二篇　机器学习进阶

第 5 章

神经网络

人工神经元是一个基于生物神经元模型的数学函数，其中每个神经元接收输入，分别对输入进行加权，相加，然后通过一个非线性函数将这个和传递给输出。这种模型的灵感很大程度上来源于神经学等多学科的具体经验。一个神经元可以看作为一个简单的神经网络模型，而神经元通过激活函数非线性组合而成的全连接神经网络模型成为神经网络模型进阶的坚实基石。

本章主要讲解神经网络中的神经元模型以及全连接神经网络模型，通过验证码图片破解过程加深对模型的应用和理解。内容分为五部分：第一部分着重介绍神经网络中的最小组成单元——神经元模型；第二部分使用单神经元模型进行验证码图片破解，通过这一过程加深对模型的应用和理解；第三部分着重分析神经元模型中使用的常用激活函数的数学定义和性能特性；第四部分着重介绍神经网络模型中的全连接神经网络模型；第五部分使用全连接神经网络模型进行验证码图片破解程序的升级，通过这一过程加深对模型的应用和理解。

5.1　神经元模型

5.1.1　神经网络的灵感来源

神经网络的灵感来自于大脑的计算机制，它由神经元的计算单元组成。人脑有数十亿个神经元，神经元是人脑中相互连接的神经细胞（示意图见图 5.1），参与处理和传递化学和电信号。树突是从其他神经元接收信息的分支，细胞核或体细胞处理从树突接收到的信息，轴突是一种神经细胞用来传递信息的电缆，突触是轴突和其他神经元树突之间的连接部位。1943 年，研究人员沃伦·麦卡洛克（Warren McCullock）和沃尔特·皮茨（Walter Pitts）发表了关于简化脑细胞的第一个概念，称为 McCullock-Pitts（MCP）神经元。他们将这种神经细胞描述为一个简单的逻辑门，输出为二进制。当多个信号到达树突时，会整合到细胞体中，如果积累的信号超过某个阈值，就会产生一个输出信号，由轴突传递。

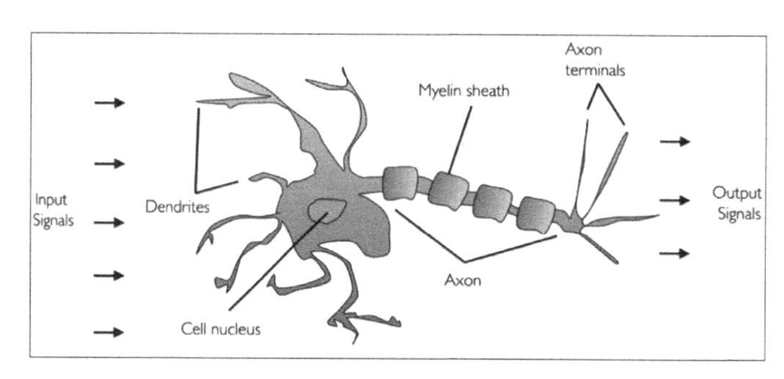

图 5.1　神经元示意图

人工神经元是一个基于生物神经元模型的数学函数，其中每个神经元接受输入，对这些输入数据进行加权，再相加，通过一个非线性函数将这个和传递给输出。然后借助感知器——一个神经网络单元（一个人工神经元）进行一定的计算来检测输入数据中的特征或商业智能。

5.1.2　感知器模型概述

感知器是由 Frank Rosenblatt 在 1957 年提出的，他提出了一种基于 MCP 神经元的感知器学习规则。感知器是一种用于二进制分类器监督学习的算法，该算法使神经元能够一次学习和处理训练集中的元素。感知器有两种类型：单层感知器和多层感知器。单层感知器只能学习线性可分的模式，如图 5.2 所示。

图 5.2　单层感知器示意图

多层感知器具有两层或两层以上的前馈神经网络，具有更强的处理能力。感知器算法通过学习输入信号得到参数矩阵的权值，从而绘制线性决策边界（见图 5.3），这使你能够区分两个线性可分的类+1 和-1。感知器学习规则表明，该算法将自动学习最优权系数，然后将输入特征与权重相乘，以确定神经元是否触发。感知器可以接收多个输入信号，如果输入信号的和超过某个阈值，它要么输出一个信号，要么不返回输出。在监督学习和分类的背景下，感知器的这一特性可以用来预测样本的类别。

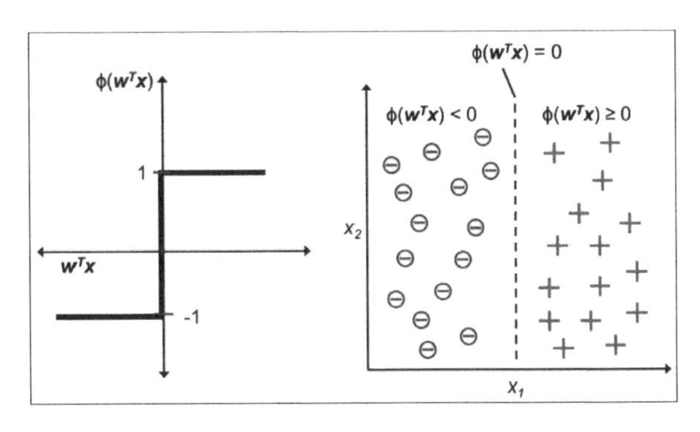

图 5.3　感知器进行回归分析

通过前面的分析我们可以写出如下神经元数学表达式，f 为 5.2 节将重点讨论的激活函数。

$$a = f(W \cdot x)$$

求解过程就是前面章节中讲到的梯度下降法，对于样本集，找到一个权重矩阵，这个权重矩阵运算后使得标签值尽可能接近，为了调节误差，使之尽可能小，因此需要求其导数，发现其下降的方向。所以，每次权重矩阵的修改应当通过求误差的偏导（梯度）来实现，比之前直接通过误差来调整具备更好的适应性。

$$w := w - \alpha \frac{dJ(w,b)}{dw}$$

$$b := b - \frac{dJ(w,b)}{db}$$

5.2　单神经元模型实例

5.1 节我们从数学定义的角度解释了神经网络中最小单元神经元的构造和计算特征。从另一个角度来说，神经元其实也是一个神经网络，是一个最简单的神经网络，就像蓝藻也是植物，草履虫也是单细胞动物一样，单神经元模型能够完整地进行神经网络的整套学习优化过程。本节将会通过一个例子来实践这一过程，展示神经网络模型构造、数据准备、模型学习优化的编程技巧，为大家后续实例的学习打好基础。

本节实例是通过神经网络模型识别验证码，是一个机器视觉的简单实例。本实例将会贯穿神经网络和卷积神经网络两章，在这两章中，一方面我们会不断升级神经网络模型，从单一神经元到全连接神经网络（其中的运算个体都是相同的单一神经元模型），再到卷积神经元模型，在这个过程中，网络对于图片输入的理解在不断加深，预测正确率也在稳步提升；另一方面，我们会在章节中穿插网络优化的一些常见方式，特别是针对过拟合问题，会在编程实践中利用部分网络优化技巧。

5.2.1 验证码概述

验证码的英文名称是Completely Automated Public Turing test to tell Computers and Humans Apart（全自动区分计算机和人类的图灵测试），可缩写为CAPTCHA，它是一种区分用户是计算机还是人的公共全自动程序，可以有效防止黑客对某一特定注册用户用特定程序以暴力破解方式不断尝试登录，如恶意破解密码、刷票、论坛灌水等。实际上用验证码是现在很多网站通用的方式，由于计算机无法解答CAPTCHA的问题，因此回答出问题的用户就可以被认为是人类。验证码这个词最早是在2002年由卡内基梅隆大学的Luis von Ahn、Manuel Blum、Nicholas J.Hopper以及IBM的John Langford提出的。卡内基梅隆大学曾试图申请此词使其成为注册商标，但该申请于2008年4月21日被拒绝。一种常用的CAPTCHA测试是让用户输入一个扭曲变形的图片上所显示的文字或数字，扭曲变形是为了避免被光学字符识别（Optical Character Recognition，OCR）之类的计算机程序自动辨识出图片上的文字或数字而失去效果。由于这个测试是由计算机来考人类的，而不是像标准图灵测试中那样由人类来考计算机，因此人们有时称CAPTCHA是一种反向图灵测试。验证码在笔者看来暗含着两方面对立的逻辑问题，一方面，对用户来说，验证码是对用户不是机器人的认证，验证码（所有验证的应用部分）的业务逻辑并不包含在用户对应用本身的使用逻辑中。我们去视频网站是去看视频，使用网购应用是为了购买自己需要的东西，而验证流程只是用来保证我们的身份。因此，从这一点出发，验证流程的设计是需要简单化的。另一方面，从对抗恶意应用的角度来说，验证码必须能使机器难以判断，或者难以完成操作，以保证应用的安全性需求。因此，验证码发展的过程中，只有遵循上述逻辑的验证码应用才能够保留并且不断演进，而对用户不友好或者很容易攻破的类型都会在发展的过程中被淘汰。例如，使用比较多的手机端混合验证的策略（短信验证码、App端扫描二维码）就使恶意程序难以完成操作，而真正的用户可以使用简单的几步没有任何歧义操作获得认证。另一个例子，目前网页应用验证码的最新一代称作No CAPTCHA，具体表现是如果你在正常操作应用，就是标准的输入用户名和密码，如果是破解程序，就会弹出验证码页面。这个应用其实也融合了机器学习的思想，通过网页埋点得到用户操作习惯的核心数据，之后利用数据矩阵通过机器学习进行模式分析，得出用户行为的模型，最后通过用户行为模型来进行验证码提示与否的预测判断。

我们在实例中使用的是图片验证码，破解方式称为端对端的验证码识别技术。端对端需要说明一下，大家都知道图形验证码中有多个字符需要辨认，很多本身是具有固定意义的。一种做法是先将图形验证码图片进行分割，分割后的图片再进行识别，将分别识别的结果最终综合起来得到验证码的值。另一种做法是将整体图片作为数据集输入程序中，最终程序得到验证码的值，这种方式就是我们提出的端对端的验证码识别方案。这种方案的主要优势在于某种验证码可能在排布和字符形态上有很大的个体差异性，但是为了满足嵌入应用的需求，应用一般会使用固定高度来约束图片控件，因此，同种验证码在总的图片像素信息上一般完全一致，图片预处理就变得非常简单。凡事有利就有弊，在图片处理上的简化必然带来在识别技术上的复杂化。当然，我们能够使用非常成熟的神经网络算法来解决这一核心问题。

5.2.2 开发实例代码详解

下面进入开发实例部分。按照我们的分析，开发实例分为两个主要组成部分：第一部分是验证码图片的生成过程；第二部分是验证码图片的破解过程。在第一部分中，我们使用 Python 的验证码生成库 captcha 作为验证码生成工具，使用该工具有两点好处：第一是免除了我们进行手工标记的工作量，第二是我们的工作保持了实验性质，没有对其他应用造成威胁。

1. 生成训练数据

首先需要使用 pip 安装 captcha 库：

```
$ pip install captcha
```

captcha 可以生成图像验证码和声音验证码，我们使用的是它生成图像验证码的功能。简单调用实例如下：

```
from captcha.image import ImageCaptcha
from PIL import Image
image = ImageCaptcha()
image.write('1234','test.png')
```

代码非常简单，实际运行结果就是在当前文件夹生成了一张显示 1234 的验证码图片，如图 5.4 所示。

图 5.4 验证码图片

有一点需要特别注意，captcha 的图片操作内部调用了 PIL（Python Imaging Library），因此需要同步引入 PIL 库。PIL 虽然不是 Python 图像处理的标准库，但是由于其 API 丰富，使用起来对应用调用特别友好，因此已经是 Python 平台图像处理中使用最多的库了。

接下来编码实现利用 captcha 生成图片验证码的数据集：

```
def gen_captcha():
    image = ImageCaptcha()
    captcha_text = gen_random_captcha_text()
    captcha = image.generate(captcha_text)
    captcha_image = np.array(Image.open(captcha))
    return captcha_text, captcha_image
```

gen_captcha 是我们定义的验证码图片和对应标签的生成程序，每调用一次生成一组验证码图片和对应的标签文本。在代码中，首先初始化 ImageCaptcha 工具类，之后调用一个 gen_random_captcha_text 方法生成字符串，接下来利用 captcha 的 API 生成验证码图片：

```
    captcha = image.generate(captcha_text)
```

与前面使用 captcha 生成验证码图片不同的是，这里并没有生成图片文件，而是图片数据的对象，因此我们可以使用 PIL 库结合 NumPy 数组操作把图片数据对象转换成三维数组：

```
captcha_image = np.array(Image.open(captcha))
```

三个维度分别表示图片的长、宽和颜色。当然，可以通过简单的函数调用和打印来印证我们的结论：

```
aptcha_text, captcha_image = gen_captcha()
    print(captcha_text)
    print(captcha_image)
    print(captcha_image.shape)
```

运行后的输出结果如下：

```
['7', '4', '5', '6']
[[[246 242 252]
  [246 242 252]
  [246 242 252]
  ...
  [246 242 252]
  [246 242 252]
  [246 242 252]]

 [[246 242 252]
  [246 242 252]
  [246 242 252]
  ...
  [246 242 252]
  [246 242 252]
  [246 242 252]]
  [246 242 252]
  [246 242 252]
  [246 242 252]]]
(60, 160, 3)
```

到这里，生成训练数据的功能就完成了。我们还留下了一个小尾巴——验证码的文本生成函数是怎么样的？其实这一部分的代码非常简单，主要是使用数组操作结合 Python 的随机数框架 random 生成 4 位数字验证码文本，大家可以自行研究一下代码，最重要的是不要忘记引入 random 库。

```
numbers = ['0','1','2','3','4','5','6','7','8','9']
def gen_random_captcha_text(char_set=numbers):
    captcha_text = []
    for i in range(4):
        captcha_text.append(random.choice(numbers))
    return captcha_text
```

2. 原始数据的预处理

有了机器学习的原始数据，接下来我们要做原始数据的预处理工作。很多读者在机器学习特别是神经网络中有一个极端的认知，就是由于神经网络在某种程度上能够自己提取特征，因此输入数据可以不经过预处理的步骤，但是截至目前还没有一种机器学习或者人工智能算法能够百分之百地解决所有的问题，正如大统一理论依然包含物理学的困惑，机器学习依然需要界定范围和进行有效的优化设计，因此数据预处理目前还是我们必须要进行的步骤。本实例的数据预处理工作比较简单，究其原因数据是由我们自己的程序控制产生的，产生数据的边界是固定的。以下是数据预处理的代码：

```
def text2vec(text):
    vector = np.zeros(40)
    for i in range(len(text)):
        idx = i*10+ord(text[i])-48
        vector[idx] = 1
    return vector

def get_batch (size = 100):
    batch_x = np.zeros([size, 9600])
    batch_y = np.zeros([size, 40])

    for i in range(size):
        text,image = gen_captcha()
        batch_x[i,:] = np.mean(image, -1).flatten()/255
        batch_y[i,:] = text2vec(text)
    return batch_x, batch_y
```

上述代码主要包含三部分，第一部分是对图片矩阵的归一化处理，对应代码如下：

```
np.mean(image, -1).flatten()/255
```

代码非常简洁，包含类灰度化、数据拍平和数据单位映射三步操作。首先观察验证码图片，发现图片的颜色对信息本身没有正增益，我们使用一个取巧的近似方法来处理图像，进行图像的灰度化，使用 mean 函数对 3 个颜色取平均值。当然，也可以使用标准的灰度化经验表达式来完成这个工作，即

```
gray = 0.2989 * r + 0.5870 * g + 0.1140 * b
```

接下来，我们需要拍平数据，Flatten 函数会帮我们处理好数据矩阵从高维度到一维的所有细节。最后，因为单一数据是 0~255 的整数，我们通过除以 255 将数据区间收缩到 0~1，便于学习算法进行分析。

第二部分是针对结果标签进行独热编码（One Hot Code）。独热编码又称为一位有效编码，主要是采用 N 位状态寄存器来对 N 个状态进行编码，每个状态都有它独立的寄存器位，并且在任意时候只有一位有效。独热编码是分类变量作为二进制向量的表示。首先，要求将分类值映射到整数值。然后，每个整数值被表示为二进制向量，除了整数的索引被标记为 1 之外，其他都是零值。这

主要针对标签数据是离散的情况，比如性别特征：["男"，"女"]，祖国特征：["中国"，"美国，"法国"]，等等。实例中主要实现下面的函数：

```
def text2vec(text):
    vector = np.zeros(40)
    for i in range(len(text)):
        idx = i*10+ord(text[i])-48
        vector[idx] = 1
    return vector
```

因为我们的验证码是 4 位数字，对于每个数字来说取值范围是 0~9，所以，一位数字对应 10 个编码位置，比如 0 对应 [1,0,0,0,0,0,0,0,0,0]，6 对应 [0,0,0,0,0,0,1,0,0,0]，整个验证码标签文本能够映射成一个 40 个元素的一维向量。

例中先初始化一个具有 40 个元素的一维向量，每个元素设置成 0，对应代码如下：

```
vector = np.zeros(40)
```

接下来，通过循环取出每个数字，判断每个数字对应向量应该将第几位元素值置为 1，注意需要加上偏置位 i*10。对应代码如下：

```
for i in range(len(text)):
        idx = i*10+ord(text[i])-48
        vector[idx] = 1
```

第三部分是将整理好的数据整理成批，支持训练时的批量梯度下降算法。对应代码如下：

```
def get_batch (size = 100):
    batch_x = np.zeros([size, 9600])
    batch_y = np.zeros([size, 40])

    for i in range(size):
        text,image = gen_captcha()
        batch_x[i,:] = np.mean(image, -1).flatten()/255
        batch_y[i,:] = text2vec(text)
    return batch_x, batch_y
```

代码中需要注意两点，因为我们需要定义一个可调整的数据批次的大小，这是因为在后续方法中批次也是训练微调整的参数之一。另外，大家可以注意一下循环和数据索引的联合使用技巧，这个技巧在机器学习数据准备环节是很常用的。

3. 模型构造

下面进入模型构造部分，实现代码如下：

```
with tf.name_scope('input') as scope:
    x = tf.placeholder(tf.float32, [None ,60*160], name='x')
    y_ = tf.placeholder(tf.float32, [None ,40], name='y_')
```

```
with tf.name_scope('net') as scope:
    w = tf.Variable(tf.random_normal([9600, 40]), name= 'w')
    b = tf.Variable(tf.random_normal([40]), name= 'b')
    y= tf.matmul(x,w) +b

with tf.name_scope('training') as scope:
    loss = tf.losses.sigmoid_cross_entropy(y_, y)
    tf.summary.scalar('loss', loss)
    optimizer  = tf.train.AdamOptimizer(learning_rate=0.001).minimize(loss)

with tf.name_scope('prediction') as scope:
    p = tf.argmax(tf.reshape(y, [-1, 4, 10]),2)
    l = tf.argmax(tf.reshape(y_, [-1, 4, 10]),2)
    pred = tf.equal(p,l)
    accuracy = tf.reduce_mean(tf.cast(pred, tf.float32))
    tf.summary.scalar('accuracy', accuracy)
```

代码通过 name_scope 将模型分隔成为不同的区域，这样做不但在代码层面增加了可读性，而且令人愉悦的是，当我们使用 TensorBoard 查看运算图的时候，这种结构能够清晰地展现出来，并且可以使用双击来打开某个特定部分的模型。有了这个架构，后续的工作就会趋于简单化，我们可以在代码开发上很好地进行增量开发，具体来说，在上面展示的代码中构造了 4 个子模型，或者称为模型区域。其中，输入模型（input）、训练模型（training）以及评价模型（prediction）在后续开发中完全可以复用，我们只需要调整网络模型（net）就可以了，这种编码架构体现了 TensorFlow 面向研究的特性。下面我们按照子模型的顺序来分析一下上述代码。

```
with tf.name_scope('input') as scope:
    x = tf.placeholder(tf.float32, [None ,60*160], name='x')
    y_ = tf.placeholder(tf.float32, [None ,40], name='y_')
```

首先是输入模型，输入分为图片输入数据和标签输入数据，两者需要和我们准备好的数据形式保持一致。我们使用 TensorFlow 的 API 的 placeholder 完成数据张量的预定义过程。具体解释大家可以参考 TensorFlow 低阶 API 的介绍。

```
with tf.name_scope('net') as scope:
    w = tf.Variable(tf.random_normal([9600, 40]), name= 'w')
    b = tf.Variable(tf.random_normal([40]), name= 'b')
    y= tf.matmul(x,w) +b
```

在网络定义阶段，我们定义了一个单一神经元，神经元模型接收图片数据的整个输入（9600 个元素的一维张量），输出是与标签一致的 40 个元素的一维张量。由于我们需要在后面的实际训练学习中更新网络中的参数集合，因此使用 Variable 来定义权重（w）和偏置（b）。在定义中，我们需要初始化这两个参数集合，使用了 TensorFlow API 中的 random_normal 函数，

tf.random_normal()函数用于从服从指定正太分布的数值中取出指定个数的值。

```
with tf.name_scope('training') as scope:
    loss = tf.losses.sigmoid_cross_entropy(y_, y)
    tf.summary.scalar('loss', loss)
    optimizer = tf.train.AdamOptimizer(learning_rate=0.001).minimize(loss)
```

在训练过程中需要定义损失函数，由于标签数据中可以出现相同的数字（比如 1111），因此判断标签中每个类别属于相互独立但互不排斥的情况。这里使用 sigmoid_cross_entropy 计算损失，对输入的 logits 先通过 Sigmoid 函数计算，再计算它们的交叉熵，但是对交叉熵的计算方式进行了优化，使得结果不至于溢出。

```
tf.summary.scalar('loss', loss)
```

通过 TensorFlow 的 summary 库可以进行优化过程的跟踪，跟踪结果会存储到对应的日志文件中，在运行过程中和运行结束后可以使用 TensorBoard 观察整个训练过程的变化趋势。在机器学习的过程中，优化器实际上是非常复杂的数学模型，但是 TensorFlow 面向用户进行了友好的封装，我们只需要指定需要优化的对象和学习速率就可以了。

```
tf.train.AdamOptimizer(learning_rate=0.001).minimize(loss)
```

我们在训练过程中不但要直观地看到损失下降，更重要的是要能够得知模型预测能力的增长。损失下降可以直观地理解为当前模型有没有达到它的性能极限，但是显然不是任何一个模型达到极限就可以解决问题。因此我们应当清晰地认识到机器学习关于模型的一个事实——任何一种模型都有它的范围和极限，针对具体问题需要设计出具有一定独特性的模型。模型预测准确率是我们需要分析监控的第一目标，而这一工作会由下面的预测子模块完成。

```
with tf.name_scope('prediction') as scope:
    p = tf.argmax(tf.reshape(y, [-1, 4, 10]),2)
    l = tf.argmax(tf.reshape(y_, [-1, 4, 10]),2)
    pred = tf.equal(p,l)
    accuracy = tf.reduce_mean(tf.cast(pred, tf.float32))
    tf.summary.scalar('accuracy', accuracy)
```

任何精确度分析的过程实际上就是模型运算值和标签值比较结果的综合。在实例中，我们使用了 3 个主要的函数：argmax、equal 和 reduce_mean，下面分别解释一下。

- tf.argmax 是 TensorFlow 用 NumPy 的 np.argmax 实现的，它能给出某个 Tensor 对象在某一维上的数据最大值所在的索引值，常用于 metric（如 acc）的计算。
  tf.argmax()函数中有一个 axis 参数（轴），该参数能够指定按照哪个维度计算。例如在矩阵的结构中，axis 可被设置为 0 或 1，0 表示按列计算，1 表示按行计算。
- tf.equal(A, B)用于对比两个矩阵或者向量相等的元素。如果是相等的，就返回 True，否则返回 False。返回的值的矩阵维度和 A 是一样的。
- tf.reduce_mean 函数用于计算张量（Tensor）沿着指定数轴（Tensor 的某一维度）上的平

均值。

因此，模块中代码的主要作用是将独热编码重新转换为验证码数字，比较两个数字是否相同，然后在整个数据集合上统计相同出现的概率。

4. 训练过程

最后进行训练过程。训练过程包括使用批量梯度下降进行学习和通过预测模块测试模型预测准确率两部分功能，代码如下：

```
with tf.Session() as sess:
    merged = tf.summary.merge_all()
    tf.global_variables_initializer().run()
    writer = tf.summary.FileWriter('./log/crack/', sess.graph)

    step = 0
    while True:
        #train
        x_train, y_train = get_batch(64)
        _, loss_ = sess.run([optimizer, loss], feed_dict={x:x_train,
y_:y_train})
        print(step, loss_)
        if step%100 ==0:
            x_test, y_test = get_batch(100)
            summary, acc = sess.run([merged, accuracy], feed_dict={x:x_test,
y_:y_test})
            print(step, acc)
            writer.add_summary(summary, step)
        step = step +1
```

代码中使用会话、初始化对象、监控对象合并等操作就不一一详述了，如果大家有疑问，可以查看2.5.1节的内容。批量梯度下降进行模型学习的过程和预测模块测试模型预测准确率都是调用对应的模型，使用的是标准的TensorFlow会话操作。唯一不同是，我们没有在每次学习后都进行模型预测，而是每学习100次后调用一次预测流程。在代码中可以尝试改变学习和预测的批次大小，可以清楚地观察到批次增大对于损失下降稳定和单次学习效率的影响。

上述代码中，其中模型学习的过程的核心代码如下：

```
x_train, y_train = get_batch(64)
        _, loss_ = sess.run([optimizer, loss], feed_dict={x:x_train,
y_:y_train})
```

预测模块测试模型预测准确率的核心代码如下：

```
x_test, y_test = get_batch(100)
            summary, acc = sess.run([merged, accuracy], feed_dict={x:x_test,
y_:y_test})
```

在终端运行实例（见图 5.5），结果使用 TensorBoard 进行可视化展示，如图 5.6 所示。我们可以清楚地看到，损失函数的值随着训练数据的增加逐渐减小到一个稳定的值，而预测准确率达到了50%。虽然这个准确率对于实际应用还有着很大的提升空间，但是作为练习，本实例中完整地展现了一个简单的神经网络的构建、学习和预测的过程，同时针对实例本身介绍了图像识别的整体流程。

```
0 32.395977
0 0.09
1 28.75614
2 25.758595
3 23.07221
4 20.065199
5 18.668772
6 15.929985
7 15.713419
8 13.919858
9 11.799208
10 11.849583
11 10.51566
12 9.640657
13 9.462906
14 8.33576
15 7.5732803
```

图 5.5　实例代码终端运行过程

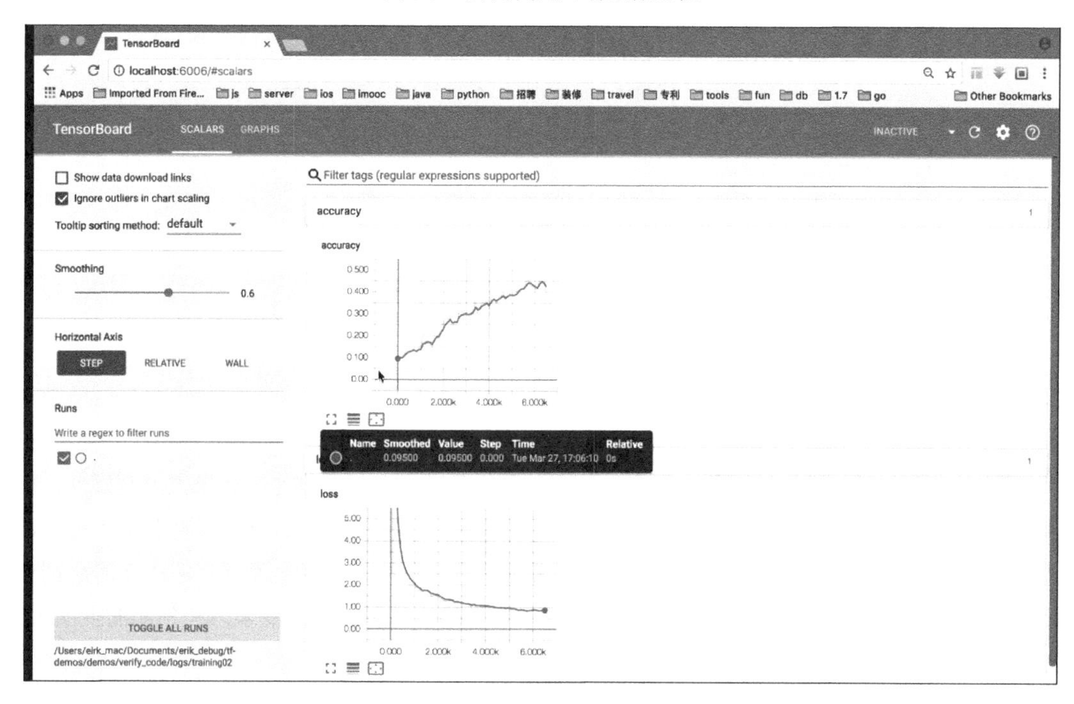

图 5.6　实例运行结果 TensorBoard 跟踪分析

5.3　激活函数

5.3.1　常用激活函数

激活函数是数据科学家根据问题陈述和期望结果的形式做出的主观决策。如果学习过程很慢，或者梯度消失或爆炸，可以尝试改变模型中的激活函数，看看问题是否可以解决。图 5.7 中列出了一些常用的激活函数。

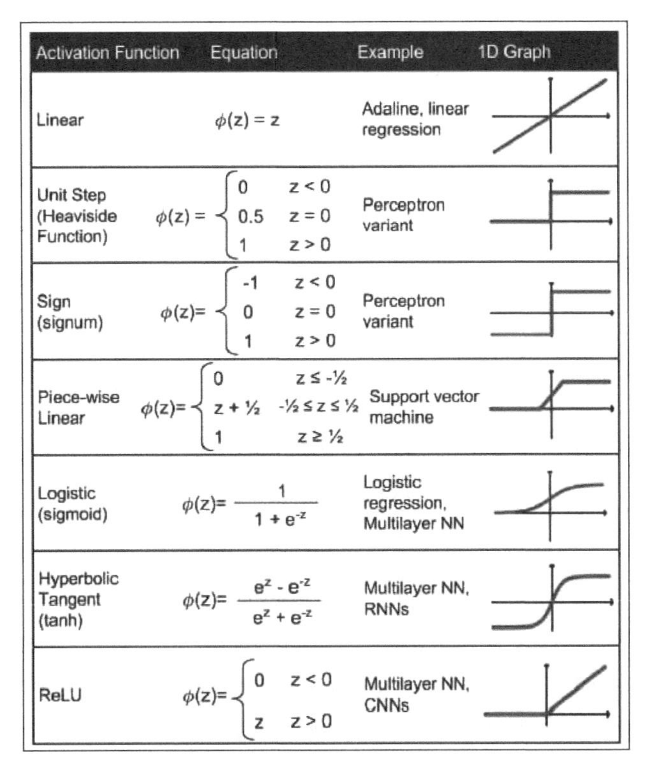

Activation Function	Equation	Example	1D Graph
Linear	$\phi(z) = z$	Adaline, linear regression	
Unit Step (Heaviside Function)	$\phi(z) = \begin{cases} 0 & z < 0 \\ 0.5 & z = 0 \\ 1 & z > 0 \end{cases}$	Perceptron variant	
Sign (signum)	$\phi(z)= \begin{cases} -1 & z < 0 \\ 0 & z = 0 \\ 1 & z > 0 \end{cases}$	Perceptron variant	
Piece-wise Linear	$\phi(z)= \begin{cases} 0 & z \le -\frac{1}{2} \\ z + \frac{1}{2} & -\frac{1}{2} \le z \le \frac{1}{2} \\ 1 & z \ge \frac{1}{2} \end{cases}$	Support vector machine	
Logistic (sigmoid)	$\phi(z)= \frac{1}{1 + e^{-z}}$	Logistic regression, Multilayer NN	
Hyperbolic Tangent (tanh)	$\phi(z)= \frac{e^{z} - e^{-z}}{e^{z} + e^{-z}}$	Multilayer NN, RNNs	
ReLU	$\phi(z)= \begin{cases} 0 & z < 0 \\ z & z > 0 \end{cases}$	Multilayer NN, CNNs	

图 5.7　常用的激活函数

本节主要关注非线性激活函数，因为对于深度学习的场景，每一层节点的输入都是上层输出的线性函数，很容易验证，无论神经网络有多少层，输出都是输入的线性组合，与没有隐藏层效果相当，因此使用非线性激活函数是保证深度神经网络的模型预测能力的关键要素。

接下来，我们一起来学习在深度学习中常用的 3 种激活函数：Sigmoid 函数、tanh 函数以及 ReLU 函数。

5.3.2 Sigmoid 函数

Sigmoid 是常用的非线性激活函数，其数学形式如下：

$$f(z) = \frac{1}{1+e^{-z}}$$

Sigmoid 的函数图像如图 5.8 所示。

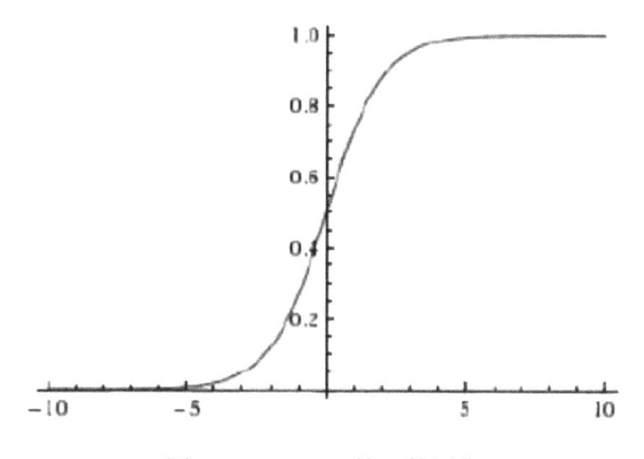

图 5.8 Sigmoid 的函数图像

特点：

Sigmoid 函数能够把输入的连续实值变换为 0 和 1 之间的输出。特别是，如果是非常大的负数，输出就是 0；如果是非常大的正数，输出就是 1。

缺点：

Sigmoid 函数曾经被使用得很多，不过近年来，使用它的人越来越少了，主要是因为它固有的一些缺点。

缺点 1：在深度神经网络中，梯度反向传递时会导致梯度爆炸和梯度消失，其中梯度爆炸发生的概率非常小，而梯度消失发生的概率比较大。首先来看 Sigmoid 函数的导数，如图 5.9 所示。

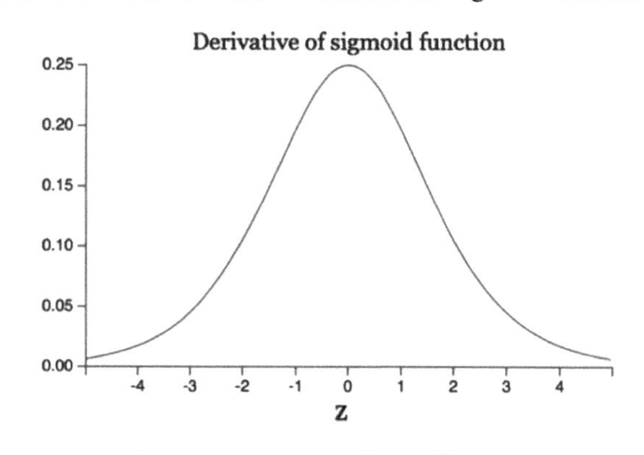

图 5.9 Sigmoid 函数的导数变化

如果初始化神经网络的权值为 [0,1] 之间的随机值，由反向传播算法的数学推导可知，梯度从后向前传播时，每传递一层梯度值都会减小为原来的 0.25 倍。如果神经网络隐层特别多，那么梯度在传过多层后将变得非常小（接近于 0），即出现梯度消失现象；如果网络权值初始化为 $(1,+\infty)$ 区间内的值，就会出现梯度爆炸情况。

缺点 2：Sigmoid 函数的输出不是零均值（Zero-Centered），这是不可取的，因为这会导致后一层神经元将得到的上一层输出的非零均值的信号作为输入。产生的一个结果就是：如果 $x>0$，那么 $f=wTx+b$ 对 w 求局部梯度都为正，这样在反向传播的过程中，w 要么都往正方向更新，要么都往负方向更新，导致有一种捆绑的效果，使得收敛缓慢。当然，如果按 Batch 去训练，那么不同的 Batch 可能得到不同的信号，所以这个问题还是可以缓解一下的。因此，非零均值这个问题虽然会产生一些不好的影响，但是和前面提到的梯度消失问题相比还是要好很多的。

缺点 3：其解析式中含有幂运算，计算机求解时相对来说比较耗时，对于规模比较大的深度网络，这会较大地增加训练时间。

5.3.3 tanh 函数

tanh 函数的解析式如下：

$$\tanh(x)=\frac{e^{x}-e^{-x}}{e^{x}+e^{-x}}$$

tanh 函数及其导数的函数图像如图 5.10 所示。

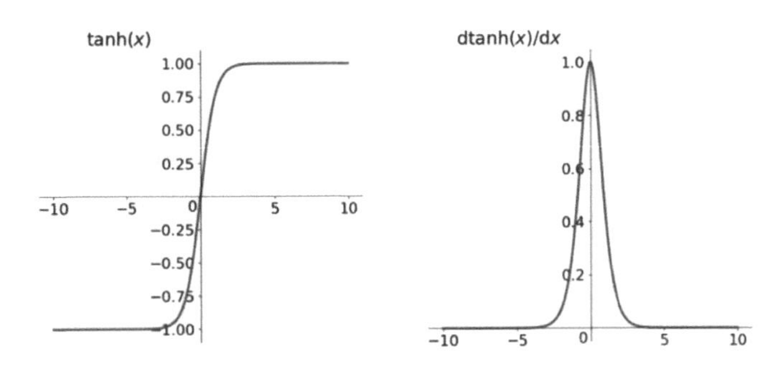

图 5.10　tanh 函数及其导数的函数图像

tanh 读作 Hyperbolic Tangent，它解决了 Sigmoid 函数非零均值输出的问题，然而梯度消失的问题和幂运算的问题仍然存在。

5.3.4 ReLU 函数以及变式

ReLU 函数数的解析式如下：

$$\text{ReLU}(x)=\max(0,x)$$

ReLU 函数及其导数的函数图像如图 5.11 所示。

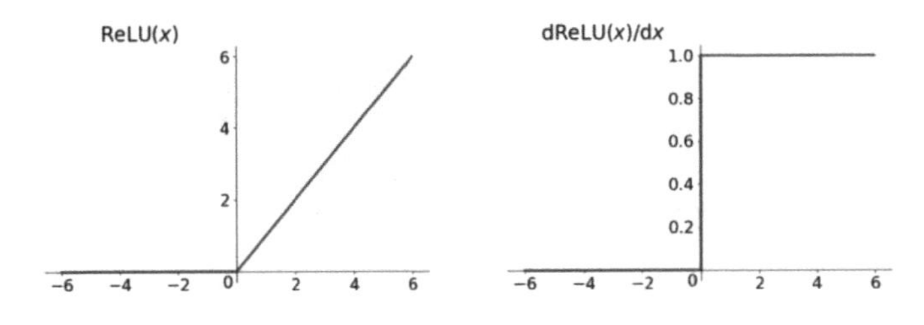

图 5.11　ReLU 函数及其导数的函数图像

ReLU 函数其实就是一个取最大值函数，注意其并不是全区间可导的，但是我们可以取次梯度（Sub-Gradient）。

ReLU 函数虽然简单，但却是近几年的重要成果，有以下几大优点：

（1）解决了梯度消失问题（在正区间）。

（2）计算速度非常快，只需要判断输入是否大于 0。

（3）收敛速度远快于 Sigmoid 和 tanh 函数。

ReLU 函数有几个需要特别注意的问题：

（1）ReLU 函数的输出不是零均值。

（2）出现 Dead ReLU Problem，指某些神经元可能永远不会被激活，导致相应的参数永远不能被更新。有两个主要原因可能导致这种情况产生：①非常不幸的参数初始化，这种情况比较少见；②学习率（Learning Rate）太高，导致在训练过程中参数更新太大，不幸使网络进入这种状态。解决方法是采用 Xavier 初始化方法，以及避免将学习率设置得太高或使用 AdaGrad 等自动调节学习率的算法。

尽管存在这两个问题，ReLU 目前仍是常用的激活函数（Activation Function），可在搭建人工神经网络的时候优先尝试。实际中，人们为了解决 Dead ReLU Problem，提出了将 ReLU 的前半段设为 αx 而非 0，通常 α=0.01。另一种直观的想法是基于参数的方法，即 ParametricRelu：f(x)=max(αx,x)，其中 α 可由反向传播算法计算出来。理论上讲，Leaky ReLU 会在 ReLU 的基础上解决 Dead ReLU 的问题，但是在实际操作中并没有完全证明 Leaky ReLU 的表现总是好于 ReLU。

Leaky ReLU 函数表达式如下：

$$f(x) = \max(\alpha x, x)$$

Leaky ReLU 函数及其导数的函数图像如图 5.12 所示。

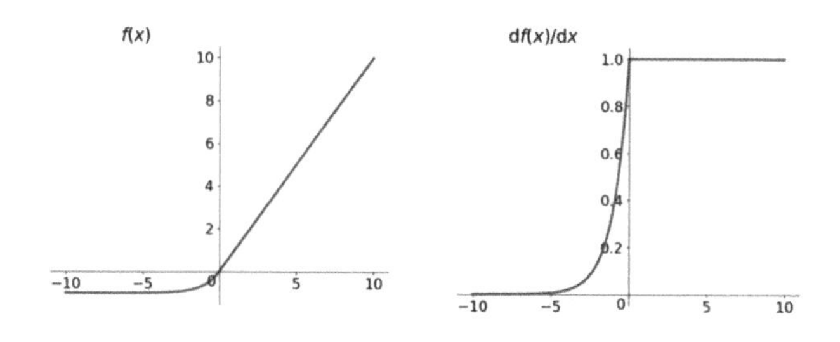

图 5.12　Leaky ReLU 函数及其导数的函数图像

5.4　全连接神经网络模型

5.4.1　多层感知器神经网络结构

在 5.1 和 5.2 节中，我们研究了感知器单层神经网络，这种网络仅限于离散分类模型。由于其基于单一神经元，所以限制了算法的计算能力。要克服单层神经络的局限性可以将感知器进行复合叠加，由此构造的神经网络我们称为多层感知器神经网络结构。多层感知器具有三个基本特征：第一，网络中每个神经元的模型包括一个非线性激活函数可微；第二，网络包含一个或多个对输入和输出节点都隐藏的层；第三，网络表现出高度的连通性，连通性的程度由网络的突触权重决定。然而，这些特征导致了我们对网络行为认识的不足。首先，由于存在分布形式的非线性和网络的高连通性，使得多层感知器的理论分析难以进行。其次，使用隐藏的神经元使得学习过程更加难以形象化。从模型表征意义上说，学习过程必须决定输入模式的哪些特征应该由隐藏的神经元来表示。因此，学习过程变得更加困难，因为搜索必须在更大可能的函数空间中进行，并且必须在输入模式的替代表示之间做出选择。一种常用的多层感知器的训练方法是反向传播算法，其中包括作为特殊情况的 LMS（Least Mean Squan，最小均方算法）算法。该训练包含两个循环进行的阶段：

（1）在正向过程中，神经网络的突触权值是固定的，输入是固定的信号通过网络一层一层地传播的，直到到达输出。因此，在这个阶段，变化仅限于激活电位和输出神经网络中的神经元。

（2）在反向过程中，通过比较系统的输出，使产生误差的错误信号传播到网络中，同样是一层一层的，但这种传播是从后向前的。在这个阶段，可根据计算输出层的调整量来推导网络的突触权重的调整方式。

虽然整个训练的机理非常简单，但是对于隐藏层实际的训练来说会有很多挑战性的状况，包括梯度消失、过拟合现象等。

“反向传播”一词的使用大约是在 1985 年以后发展起来的，当时这个术语是通过出版具有开创性的《平行分布》一书而普及的（Rumelhart 和 McClelland，1986）。这本书介绍了反向传播算法，是 20 世纪 80 年代中期神经网络发展的一个里程碑，它提供了一个具有计算效率的多层感知器的训练方法。

图 5.13 显示了一个隐藏了两个感知器的多层感知器的架构图层和输出层，为多层感知器的描述奠定了基础。

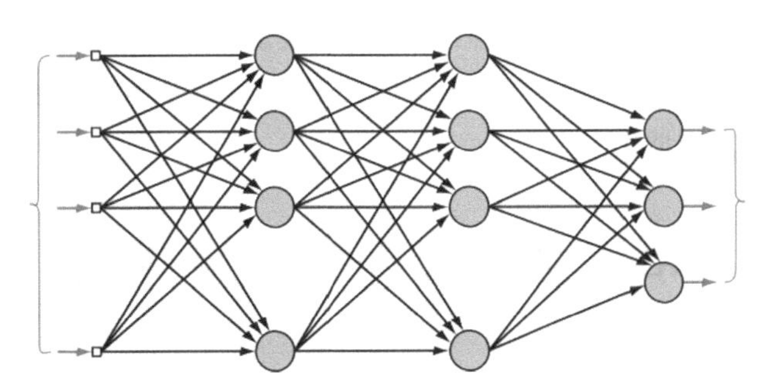

图 5.13　多层感知器示意图

多层感知器中实际运行着两种信号，这两种信号的交替过程就是网络最终确认的过程。

（1）函数信号。函数信号是输入信号（刺激）网络的输入端，通过神经元一个一个地向前传播，并作为输出信号出现在网络的输出端。我们一般假定在网络的输出端会得到一个能够预测问题结果的函数。在计算过程中，每个神经元函数信号通过构造的网络，其信号计算量作用于该神经元的输入和相关权重的最终函数。因此，这个函数信号也称为输入信号。

（2）误差信号。误差信号产生于网络的输出神经元，通过网络向后（逐层）传播，我们称它为“误差信号”是因为网络中的每个神经元计算过程中会产生和最优解的误差偏移。

输出神经元构成网络的输出层，剩下的神经元构成了网络的隐藏层。因此，隐藏的单位不是其中的一部分网络的输出或输入，而是它们被指定为“隐藏的”。第一隐层由感知单元（源节点）构成的输入层馈电，第一隐层的结果作为输出依次应用于下一隐层，对网络的其他部分也是如此。多层感知器的每个隐藏或输出神经元都被设计用来执行以下两个计算：

（1）计算每个神经元输出的函数信号，用输入信号的连续非线性函数表示与神经元相关的突触重量。

（2）梯度向量估计值的计算。换句话说，就是梯度与神经元输入权值相关的误差曲面，这是后面进行的反向传播过程所需要的。

5.4.2 BP 算法

BP（Back Propagation Algorithm，反向传播算法）算法就是反向传播算法，它将误差进行反向传播，从而获取更高的学习效率。这很像烽火台，如果前线战败了，消息就通过烽火台传递回指挥部，指挥部去反思问题，最终改变策略。但这带来一个问题——中间层的误差怎么计算？为解决这个问题，可以简单地将权重和残差的乘积返回给上一层节点，参见下述公式：

$$\delta_i^{(l)}=(\sum_{j=1}^{S_{i+1}}W_{ji}^{(l)}\delta_i^{(l+1)})f'(z_i^{(l)})$$

类似 BP 算法，隐藏神经元的功能也需要我们在抽象意义上进行分析。由于隐藏神经元作为特征检测器在深度学习模型中使用，因此它在多层感知器的学习和预测过程中发挥了关键作用。随着学习过程跨越多层感知器，隐藏的神经元开始逐渐“发现”那些在描述训练数据过程中对于表征结果标签更加有效的特征。它们是通过一个非线性变换来实现的，即将输入数据输入一个特征空间的新空间中，在这个新的空间里，类似对模式分类任务的兴趣可能更容易与之分离，而在原来的输入数据空间中则可能是不同的情况。

5.5 全连接神经网络实例

在 5.2 节使用单一神经元模型对验证码进行了预测，在实例中完成了一个简单的神经网络的构建、学习和预测的过程，也针对实例本身介绍了图像识别的整体流程。主要存在的问题在于模型的预测能力不强，没有达到实际使用的要求，一般图像识别要求模型在特定数据集合上达到 90%以上，因此我们需要利用对神经网络模型的理解升级模型。在 5.4 节介绍了全连接神经网络，在本节的实例中将单一神经元模型升级为全连接神经网络模型。

5.2 节的实例通过 name_scope 将模型分隔成为不同的区域，原始的网络模型（net）代码如下：

```
with tf.name_scope('net') as scope:
    w = tf.Variable(tf.random_normal([9600, 40]), name= 'w')
    b = tf.Variable(tf.random_normal([40]), name= 'b')
    y= tf.matmul(x,w) +b
```

在单神经元模型中，模型中只经过一次矩阵运算，运算结果就传递到了输出层中，从直观上可以理解为机器学习破解程序只是整张地观察了验证码图片。我们在模型中增加多层隐藏神经元，具体代码如下：

```
with tf.name_scope('net') as scope:
    with tf.name_scope('hidden') as scope:
        w1 = tf.Variable(tf.random_normal([9600, 1024]), name= 'w1')
        b1 = tf.Variable(tf.random_normal([1024]), name= 'b1')
        a1 = tf.nn.relu(tf.matmul(x,w1) +b1)
        for i in range(10):
            w2 = tf.Variable(tf.random_normal([1024, 1024]), name= 'w2')
            b2 = tf.Variable(tf.random_normal([1024]), name= 'b2')
            a2 = tf.nn.relu(tf.matmul(a1,w2) +b2)
            a1 = a2
with tf.name_scope('output') as scope:
        w = tf.Variable(tf.random_normal([1024, 40]), name= 'w')
        b = tf.Variable(tf.random_normal([40]), name= 'b')
        y=  tf.matmul(a2,w) +b
```

整个神经网络模型架构对比 5.2 节单一神经元模型增加了用于中间处理的隐藏神经元层次结构，相关代码定义为隐藏层（tf.name_scope('hidden')）。在第一层中接收输入图片的张量数据，进行运算和激活，为了保证网络的运算效率，激活函数选择 ReLU，这样就构建了一个隐藏神经网络层。其他层的构造与这一层相似，我们使用一个循环来控制，创建多个隐藏层。

由于网络层新的输出张量大小为 1 024，因此最终输出层（output）需要进行微调整，将权重矩阵中的 9 600 减小到 1 024。更新后的代码如下：

```
with tf.name_scope('output') as scope:
        w = tf.Variable(tf.random_normal([1024, 40]), name= 'w')
        b = tf.Variable(tf.random_normal([40]), name= 'b')
        y= tf.matmul(a2,w) +b
```

在终端运行实例代码，通过我们构造的训练数据集合不断更新整个网络的权重和偏置，训练和预测结果使用 TensorBoard 进行可视化展示，如图 5.14 所示。可以清楚地看到，损失函数的值随着训练数据的增加逐渐减小到一个稳定的值的过程与单神经元模型是完全一致的，而预测准确率获得了一定的提升。另外，使用循环可以进行输入层数的调整，从结果可以清晰地看到随着全连接神经网络深度的增加，在训练过程中，同样的数据量可以更快地提升模型的预测准确率，但是模型最终的预测准确率的提升相对比较小。从 TensorBoard 准确率的折线图来看，当训练次数达到 1 000 次时，具有 10 层隐藏层的全连接模型已经基本达到模型预测能力的最大值，而只有一层隐藏层的模型需要 5 000 次以上的训练次数。从直观上来理解，模型虽然多次分析网络，但是只能提取综合特征，因此模型的理解力很快就达到了瓶颈。为了能够真正完成验证码破解的要求，我们将在第 6

章中祭出神经网络模型中的“大杀器”——卷积神经网络，完成端对端验证码识别程序的最终版本。

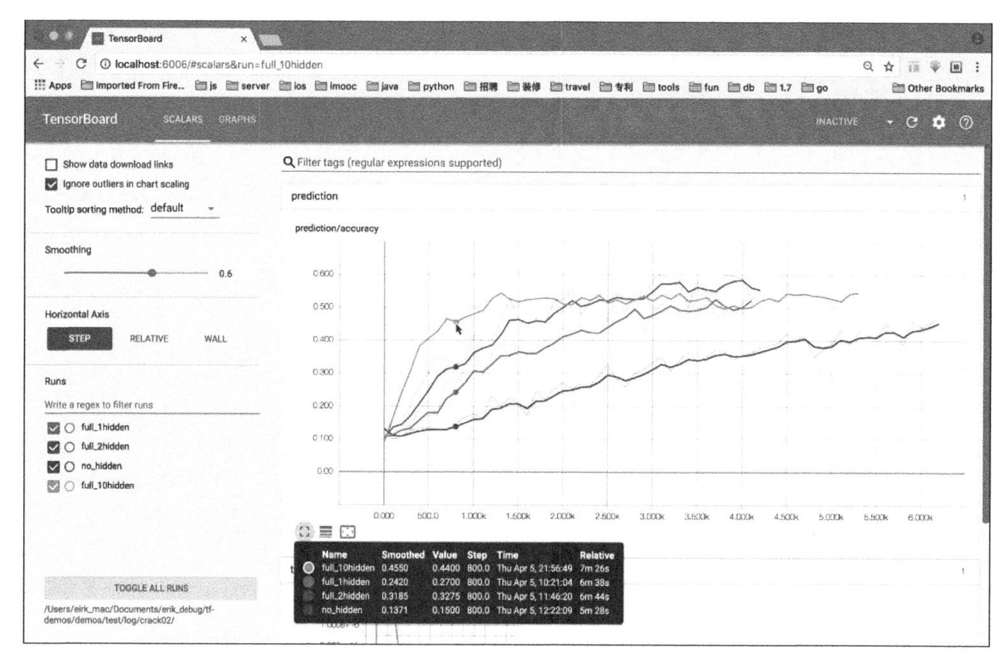

图 5.14 训练和预测结果使用 TensorBoard 进行可视化展示

参考文献

[1] Rosenblatt F. The perceptron: a probabilistic model for information storage and organization in the brain.[M]// Neurocomputing: foundations of research. 1988.

[2] Widrow B, Lehr M A. 30 years of adaptive neural networks: perceptron, Madaline, and backpropagation[J]. Proceedings of the IEEE, 1990, 78(9):1415-1442.

[3] Gardner M W. Artificial neural networks (the multiplayer perceptron)-a review of applications in the atmospheric sciences[J]. Atmospheric Environment, 1998, 32.

[4] Setiono R, Liu H. Improving backpropagation learning with feature selection[J]. Applied Intelligence: The International Journal of Artificial, Intelligence, Neural Networks, and Complex Problem-Solving Technologies, 1996, 6(2):129-139.

[5] Lu H, Setiono R, Liu H. NeuroRule: A Connectionist Approach to Data Mining[J]. 2017.

[6] Banerjee M, Mitra S, Pal S K. Rough fuzzy MLP: knowledge encoding and classification[J]. IEEE Transactions on Neural Networks, 1998, 9(6):1203-1216.

[7] Pal S K, Mitra S. Fuzzy versions of Kohonen\"s net and MLP-based classification: Performance evaluation for certain nonconvex decision regions[J]. Information Sciences, 1994, 76(3-4):297-337.

[8] Robert Fullér. Fuzzy neural networks[M]// Introduction to Neuro-Fuzzy Systems. Physica-Verlag HD, 2000.

第6章

卷积神经网络

卷积神经网络无疑是近年来最成功的机器学习算法，通过机器视觉应用确立了其算法设计的预测准确性优势后，借助与其他模型的前后置组合或者对计算层流程的再设计，使得卷积神经网络在自然语言处理、机器听觉等方面也大放异彩，硕果累累。

本章主要讲解神经网络中的神经元模型以及全连接神经网络模型，通过验证码图片破解过程加深对模型的应用和理解。内容分为三部分，第一部分着重介绍卷积神经网络模型的架构，特别着重分析模型中重要的卷积操作和池化操作；第二部分使用卷积神经网络模型进行验证码图片破解程序的升级，通过这一过程加深对模型的应用和理解；第三部分着重分析机器学习中常见的过拟合和欠拟合问题，通过实例得到对应的处理方式。

6.1　卷积神经网络概述

卷积神经网络作为一种非常有前途的深度学习的网络构建形式，近年来在人工智能应用领域取得了长足的进展，也获得了特殊的地位。卷积层的设计虽然最初是针对图像处理的特征提取问题，但是随着发展，实际上它已经进入深度学习的所有子领域并且在很大程度上取得了巨大成功。

6.1.1　卷积神经网络架构

全连接神经网络和卷积神经网络的根本区别在于连续层之间的连接模式。前者是在完全连接的情况下，每个神经元都连接到上一层中的所有神经元，而后者与之不同，在神经网络的卷积层中，每一个神经元都连接到前一层中一小部分神经元，更重要的是，所有单元都以相同的方式连接到上一层，具有完全相同的权重和结构。这种部分传导运算被称作卷积运算，但简而言之，它对我们的全部意义就是在图像上应用一个小的权重窗口（也称过滤器），因此神经网络结构称为卷积神经网络。

通常认为，卷积神经网络算法的核心架构卷积层的设计形成过程来自不同学派的理论基础，目前在人工智能领域有三种主要的观点：第一种观点是模型背后所谓的神经科学灵感；第二种观点是图像科学对图像本质的洞察；第三种观点是学习理论。在深入研究实际的卷积神经网络的数学运算和训练技巧之前，我们将逐一介绍这些理论。

一般来说，学术界将神经网络，特别是卷积神经网络描述为受生物学启发的计算模型是一种很流行的说法。其中的一些学派甚至声称这些模拟了大脑执行计算的方式，虽然从表面上看有误导作用，但生物学上的类比还是有一定意义的。诺贝尔奖得主神经生理学家 Hubel 和 Wiesel 早在 20 世纪 60 年代就发现，大脑视觉处理的第一阶段包括对视野的所有部分应用相同的局部过滤器（例如边缘探测器）。神经科学界目前的理解是，随着视觉处理的进行，输入的图像信息的像素特征逐渐被部分整合，这个过程是分层进行的，卷积神经网络遵循同样的模式。随着网络的深入，我们在每个卷积层都能看到图像中越来越大的部分。常见的情况是，之后会出现完全连接的层，在生物启发的类比中，这些层将会作为处理全局信息的更高层次的视觉处理。

第二种观点更加强调卷积神经网络在图像处理领域取得的出色成果，通过更多事实工程为导向的实际机器视觉案例分析得出卷积神经网络源于图像的本质及其内容。具体来说，当在图像中寻找对象时，比如猫的脸，我们通常希望能够探测到它，而不管它在图像中的位置如何。这反映了自然图像的特性，即相同的内容可以在图像的不同位置找到，这个性质被称为不变性，这种类型的不变性也可以在（小）旋转、改变光照条件等方面得到预期。相应的，当构建一个对象识别系统时，它应该对平移保持不变。简单地说，图像的不同部分执行相同的精确计算是有意义的。因此，在这种观点中，卷积神经网络层遍历所有空间区域计算图像的相同特征是很有实际工程意义的。

最后一种观点，卷积结构可以看作一种正则化机制。在这种机制中，卷积层在结构上类似全连接层，只是增加了对整个空间的搜索权重矩阵（一般需要指定特定大小），通过限制搜索矩阵的卷积核的固定大小的变化减少卷积运算操作的自由度大小。由于采用局部连接和权值共享，保持网络深层结构的同时又大大减少了网络参数，使卷积神经网络模型既具有良好的泛化能力又较容易训练。

在卷积神经网络的运算中有两种标志性的数学变换操作，它们分别是卷积操作和池化操作。接下来，我们分别介绍这两种操作的具体过程和直观意义。

6.1.2 卷积操作

首先，我们来一起学习一下卷积操作。从算法的名字上就能够清楚地感受到这种操作对于卷积神经网络算法的意义。我们可以思考一个机器视觉的问题：人类为什么能够很快地通过看物体识别出物体的类别。有一种比较可靠的解释可以通俗地表述为，观察者主要是看到了物体的某些抽象化的标志特征，并且能够组合这些标志特征形成更加确定的复杂特征，一旦标志特征的信号强度超过观察者大脑对物体认知的阈值，我们就完成了物体的视觉识别过程。这种解释提供了我们升级神经网络的前进方向。

从本质上来说，每张图像都可以表示为像素值的矩阵。通道常用于表示图像的某种组成，一个标准数字相机拍摄的图像有 3 个通道：红、绿和蓝，你可以把它们看作互相堆叠在一起的二维矩阵（每一个通道代表一种颜色），每个通道的像素值在 0~255 的范围内。灰度图像只有一个通道。

本节仅考虑灰度图像，这样就只有一个二维的矩阵来表示图像。矩阵中各个像素的值在 0~255 的范围内（0 表示黑色，255 表示白色）。卷积的主要目的是为了从输入图像中提取特征。卷积可以通过从输入的一小块数据中学到图像的特征，并可以保留像素间的空间关系。在这里不会详细讲解卷积的数学细节，但我们可以试着理解卷积是如何处理图像的。正如前面所说的，每张图像都可以看作像素值的矩阵。考虑一个 5×5 的图像，它的像素值仅为 0 或者 1（注意，对于灰度图像而言，像素值的范围是 0~255，图 6.1 所示的像素值为 0 和 1 的示意矩阵仅为特例）。

同时，考虑另一个 3×3 的矩阵，如图 6.2 所示。

图 6.1　图像矩阵

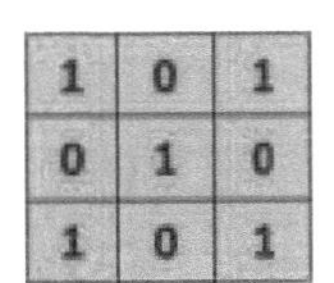

图 6.2　卷积核矩阵

接下来，5×5 的图像和 3×3 的矩阵的卷积可以按图 6.3 所示的过程进行计算。

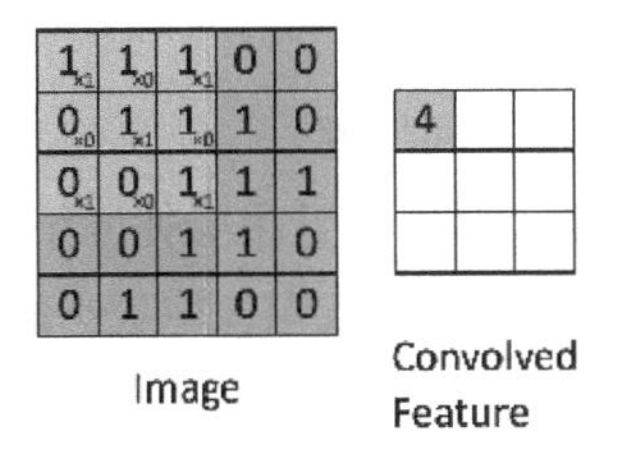

图 6.3　卷积过程示意图

下面详细说明图 6.3 的计算是怎样完成的。我们用橙色的矩阵在原始图像（绿色）上滑动，每次滑动一个像素（也叫作步长），在每个位置上计算对应元素的乘积（两个矩阵间），并把乘积的和作为最后的结果，得到输出矩阵（粉色）中的每一个元素的值。注意，3×3 的矩阵每个步长中仅可以“看到”输入图像的一部分。

在卷积神经网络的术语中，3×3 的矩阵叫作滤波器（Filter）、核（Kernel）或者特征检测器（Feature Detector），通过在图像上滑动滤波器并计算点乘得到的矩阵叫作卷积特征（Convolved Feature）、激活图（Activation Map）或者特征图（Feature Map）。

滤波器在原始输入图像上的作用是特征检测器，需要注意的是不同值的滤波器会生成不同的特征图，通过在卷积操作前修改滤波矩阵的数值，可以进行诸如边缘检测、锐化和模糊等操作，这表明不同的滤波器可以从图中检测到不同的特征，比如边缘、曲线等。

6.1.3　池化操作

本节分析池化操作的一些要点。空间池化（Spatial Pooling，也叫作亚采样或者下采样）降低了各个特征图的维度，但可以保持大部分重要的信息。空间池化有最大化、平均化、加和等方式。

对于最大池化（Max Pooling），我们可以定义一个空间邻域（比如 2×2 的窗口），并从窗口内的修正特征图中取出最大的元素。除了取最大元素外，也可以取平均池化（Average Pooling），即取窗口元素和的平均值，或者直接对窗口内的元素求和。实际上，最大池化被证明效果更好一些。

图 6.4 展示了使用 2×2 窗口在修正特征图（在卷积 + ReLU 操作后得到）中使用最大池化的例子。

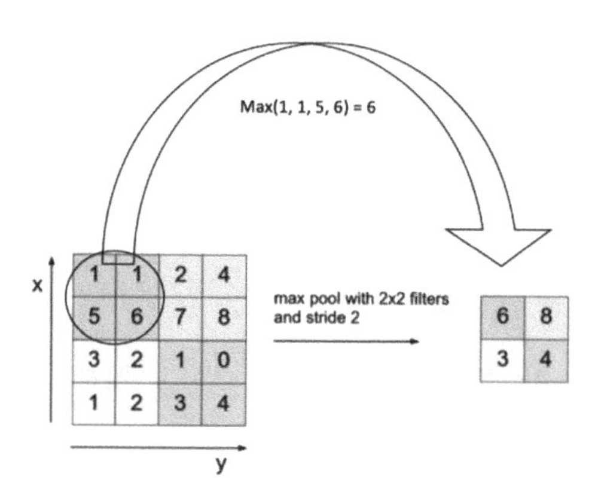

图 6.4　最大池化示意图

池化函数可以逐渐降低输入表示的空间尺度。

池化使输入表示（特征维度）变得更小，并且也会使网络中的参数和计算的数量更加可控地减小，因此可以控制过拟合。池化使网络对于输入图像中更小的变化、冗余和变换变得具有不变性，输入的微小冗余将不会改变池化的输出，因为我们在局部邻域中使用了最大化/平均值的操作。池化可以帮助我们最大程度上获取图像的尺度不变性。可以说，池化的功能非常强大，我们可以利用其检测图像中的物体，无论物体的位置在哪里。

6.1.4　卷积神经网络的特点

总结来说，卷积神经网络是 BP 神经网络的改进，与 BP 神经网络类似，都采用了前向传播计算输出值，反向传播调整权重和偏置。卷积神经网络与标准的 BP 神经网络最大的不同是：卷积神经网络中相邻层之间的神经单元并不是全连接的，而是部分连接的，也就是某个神经单元的感知区域来自于上层的部分神经单元，而不是像 BP 神经网络那样与所有的神经单元相连接。卷积神经网络有 3 个重要的思想架构：局部区域感知、权重共享和空间或时间上的采样。

局部区域感知能够发现数据的一些局部特征，比如图片上的一个角、一段弧，这些基本特征是构成动物视觉的基础；而 BP 神经网络中，所有的像素点是一堆混乱的点，相互之间的关系没有被挖掘。

卷积神经网络中每一层由多个 Map 组成，每个 Map 由多个神经单元组成，同一个 Map 的所有神经单元共用一个卷积核（权重），卷积核往往代表一个特征，比如某个卷积和代表一段弧，把这个卷积核在整个图片上滚一下，卷积值较大的区域就很有可能是一段弧。注意，卷积核其实就是权重，我们并不需要单独去计算一个卷积，而是一个固定大小的权重矩阵去匹配整个图像，这个操作与卷积类似，因此我们称之为卷积神经网络。实际上，BP 神经网络也可以看作是一种特殊的卷积神经网络，只是这个卷积核就是某层的所有权重，即感知区域是整个图像。卷积神经网络的权重共享策略减少了需要训练的参数，使得训练出来的模型的泛化能力更强。

采样的目的主要是混淆特征的具体位置，因为某个特征找出来后，它的具体位置已经不重要了，我们只需要这个特征与其他的相对位置，比如一个 8，当我们得到了上面一个 o 时，不需要知道它在图像上的具体位置，只需要知道它下面又是一个 o，就可以知道是一个 8 了，因为图片中“8”在图片中偏左或者偏右都不影响我们认识它。这种混淆具体位置的策略能对变形和扭曲的图像进行识别。

卷积神经网络的这 3 个特点使其对输入数据在空间（主要针对图像数据）和时间（主要针对时间序列数据，参考时延神经网络（Time-Delay Neural Network，TDNN）上的扭曲有很强的鲁棒性。卷积神经网络一般采用卷积层与采样层交替设置，即一层卷积层接一层采样层，采样层后接一层卷积……这样卷积层提取出特征，再进行组合形成更抽象的特征，最后形成对图片对象的描述特征，卷积神经网络后面还可以跟全连接层，全连接层跟 BP 神经网络一样。

6.2　实例 1：验证码识别

前文我们理解了卷积神经网络的核心架构，包括卷积层的工作原理和池化层的工作原理，本实例将会再次升级验证码识别的神经网络模型，使用卷积神经网络进行验证码图片的多次特征提取，最终达到模型能够很好地理解验证码图片的结果（模型预测准确率达到 90%+）。

6.2.1　神经网络的具体设计

下面首先分析整个验证码识别的卷积神经网络的架构，我们设计通过多层（3 层）卷积和池化操作来提取验证码图片输入的图像特征，对应图 6.5，在输入层之后串行 3 个层（卷积 1、卷积 2 和卷积 3）。卷积完成后的输出结果需要通过一个全连接层来调整输出数据的规模，最终通过输出层（广义上也可以认为是一个全连接层）输出预测结果。

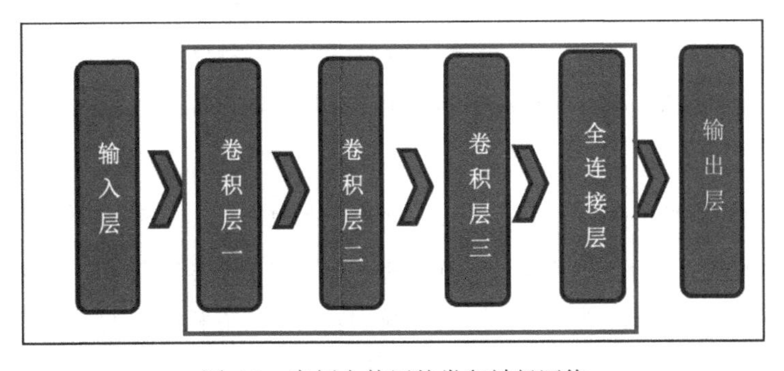

图 6.5　实例中使用的卷积神经网络

根据我们的分析，按照模型架构只需要更新神经网络模型部分（tf.name_scope('net')）即可，更新代码如下：

```
with tf.name_scope('net') as scope:
    w_alpha = 0.01
    b_alpha = 0.1
    x_ = tf.reshape(x, shape=[-1, 60, 160, 1])
    with tf.name_scope('conv1') as scope:
        w_c1 = tf.Variable(w_alpha*tf.random_normal([3, 3, 1, 32]))
        b_c1 = tf.Variable(b_alpha*tf.random_normal([32]))
        conv1 = tf.nn.conv2d(x_, w_c1, strides=[1, 1, 1, 1], padding='SAME')
        conv1 = tf.nn.relu(tf.nn.bias_add(conv1, b_c1))
        conv1 = tf.nn.max_pool(conv1, ksize=[1, 2, 2, 1], strides=[1, 2, 2, 1],
padding='SAME')
    with tf.name_scope('conv2') as scope:
        w_c2 = tf.Variable(w_alpha*tf.random_normal([3, 3, 32, 64]))
        b_c2 = tf.Variable(b_alpha*tf.random_normal([64]))
        conv2 = tf.nn.conv2d(conv1, w_c2, strides=[1, 1, 1, 1], padding='SAME')
        conv2 = tf.nn.relu(tf.nn.bias_add(conv2, b_c2))
        conv2 = tf.nn.max_pool(conv2, ksize=[1, 2, 2, 1], strides=[1, 2, 2, 1],
padding='SAME')
    with tf.name_scope('conv3') as scope:
        w_c3 = tf.Variable(w_alpha*tf.random_normal([3, 3, 64, 64]))
        b_c3 = tf.Variable(b_alpha*tf.random_normal([64]))
        conv3 = tf.nn.conv2d(conv2, w_c3, strides=[1, 1, 1, 1], padding='SAME')
        conv3 = tf.nn.relu(tf.nn.bias_add(conv3, b_c3))
        conv3 = tf.nn.max_pool(conv3, ksize=[1, 2, 2, 1], strides=[1, 2, 2, 1],
padding='SAME')
    with tf.name_scope('dense') as scope:
        w_d = tf.Variable(w_alpha*tf.random_normal([8*20*64, 1024]))
        b_d = tf.Variable(b_alpha*tf.random_normal([1024]))
        dense = tf.reshape(conv3, [-1, w_d.get_shape().as_list()[0]])
        dense = tf.nn.relu(tf.add(tf.matmul(dense, w_d), b_d))
```

代码中定义了卷积层为conv1、conv2和conv3，以及全连接层dense。全连接层与第5章的全连接层的结构是完全一致的。唯一需要注意的是全连接层的输入需要进行数据预处理（数据拍平），对应代码如下：

```
dense = tf.reshape(conv3, [-1, w_d.get_shape().as_list()[0]])
```

接下来，我们需要认真分析卷积层的工作原理，由于在3个卷积层的工作方法是一致的，只是卷积的权重和偏置矩阵的形式有些许不同，因此我们以第一层卷积网络为核心模型，使用分治的思想进行详细分析。在计算机领域分析问题经常使用的是分治法，在计算机科学中，分治法是一种很重要的算法。分治法字面上的解释是“分而治之”，就是把一个复杂的问题分成两个或更多的相同或相似的子问题，再把子问题分成更小的子问题，重复这个过程直到最后子问题可以简单地直接求解，求解完成后通过合并子问题的解得到原问题的完整解答。为了清晰地解释卷积神经网络模型进行验证码识别的实例，我们使用分治的思想，首先分别解释实例中卷积代码和池化代码的运作流程和特点，然后以一次卷积池化过程为蓝本分析训练过程特征的变化趋势和预测性能的提升，最终进行完整模型的训练和预测准确率的跟踪显示。在模型分析的过程中，我们将使用 TensorBoard 的图片（Image）功能，图片功能能够保存并展示卷积学习训练过程中通过卷积、激活和池化功能对验证码图片矩阵的特征提取过程，另外我们可以比较清晰地看到随着训练数据的注入，模型对于提取特征逐渐强化的过程。

6.2.2　卷积过程分析

卷积过程的核心代码如下：

```
conv1 = tf.nn.conv2d(x_, w_c1, strides=[1, 1, 1, 1], padding='SAME')
```

核心函数是 tf.nn.conv2d，这个函数是 TensorFlow 对卷积操作的实用封装。函数定义如下：

```
    tf.nn.conv2d(input, filter, strides, padding, use_cudnn_on_gpu=None,
name=None)
```

除去 name 参数用以指定该操作的 name 外，与方法有关的参数一共有 5 个：

第 1 个参数 input：指需要进行卷积的输入图像，要求是一个张量，具有[batch, in_height, in_width, in_channels]这样的 shape，具体含义是[训练时一个 batch 的图片数量，图片高度，图片宽度，图像通道数]，注意这是一个 4 维的张量，要求类型为 float32 或 float64。

第 2 个参数 filter：相当于卷积神经网络中的卷积核，要求是一个张量，具有[filter_height, filter_width, in_channels, out_channels]这样的 shape，具体含义是[卷积核的高度，卷积核的宽度，图像通道数，卷积核个数]，要求类型与参数 input 相同，有一个地方需要注意，第 3 维 in_channels 就是参数 input 的第 4 维。

第 3 个参数 strides：卷积时在图像每一维的步长，这是一个一维的向量，长度为 4。

第 4 个参数 padding：string 类型的量，只能是 SAME、VALID 其中之一，这个值决定了不同的卷积方式。

第 5 个参数：use_cudnn_on_gpu:bool 类型，是否使用 cuDNN 加速，默认为 True。

结果返回一个张量，这个输出就是我们常说的特征图（Feature Map），shape 仍然是[batch, height, width, channels]这种形式。

我们在实例中使用了[3, 3, 1, 32]的卷积核，因为卷积核矩阵会随着训练的进行不但优化，因此我们可以直观地理解卷积核也是权重矩阵的一种。

```
w_c1 = tf.Variable(w_alpha*tf.random_normal([3, 3, 1, 32]))
```

卷积结果和原始图像输入关系相对难以理解和想象，因此我们利用前文提到过的 TensorBoard，图像输出功能将卷积结果直接映射成图像来直观地展示分析。在分析之前，我们仍然需要通过

tf.summary 指定存储格式。

```
image = tf.reshape(conv1, shape=[-1, 60, 160, channels])
        tf.summary.image("image1-1", image)
```

存储之前的数据转换是必需的，shape 的最后一位是和卷积的通道数相对应的，例如实例中使用 32，这里就是 32。实例中尝试存储 1、2、32 通道的卷积结果，重构为图像，如图 6.6 所示。我们可以看到随着使用通道的维度上升，验证码本身的图片特征被提取为更加抽象的图像特征。

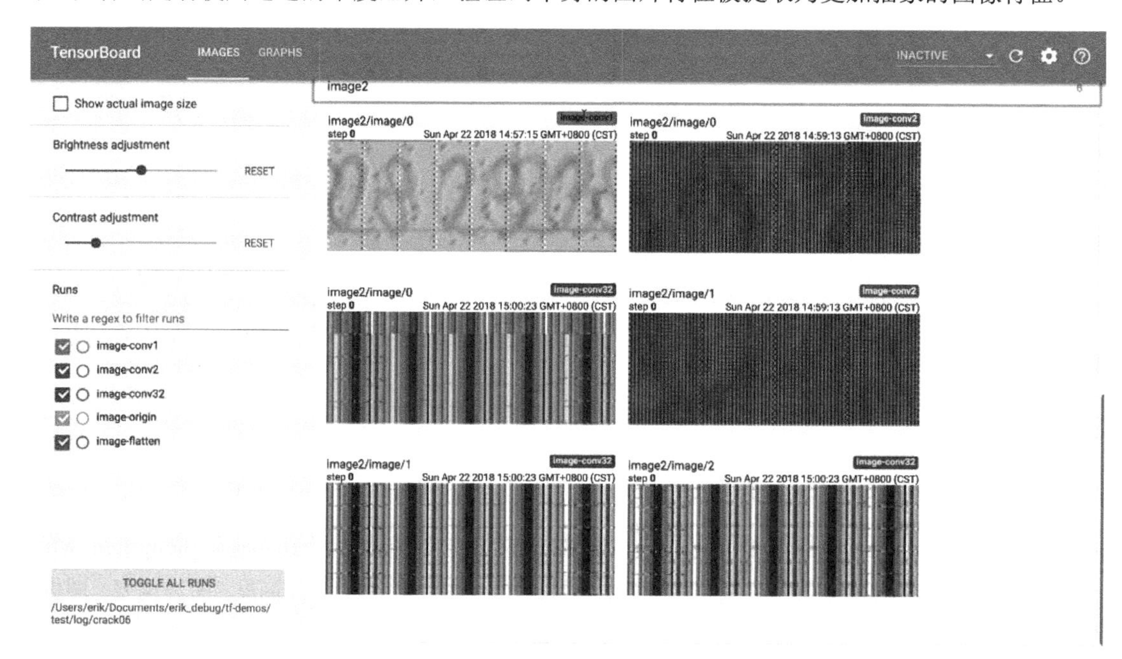

图 6.6 使用 1、2、32 通道的卷积结果

6.2.3 池化过程分析

池化过程的核心代码如下：

```
conv1 = tf.nn.max_pool(conv1, ksize=[1, 2, 2, 1], strides=[1, 2, 2, 1], padding='SAME')
tf.nn.max_pool(value, ksize, strides, padding, name=None)
```

参数是 4 个，和卷积很类似：

第一个参数 value：需要池化的输入，一般池化层接在卷积层后面，所以输入通常是特征图（Feature Map），依然是[batch, height, width, channels]这样的 shape。

第二个参数 ksize：池化窗口的大小，取一个四维向量，一般是[1, height, width, 1]，因为我们不想在 batch 和 channels 上做池化，所以这两个维度设为 1。

第三个参数 strides：和卷积类似，窗口在每一个维度上滑动的步长，一般也是[1, stride,stride, 1]。

第四个参数 padding：和卷积类似，可以取 VALID 或者 SAME。

返回一个张量，类型不变，shape 仍然是[batch, height, width, channels]这种形式。

我们使用 summary 继续进行输出特征图的跟踪，需要注意的是池化操作实际上是对特征图的亚采样，最终输出的结果是原始特征图的子集，因此数据转换时需要输入正确的 shape 矩阵，在当前实例中，实现代码如下：

```
image = tf.reshape(conv1, shape=[-1, 30, 80, channels])
        tf.summary.image("image1-3", image)
```

图 6.7 展示了 32 维池化的结果，直观看来图像中的黑色部分明显增多，并且整体数据规模减小为原来的 1/4。当我们运行程序时，运行效率会有很大的提升。

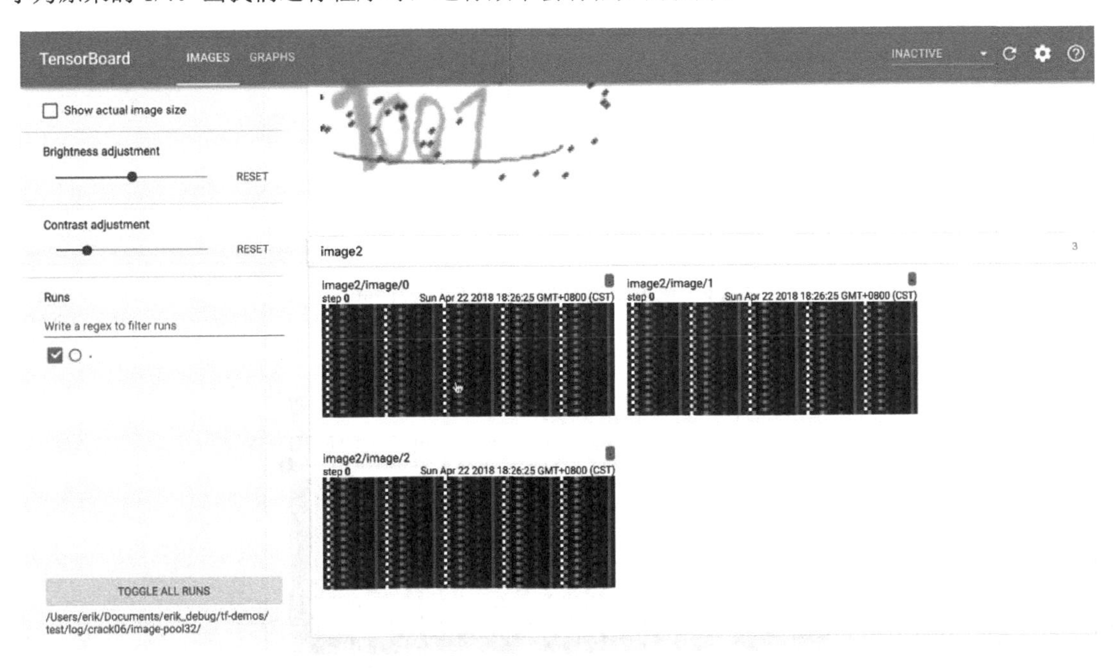

图 6.7　32 维池化的结果

6.2.4　完整学习过程分析

在单次完整卷积池化操作中，通过卷积、ReLU 激活、池化三个步骤得到图 6.8 所示的结果，可以清楚地看到预测准确率有了质的提升。同时，通过 TensorBoard 可以看到卷积、激活和池化对特征图的作用过程。

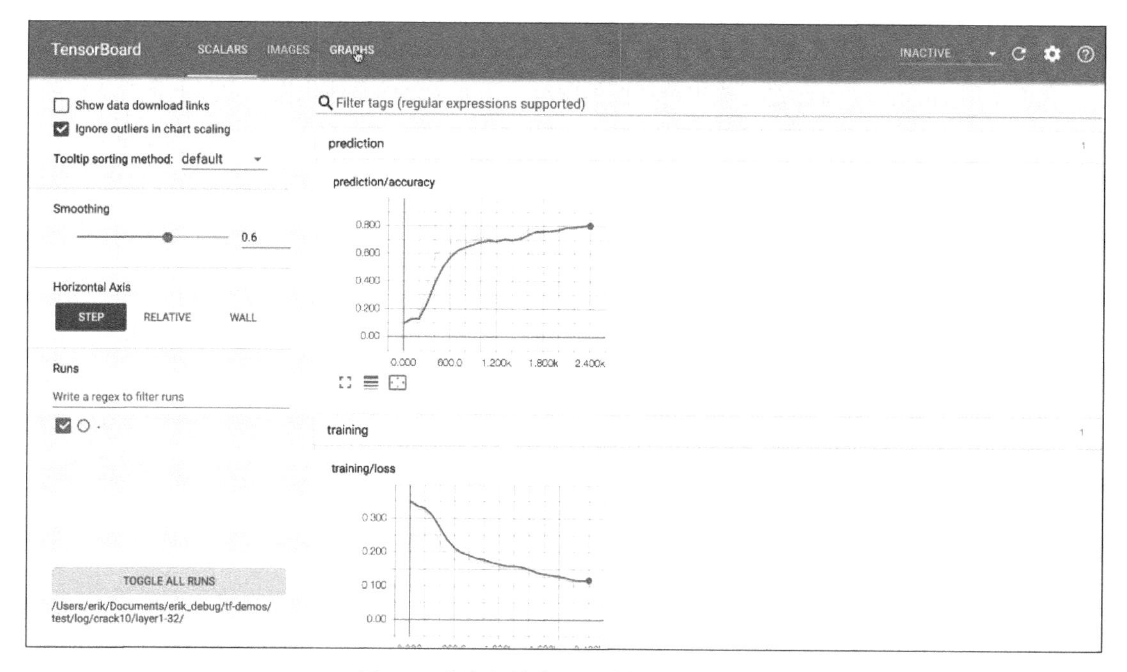

图 6.8　单次完整卷积池化操作预测精度

卷积后的图像结果如图 6.9 所示。

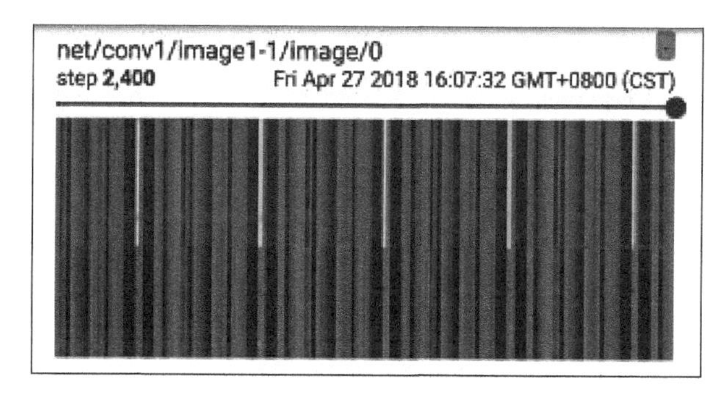

图 6.9　卷积后的图像结果

卷积并且 ReLU 激活后的图像结果如图 6.10 所示。

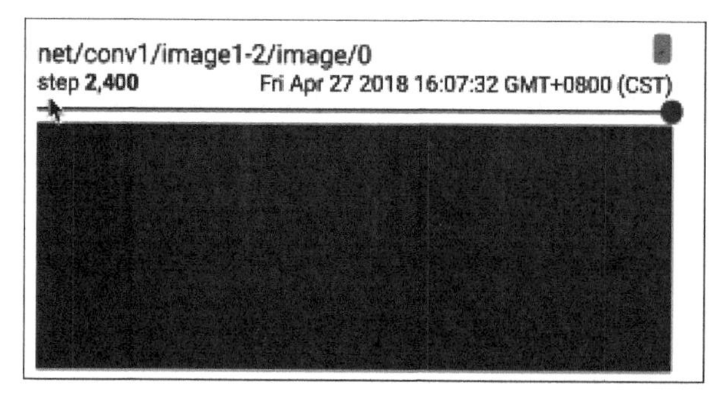

图 6.10　卷积并且 ReLU 激活后的图像结果

卷积、ReLU 激活和池化后的图像结果如图 6.11 所示。

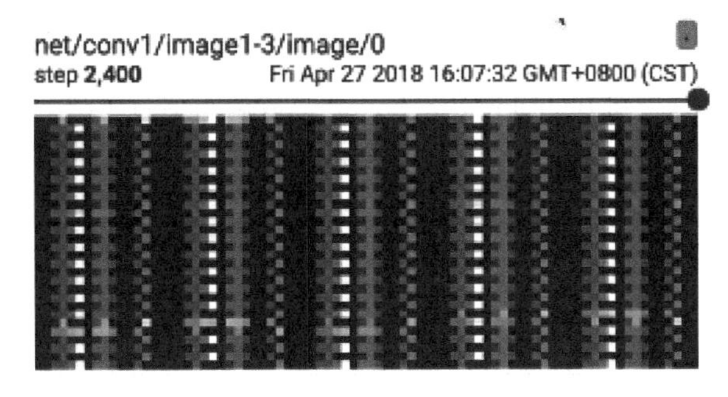

图 6.11 卷积、ReLU 激活和池化后的图像结果

当我们使用 3 层完整的卷积神经网络进行验证码预测时，预测准确率超过了 95%，已经满足验证码端对端识别的要求，如图 6.12 所示。

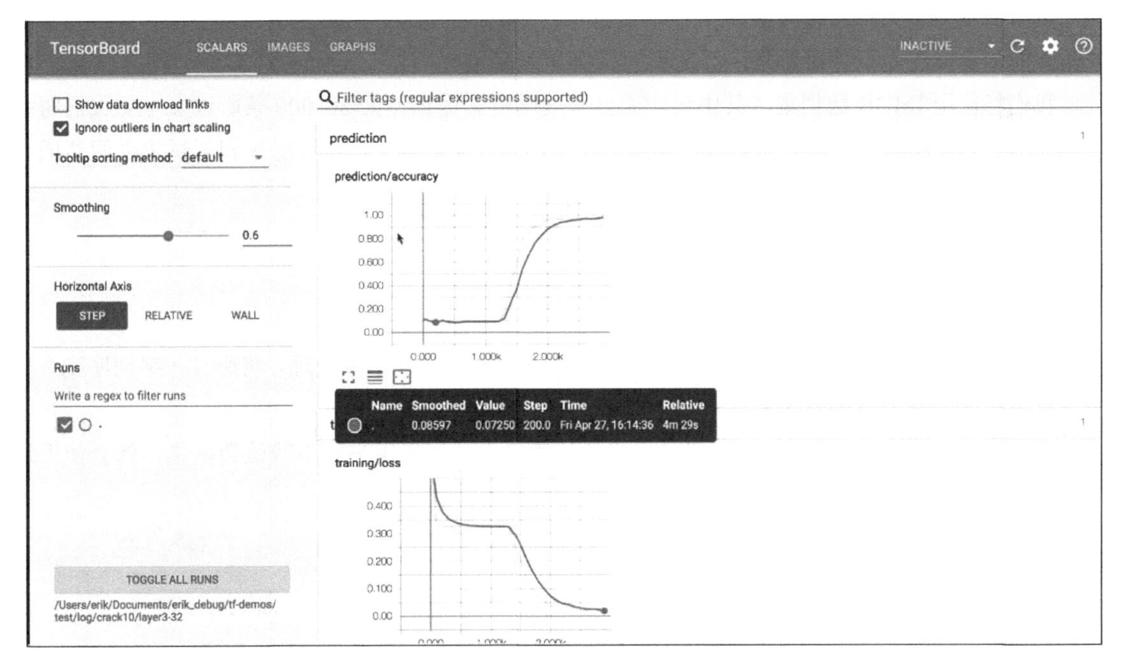

图 6.12 3 层完整卷积池化操作预测精度

6.3 实例 2：过拟合和欠拟合

在很多机器学习，特别是神经网络的实际开发实践中，我们有机会了解在训练周期达到一定次数后，模型在验证数据上的准确率会达到峰值，然后便开始下降。

也就是说，模型会过拟合训练数据。按照我们之前的介绍，模型在训练数据集上取得很好的预测结果和保持在训练和测试数据集合上的能力一致性同等重要。尽管通常可以在训练集上实现很高的准确率，但我们真正想要的是开发出能够很好地泛化到测试数据（或之前未见过的数据）的模型。

与过拟合相对的是欠拟合。当测试数据仍存在改进空间时便会发生欠拟合。出现这种情况的原因有很多，如模型不够强大、过于正则化，或者根本没有训练足够长的时间等。这意味着网络未学习训练数据中的相关模式。

如果训练时间过长，模型将开始过拟合，并从训练数据中学习无法泛化到测试数据的模式。我们需要在这两者之间实现平衡。了解如何训练适当的周期次数是一项很实用的技能。接下来我们将介绍这一技能。

为了防止发生过拟合，最好的解决方案是使用更多的训练数据，用更多数据进行训练的模型自然能够更好地泛化。若无法采用这种解决方案，则次优解决方案是使用正则化等技术，这些技术会限制模型可以存储的信息的数量和类型。如果网络只能记住少量模式，那么优化过程将迫使它专注于最突出的模式，因为这些模式更有机会更好地泛化。

在本节中，我们将探索两种常见的正则化技术：权重正则化和丢弃，并使用它们改进 IMDB 影评分类模型。

本节内容会将文本形式的影评分为“正面”或“负面”影评。这是一个二元分类（又称为两类分类）的示例，也是一种重要且广泛适用的机器学习问题。

我们将使用 IMDB 数据集，其中包含来自互联网电影数据库的 50 000 条影评文本。我们将这些影评拆分为训练集（25 000 条影评）和测试集（25 000 条影评），当训练集和测试集之间达成了平衡时，意味着它们包含相同数量的正面和负面影评。

6.3.1 下载 IMDB 数据集

TensorFlow 中包含 IMDB 数据集。我们已对该数据集进行了预处理，将影评（字词序列）转换为整数序列，其中每个整数表示字典中的一个特定字词。

以下代码会将 IMDB 数据集下载到用户的计算机上（如果用户已下载该数据集，就会使用缓存副本）：

```
    imdb = keras.datasets.imdb

(train_data, train_labels), (test_data, test_labels) =
imdb.load_data(num_words=10000)

    Downloading data from
https://storage.googleapis.com/tensorflow/tf-keras-datasets/imdb.npz
    17465344/17464789 [==============================] - 0s 0us/step
```

参数 num_words=10000 会保留训练数据中出现频次在前 10 000 位的字词。为确保数据规模处于可管理的水平，罕见字词将被舍弃。

我们花点时间来了解一下数据的格式。该数据集已经过预处理：每个样本都是一个整数数组，表示影评中的字词；每个标签都是整数值 0 或 1，其中 0 表示负面影评，1 表示正面影评。

```
    print("Training entries: {}, labels: {}".format(len(train_data),
len(train_labels)))
    Training entries: 25000, labels: 25000
```

影评文本已转换为整数，其中每个整数都表示字典中的一个特定字词。第一条影评如下：

```
print(train_data[0])

[1, 14, 22, 16, 43, 530, 973, 1622, 1385, 65, 458, 4468, 66, 3941, 4, 173, 36,
256, 5, 25, 100, 43, 838, 112, 50, 670, 2, 9, 35, 480, 284, 5, 150, 4, 172, 112,
167, 2, 336, 385, 39, 4, 172, 4536, 1111, 17, 546, 38, 13, 447, 4, 192, 50, 16,
6, 147, 2025, 19, 14, 22, 4, 1920, 4613, 469, 4, 22, 71, 87, 12, 16, 43, 530, 38,
76, 15, 13, 1247, 4, 22, 17, 515, 17, 12, 16, 626, 18, 2, 5, 62, 386, 12, 8, 316,
8, 106, 5, 4, 2223, 5244, 16, 480, 66, 3785, 33, 4, 130, 12, 16, 38, 619, 5, 25,
124, 51, 36, 135, 48, 25, 1415, 33, 6, 22, 12, 215, 28, 77, 52, 5, 14, 407, 16,
82, 2, 8, 4, 107, 117, 5952, 15, 256, 4, 2, 7, 3766, 5, 723, 36, 71, 43, 530, 476,
26, 400, 317, 46, 7, 4, 2, 1029, 13, 104, 88, 4, 381, 15, 297, 98, 32, 2071, 56,
26, 141, 6, 194, 7486, 18, 4, 226, 22, 21, 134, 476, 26, 480, 5, 144, 30, 5535,
18, 51, 36, 28, 224, 92, 25, 104, 4, 226, 65, 16, 38, 1334, 88, 12, 16, 283, 5,
16, 4472, 113, 103, 32, 15, 16, 5345, 19, 178, 32]
```

影评的长度可能会有所不同。以下代码显示了第一条和第二条影评中的字词数。由于神经网络的输入必须具有相同长度，因此我们稍后需要解决此问题。

```
len(train_data[0]), len(train_data[1])
(218, 189)
```

了解如何将整数转换回文本可能很有用。在以下代码中，我们将创建一个辅助函数来查询包含整数到字符串映射的字典对象。

```
# A dictionary mapping words to an integer index
word_index = imdb.get_word_index()

# The first indices are reserved
word_index = {k:(v+3) for k,v in word_index.items()}
word_index["<PAD>"] = 0
word_index["<START>"] = 1
word_index["<UNK>"] = 2 # unknown
word_index["<UNUSED>"] = 3

reverse_word_index = dict([(value, key) for (key, value) in word_index.items()])

def decode_review(text):
  return ' '.join([reverse_word_index.get(i, '?') for i in text])

  Downloading data from
https://storage.googleapis.com/tensorflow/tf-keras-datasets/imdb_word_index.js
on
  1646592/1641221 [==============================] - 0s 0us/step
```

现在，我们可以使用 decode_review 函数显示第一条影评的文本：

```
decode_review(train_data[0])

" this film was just brilliant casting location scenery story direction
everyone's really suited the part they played and you could just imagine being there
robert  is an amazing actor and now the same being director  father came from the
same scottish island as myself so i loved the fact there was a real connection with
this film the witty remarks throughout the film were great it was just brilliant
so much that i bought the film as soon as it was released for  and would recommend
it to everyone to watch and the fly fishing was amazing really cried at the end
it was so sad and you know what they say if you cry at a film it must have been
good and this definitely was also  to the two little boy's that played the  of norman
and paul they were just brilliant children are often left out of the  list i think
because the stars that play them all grown up are such a big profile for the whole
film but these children are amazing and should be praised for what they have done
don't you think the whole story was so lovely because it was true and was someone's
life after all that was shared with us all"
```

影评（整数数组）必须转换为张量，然后才能馈送到神经网络中。我们可以通过以下两种方法实现这种转换：

（1）对数组进行独热编码，将它们转换为由 0 和 1 构成的向量。例如，序列[3,5]将变成一个 10 000 维的向量，除索引 3 和 5 转换为 1 之外，其余全转换为 0。然后，将它作为网络的第一层，一个可以处理浮点向量数据的密集层。不过，这种方法会占用大量内存，需要一个大小为 num_words * num_reviews 的矩阵。

（2）填充数组，使它们具有相同的长度，然后创建一个形状为 max_length * num_reviews 的整数张量。我们可以使用一个能够处理这种形状的嵌入层作为网络中的第一层。

在本书中，我们将使用第二种方法。

由于影评的长度必须相同，因此我们将使用 pad_sequences 函数将长度标准化：

```
train_data = keras.preprocessing.sequence.pad_sequences(train_data,
                                                         value=word_index["<PAD>"],
                                                         padding='post',
                                                         maxlen=256)

test_data = keras.preprocessing.sequence.pad_sequences(test_data,
                                                        value=word_index["<PAD>"],
                                                        padding='post',
                                                        maxlen=256)
```

现在，我们来看看样本的长度：

```
len(train_data[0]), len(train_data[1])
```

```
(256, 256)
```

并检查（现已填充的）第一条影评：

```
print(train_data[0])

[   1   14   22   16   43  530  973 1622 1385   65  458 4468   66 3941
    4  173   36  256    5   25  100   43  838  112   50  670    2    9
   35  480  284    5  150    4  172  112  167    2  336  385   39    4
  172 4536 1111   17  546   38   13  447    4  192   50   16    6  147
 2025   19   14   22    4 1920 4613  469    4   22   71   87   12   16
   43  530   38   76   15   13 1247    4   22   17  515   17   12   16
  626   18    2    5   62  386   12    8  316    8  106    5    4 2223
 5244   16  480   66 3785   33    4  130   12   16   38  619    5   25
  124   51   36  135   48   25 1415   33    6   22   12  215   28   77
   52    5   14  407   16   82    2    8    4  107  117 5952   15  256
    4    2    7 3766    5  723   36   71   43  530  476   26  400  317
   46    7    4    2 1029   13  104   88    4  381   15  297   98   32
 2071   56   26  141    6  194 7486   18    4  226   22   21  134  476
   26  480    5  144   30 5535   18   51   36   28  224   92   25  104
    4  226   65   16   38 1334   88   12   16  283    5   16 4472  113
  103   32   15   16 5345   19  178   32    0    0    0    0    0    0
    0    0    0    0    0    0    0    0    0    0    0    0    0    0
    0    0    0    0    0    0    0    0    0    0    0    0    0    0
    0    0    0    0]
```

6.3.2 构建模型

神经网络通过堆叠层创建而成，这需要做出两个架构方面的决策：

（1）在模型中使用多少个层？

（2）针对每个层使用多少个隐藏单元？

在本示例中，输入数据由字词-索引数组构成，要预测的标签是 0 或 1。接下来，我们为此问题构建一个模型：

```
  # input shape is the vocabulary count used for the movie reviews (10,000 words)
vocab_size = 10000

model = keras.Sequential()
model.add(keras.layers.Embedding(vocab_size, 16))
model.add(keras.layers.GlobalAveragePooling1D())
model.add(keras.layers.Dense(16, activation=tf.nn.relu))
model.add(keras.layers.Dense(1, activation=tf.nn.sigmoid))
```

```
model.summary()

  ______________________________________________________________________
  Layer (type)                Output Shape            Param #
  ======================================================================
  embedding (Embedding)        (None, None, 16)          160000
  ______________________________________________________________________
  global_average_pooling1d (Gl (None, 16)                 0
  ______________________________________________________________________
  dense (Dense)               (None, 16)              272
  ______________________________________________________________________
  dense_1 (Dense)              (None, 1)               17
  ======================================================================
  Total params: 160,289
  Trainable params: 160,289
  Non-trainable params: 0
  ______________________________________________________________________
```

按顺序堆叠各个层以构建分类器：

第一层是 Embedding 层。该层会在整数编码的词汇表中查找每个字词-索引的嵌入向量。模型在接受训练时会学习这些向量，然后这些向量会向输出数组添加一个维度。生成的维度为：(batch, sequence, embedding)。

接下来，一个 GlobalAveragePooling1D 层通过对序列维度求平均值，针对每个样本返回一个长度固定的输出向量。这样，模型便能够以尽可能简单的方式处理各种长度的输入。

该长度固定的输出向量会传入一个全连接（Dense）层（包含 16 个隐藏单元）。

最后一层与单个输出节点密集连接。应用 Sigmoid 激活函数后，结果是介于 0~1 的浮点值，表示概率或置信水平。

上述模型在输入和输出之间有两个中间层（也称为隐藏层）。输出（单元、节点或神经元）的数量是对应层的法空间向量的维度。换句话说，该数值表示学习内部表示法时网络所允许的自由度。

如果模型具有更多隐藏单元（更高维度的表示空间）和/或更多层，就说明网络可以学习更复杂的表示法。不过，这会使网络耗费更多计算资源，并且可能导致学习不必要的模式（可以优化在训练数据上的表现，但不会优化在测试数据上的表现）。这称为过拟合，我们稍后会加以探讨。

模型在训练时需要一个损失函数和一个优化器。由于这是一个二元分类问题且模型会输出一个概率（应用 S 型激活函数的单个单元层），因此我们将使用 binary_crossentropy 损失函数。

该函数并不是唯一的损失函数，例如可以选择 mean_squared_error。但一般来说，binary_crossentropy 更适合处理概率问题，它可以测量概率分布之间的“差距”，在本例中则为实际分布和预测之间的“差距”。

稍后，在探索回归问题（比如预测房价）时，我们将了解如何使用均方误差的损失函数。

现在，配置模型以使用优化器和损失函数：

```
model.compile(optimizer=tf.train.AdamOptimizer(),
         loss='binary_crossentropy',
         metrics=['accuracy'])
```

6.3.3 训练模型

在训练时，我们需要检查模型处理从未见过的数据的准确率。从原始训练数据中分离出 10 000 个样本，创建一个验证集（为什么现在不使用测试集？这是因为我们的目标是仅使用训练数据开发和调整模型，然后仅使用一次测试数据评估准确率）。

```
x_val = train_data[:10000]
partial_x_train = train_data[10000:]

y_val = train_labels[:10000]
partial_y_train = train_labels[10000:]
```

用有 512 个样本的小批次训练模型 40 个周期。这将对 x_train 和 y_train 张量中的所有样本进行 40 次迭代。在训练期间，监控模型在验证集的 10 000 个样本上的损失和准确率：

```
history = model.fit(partial_x_train,
            partial_y_train,
            epochs=40,
            batch_size=512,
            validation_data=(x_val, y_val),
            verbose=1)

Train on 15000 samples, validate on 10000 samples
Epoch 1/40
15000/15000 [==============================] - 1s 53us/step - loss: 0.6916 -
acc: 0.5907 - val_loss: 0.6894 - val_acc: 0.6462
Epoch 2/40
15000/15000 [==============================] - 1s 41us/step - loss: 0.6849 -
acc: 0.7349 - val_loss: 0.6802 - val_acc: 0.7464
Epoch 3/40
15000/15000 [==============================] - 1s 40us/step - loss: 0.6711 -
acc: 0.7663 - val_loss: 0.6632 - val_acc: 0.7627
Epoch 4/40
15000/15000 [==============================] - 1s 40us/step - loss: 0.6472 -
acc: 0.7761 - val_loss: 0.6373 - val_acc: 0.7676
Epoch 5/40
15000/15000 [==============================] - 1s 39us/step - loss: 0.6133 -
acc: 0.7973 - val_loss: 0.6016 - val_acc: 0.7887
Epoch 6/40
15000/15000 [==============================] - 1s 38us/step - loss: 0.5716 -
```

```
acc: 0.8146 - val_loss: 0.5617 - val_acc: 0.8029
    Epoch 7/40
    15000/15000 [==============================] - 1s 38us/step - loss: 0.5259 -
acc: 0.8329 - val_loss: 0.5202 - val_acc: 0.8204
    Epoch 8/40
    15000/15000 [==============================] - 1s 37us/step - loss: 0.4804 -
acc: 0.8466 - val_loss: 0.4805 - val_acc: 0.8352
    Epoch 9/40
    15000/15000 [==============================] - 1s 40us/step - loss: 0.4387 -
acc: 0.8589 - val_loss: 0.4454 - val_acc: 0.8442
    Epoch 10/40
    15000/15000 [==============================] - 1s 40us/step - loss: 0.4013 -
acc: 0.8733 - val_loss: 0.4158 - val_acc: 0.8495
    Epoch 11/40
    15000/15000 [==============================] - 1s 41us/step - loss: 0.3697 -
acc: 0.8810 - val_loss: 0.3934 - val_acc: 0.8542
    Epoch 12/40
    15000/15000 [==============================] - 1s 39us/step - loss: 0.3432 -
acc: 0.8869 - val_loss: 0.3712 - val_acc: 0.8627
    Epoch 13/40
    15000/15000 [==============================] - 1s 39us/step - loss: 0.3192 -
acc: 0.8941 - val_loss: 0.3556 - val_acc: 0.8679
    Epoch 14/40
    15000/15000 [==============================] - 1s 40us/step - loss: 0.2993 -
acc: 0.8997 - val_loss: 0.3416 - val_acc: 0.8721
    Epoch 15/40
    15000/15000 [==============================] - 1s 38us/step - loss: 0.2820 -
acc: 0.9046 - val_loss: 0.3307 - val_acc: 0.8736
    Epoch 16/40
    15000/15000 [==============================] - 1s 39us/step - loss: 0.2672 -
acc: 0.9076 - val_loss: 0.3218 - val_acc: 0.8753
    Epoch 17/40
    15000/15000 [==============================] - 1s 39us/step - loss: 0.2528 -
acc: 0.9143 - val_loss: 0.3143 - val_acc: 0.8774
    Epoch 18/40
    15000/15000 [==============================] - 1s 40us/step - loss: 0.2404 -
acc: 0.9192 - val_loss: 0.3080 - val_acc: 0.8817
    Epoch 19/40
    15000/15000 [==============================] - 1s 41us/step - loss: 0.2289 -
acc: 0.9227 - val_loss: 0.3028 - val_acc: 0.8816
    Epoch 20/40
    15000/15000 [==============================] - 1s 37us/step - loss: 0.2189 -
acc: 0.9249 - val_loss: 0.2987 - val_acc: 0.8829
```

```
  Epoch 21/40
  15000/15000 [==============================] - 1s 39us/step - loss: 0.2089 -
acc: 0.9293 - val_loss: 0.2951 - val_acc: 0.8834
  Epoch 22/40
  15000/15000 [==============================] - 1s 41us/step - loss: 0.1999 -
acc: 0.9323 - val_loss: 0.2925 - val_acc: 0.8833
  Epoch 23/40
  15000/15000 [==============================] - 1s 41us/step - loss: 0.1916 -
acc: 0.9351 - val_loss: 0.2907 - val_acc: 0.8840
  Epoch 24/40
  15000/15000 [==============================] - 1s 40us/step - loss: 0.1834 -
acc: 0.9402 - val_loss: 0.2882 - val_acc: 0.8849
  Epoch 25/40
  15000/15000 [==============================] - 1s 41us/step - loss: 0.1761 -
acc: 0.9431 - val_loss: 0.2868 - val_acc: 0.8853
  Epoch 26/40
  15000/15000 [==============================] - 1s 40us/step - loss: 0.1688 -
acc: 0.9459 - val_loss: 0.2864 - val_acc: 0.8857
  Epoch 27/40
  15000/15000 [==============================] - 1s 38us/step - loss: 0.1628 -
acc: 0.9490 - val_loss: 0.2861 - val_acc: 0.8859
  Epoch 28/40
  15000/15000 [==============================] - 1s 42us/step - loss: 0.1562 -
acc: 0.9520 - val_loss: 0.2854 - val_acc: 0.8866
  Epoch 29/40
  15000/15000 [==============================] - 1s 39us/step - loss: 0.1503 -
acc: 0.9536 - val_loss: 0.2854 - val_acc: 0.8866
  Epoch 30/40
  15000/15000 [==============================] - 1s 40us/step - loss: 0.1451 -
acc: 0.9556 - val_loss: 0.2861 - val_acc: 0.8864
  Epoch 31/40
  15000/15000 [==============================] - 1s 41us/step - loss: 0.1390 -
acc: 0.9585 - val_loss: 0.2870 - val_acc: 0.8872
  Epoch 32/40
  15000/15000 [==============================] - 1s 41us/step - loss: 0.1342 -
acc: 0.9604 - val_loss: 0.2883 - val_acc: 0.8867
  Epoch 33/40
  15000/15000 [==============================] - 1s 40us/step - loss: 0.1287 -
acc: 0.9631 - val_loss: 0.2896 - val_acc: 0.8869
  Epoch 34/40
  15000/15000 [==============================] - 1s 40us/step - loss: 0.1242 -
acc: 0.9647 - val_loss: 0.2911 - val_acc: 0.8856
  Epoch 35/40
```

```
    15000/15000 [==============================] - 1s 41us/step - loss: 0.1200 -
acc: 0.9656 - val_loss: 0.2925 - val_acc: 0.8873
    Epoch 36/40
    15000/15000 [==============================] - 1s 39us/step - loss: 0.1150 -
acc: 0.9681 - val_loss: 0.2949 - val_acc: 0.8851
    Epoch 37/40
    15000/15000 [==============================] - 1s 40us/step - loss: 0.1110 -
acc: 0.9694 - val_loss: 0.2972 - val_acc: 0.8852
    Epoch 38/40
    15000/15000 [==============================] - 1s 40us/step - loss: 0.1075 -
acc: 0.9697 - val_loss: 0.2993 - val_acc: 0.8840
    Epoch 39/40
    15000/15000 [==============================] - 1s 40us/step - loss: 0.1030 -
acc: 0.9715 - val_loss: 0.3016 - val_acc: 0.8826
    Epoch 40/40
    15000/15000 [==============================] - 1s 40us/step - loss: 0.0991 -
acc: 0.9731 - val_loss: 0.3042 - val_acc: 0.8842
```

我们来看看模型的表现如何。模型会返回两个值：损失（表示误差的数字，越低越好）和准确率。

```
    results = model.evaluate(test_data, test_labels)

print(results)

    25000/25000 [==============================] - 1s 35us/step
    [0.32395469411849975, 0.87264]
```

使用这种相当简单的方法可实现约 87%的准确率。如果采用更高级的方法，那么模型的准确率应该会接近 95%。

model.fit()返回一个 history 对象，该对象包含一个字典，其中包括训练期间发生的所有情况：

```
    history_dict = history.history
history_dict.keys()

    dict_keys(['loss', 'val_loss', 'val_acc', 'acc'])
```

一共有 4 个条目：每个条目对应训练和验证期间的一个受监控的指标。可以使用这些指标绘制训练损失与验证损失图表以进行对比，并绘制训练准确率与验证准确率图表（见图 6.13）：

```
    import matplotlib.pyplot as plt

acc = history.history['acc']
val_acc = history.history['val_acc']
loss = history.history['loss']
val_loss = history.history['val_loss']
```

```
epochs = range(1, len(acc) + 1)

# "bo" is for "blue dot"
plt.plot(epochs, loss, 'bo', label='Training loss')
# b is for "solid blue line"
plt.plot(epochs, val_loss, 'b', label='Validation loss')
plt.title('Training and validation loss')
plt.xlabel('Epochs')
plt.ylabel('Loss')
plt.legend()

plt.show()

   plt.clf()   # clear figure
acc_values = history_dict['acc']
val_acc_values = history_dict['val_acc']

plt.plot(epochs, acc, 'bo', label='Training acc')
plt.plot(epochs, val_acc, 'b', label='Validation acc')
plt.title('Training and validation accuracy')
plt.xlabel('Epochs')
plt.ylabel('Accuracy')
plt.legend()

plt.show()
```

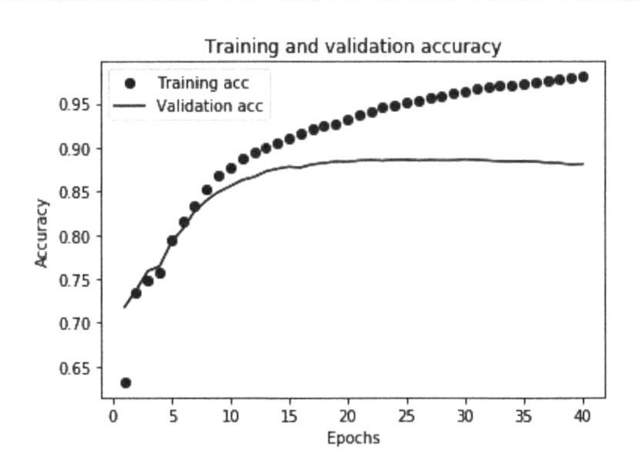

图 6.13　训练准确率与验证准确率图表

在该图表中，圆点表示训练损失和准确率，实线表示验证损失和准确率。

我们注意到，训练损失随着周期数的增加而降低，训练准确率随着周期数的增加而提高。在使用梯度下降法优化模型时，这属于正常现象，该方法应在每次迭代时尽可能降低目标值。

验证损失和准确率的变化情况并非如此，它们似乎在大约 20 个周期后达到峰值。这是一种过拟合现象：模型在训练数据上的表现要优于在从未见过的数据上的表现。在此之后，模型会过度优化和学习特定于训练数据的表示法，而无法泛化到测试数据。

对于这种特殊情况，我们可以在大约 20 个周期后停止训练，防止出现过拟合。稍后将了解如何使用回调自动执行此操作。

6.3.4 过拟合过程实践

接下来修改我们使用的嵌入编码模型，对句子进行多热编码。该模型将很快过拟合训练集，将用来演示何时发生过拟合，以及如何防止过拟合。

对列表进行多热编码意味着将它们转换为由 0 和 1 组成的向量。例如，将序列[3,5]转换为一个 10 000 维的向量（除索引 3 和 5 转换为 1 之外，其余全为 0）。

```
NUM_WORDS = 10000

(train_data, train_labels), (test_data, test_labels) =
keras.datasets.imdb.load_data(num_words=NUM_WORDS)

def multi_hot_sequences(sequences, dimension):
    # Create an all-zero matrix of shape (len(sequences), dimension)
    results = np.zeros((len(sequences), dimension))
    for i, word_indices in enumerate(sequences):
        results[i, word_indices] = 1.0  # set specific indices of results[i] to 1s
    return results

train_data = multi_hot_sequences(train_data, dimension=NUM_WORDS)
test_data = multi_hot_sequences(test_data, dimension=NUM_WORDS)

    Downloading data from
https://storage.googleapis.com/tensorflow/tf-keras-datasets/imdb.npz
    17465344/17464789 [==============================] - 0s 0us/step
```

我们来看看生成的其中一个多热向量。字词索引按频率排序，因此索引 0 附近应该有更多的 1 值（见图 6.14）：

```
plt.plot(train_data[0])
[]
```

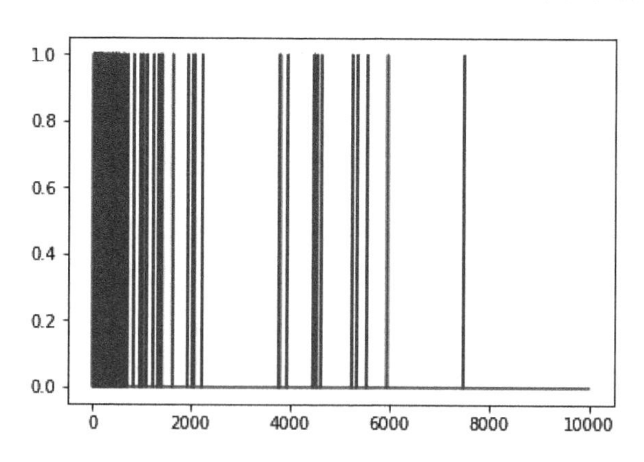

图 6.14 字词索引按频率排序

要防止过拟合，最简单的方法是缩小模型，即减少模型中可学习参数的数量（由层数和每层的单元数决定）。在深度学习中，模型中可学习参数的数量通常称为模型的“容量”。直观而言，参数越多的模型“记忆容量”越大，因此能够轻松学习训练样本与其目标之间的字典式完美映射（无任何泛化能力的映射），但如果要对之前未见过的数据做出预测，这种映射就毫无用处。

深度学习模型往往善于与训练数据拟合，但真正的挑战是泛化，而非拟合。另一方面，如果网络的记忆资源有限，便无法轻松学习映射。为了最小化损失，必须学习具有更强预测能力的压缩表示法。同时，如果模型太小，就会难以与训练数据拟合。我们需要在“太多容量”和“容量不足”这两者之间实现平衡。

遗憾的是，并没有什么神奇公式可用来确定合适的模型大小或架构（由层数或每层的合适大小决定），我们需要尝试一系列不同的网络模型架构。要找到合适的模型大小，最好先使用相对较少的层和参数，然后开始增加层的大小或添加新的层，直到看到返回的验证损失不断减小为止。下面在影评分类网络上尝试这个方法。我们将仅使用 Dense 层创建一个简单的基准模型，然后创建更小和更大的版本，并比较这些版本。

创建基准模型的代码如下：

```
    baseline_model = keras.Sequential([
    # 'input_shape' is only required here so that '.summary' works.
    keras.layers.Dense(16, activation=tf.nn.relu, input_shape=(NUM_WORDS,)),
    keras.layers.Dense(16, activation=tf.nn.relu),
    keras.layers.Dense(1, activation=tf.nn.sigmoid)
])

baseline_model.compile(optimizer='adam',
                       loss='binary_crossentropy',
                       metrics=['accuracy', 'binary_crossentropy'])

baseline_model.summary()

    _________________________________________________________________
    Layer (type)                 Output Shape              Param #
    =================================================================
    dense (Dense)                (None, 16)                160016
    _________________________________________________________________
    dense_1 (Dense)              (None, 16)                272
    _________________________________________________________________
    dense_2 (Dense)              (None, 1)                 17
    =================================================================
    Total params: 160,305
    Trainable params: 160,305
    Non-trainable params: 0
    _________________________________________________________________
```

```
baseline_history = baseline_model.fit(train_data,
                                      train_labels,
                                      epochs=20,
                                      batch_size=512,
                                      validation_data=(test_data, test_labels),
                                      verbose=2)

Train on 25000 samples, validate on 25000 samples
Epoch 1/20
 - 4s - loss: 0.4765 - acc: 0.8168 - binary_crossentropy: 0.4765 - val_loss:
0.3289 - val_acc: 0.8784 - val_binary_crossentropy: 0.3289
Epoch 2/20
 - 3s - loss: 0.2437 - acc: 0.9122 - binary_crossentropy: 0.2437 - val_loss:
0.2828 - val_acc: 0.8878 - val_binary_crossentropy: 0.2828
Epoch 3/20
 - 3s - loss: 0.1782 - acc: 0.9376 - binary_crossentropy: 0.1782 - val_loss:
0.2912 - val_acc: 0.8845 - val_binary_crossentropy: 0.2912
Epoch 4/20
 - 3s - loss: 0.1416 - acc: 0.9506 - binary_crossentropy: 0.1416 - val_loss:
0.3214 - val_acc: 0.8790 - val_binary_crossentropy: 0.3214
Epoch 5/20
 - 3s - loss: 0.1173 - acc: 0.9608 - binary_crossentropy: 0.1173 - val_loss:
0.3502 - val_acc: 0.8735 - val_binary_crossentropy: 0.3502
Epoch 6/20
 - 3s - loss: 0.0964 - acc: 0.9697 - binary_crossentropy: 0.0964 - val_loss:
0.3860 - val_acc: 0.8688 - val_binary_crossentropy: 0.3860
Epoch 7/20
 - 3s - loss: 0.0795 - acc: 0.9768 - binary_crossentropy: 0.0795 - val_loss:
0.4404 - val_acc: 0.8605 - val_binary_crossentropy: 0.4404
Epoch 8/20
 - 3s - loss: 0.0662 - acc: 0.9814 - binary_crossentropy: 0.0662 - val_loss:
0.4748 - val_acc: 0.8612 - val_binary_crossentropy: 0.4748
Epoch 9/20
 - 3s - loss: 0.0540 - acc: 0.9871 - binary_crossentropy: 0.0540 - val_loss:
0.5181 - val_acc: 0.8584 - val_binary_crossentropy: 0.5181
Epoch 10/20
 - 3s - loss: 0.0441 - acc: 0.9906 - binary_crossentropy: 0.0441 - val_loss:
0.5704 - val_acc: 0.8554 - val_binary_crossentropy: 0.5704
Epoch 11/20
 - 3s - loss: 0.0376 - acc: 0.9919 - binary_crossentropy: 0.0376 - val_loss:
0.6164 - val_acc: 0.8547 - val_binary_crossentropy: 0.6164
Epoch 12/20
 - 3s - loss: 0.0289 - acc: 0.9951 - binary_crossentropy: 0.0289 - val_loss:
```

```
0.6639 - val_acc: 0.8528 - val_binary_crossentropy: 0.6639
  Epoch 13/20
   - 3s - loss: 0.0225 - acc: 0.9967 - binary_crossentropy: 0.0225 - val_loss:
0.7050 - val_acc: 0.8518 - val_binary_crossentropy: 0.7050
  Epoch 14/20
   - 3s - loss: 0.0174 - acc: 0.9982 - binary_crossentropy: 0.0174 - val_loss:
0.7472 - val_acc: 0.8502 - val_binary_crossentropy: 0.7472
  Epoch 15/20
   - 3s - loss: 0.0132 - acc: 0.9990 - binary_crossentropy: 0.0132 - val_loss:
0.7909 - val_acc: 0.8502 - val_binary_crossentropy: 0.7909
  Epoch 16/20
   - 3s - loss: 0.0103 - acc: 0.9994 - binary_crossentropy: 0.0103 - val_loss:
0.8295 - val_acc: 0.8500 - val_binary_crossentropy: 0.8295
  Epoch 17/20
   - 3s - loss: 0.0080 - acc: 0.9998 - binary_crossentropy: 0.0080 - val_loss:
0.8620 - val_acc: 0.8490 - val_binary_crossentropy: 0.8620
  Epoch 18/20
   - 3s - loss: 0.0063 - acc: 0.9999 - binary_crossentropy: 0.0063 - val_loss:
0.8932 - val_acc: 0.8484 - val_binary_crossentropy: 0.8932
  Epoch 19/20
   - 3s - loss: 0.0051 - acc: 1.0000 - binary_crossentropy: 0.0051 - val_loss:
0.9225 - val_acc: 0.8490 - val_binary_crossentropy: 0.9225
  Epoch 20/20
   - 3s - loss: 0.0042 - acc: 1.0000 - binary_crossentropy: 0.0042 - val_loss:
0.9477 - val_acc: 0.8478 - val_binary_crossentropy: 0.9477
```

创建一个隐藏单元更少的模型，然后与刚刚创建的基准模型进行比较：

```
  smaller_model = keras.Sequential([
   keras.layers.Dense(4, activation=tf.nn.relu, input_shape=(NUM_WORDS,)),
   keras.layers.Dense(4, activation=tf.nn.relu),
   keras.layers.Dense(1, activation=tf.nn.sigmoid)
])

smaller_model.compile(optimizer='adam',
           loss='binary_crossentropy',
           metrics=['accuracy', 'binary_crossentropy'])

smaller_model.summary()

  _________________________________________________________________
  Layer (type)                 Output Shape              Param #
  =================================================================
  dense_3 (Dense)              (None, 4)                 40004
```

```
_________________________________________________________________
dense_4 (Dense)              (None, 4)                 20
_________________________________________________________________
dense_5 (Dense)              (None, 1)                 5
=================================================================
Total params: 40,029
Trainable params: 40,029
Non-trainable params: 0
_________________________________________________________________
```

使用相同的数据训练该模型：

```
smaller_history = smaller_model.fit(train_data,
                                    train_labels,
                                    epochs=20,
                                    batch_size=512,
                                    validation_data=(test_data, test_labels),
                                    verbose=2)

Train on 25000 samples, validate on 25000 samples
Epoch 1/20
 - 3s - loss: 0.6094 - acc: 0.6707 - binary_crossentropy: 0.6094 - val_loss: 
0.5141 - val_acc: 0.8269 - val_binary_crossentropy: 0.5141
Epoch 2/20
 - 3s - loss: 0.4043 - acc: 0.8810 - binary_crossentropy: 0.4043 - val_loss: 
0.3588 - val_acc: 0.8788 - val_binary_crossentropy: 0.3588
Epoch 3/20
 - 3s - loss: 0.2764 - acc: 0.9154 - binary_crossentropy: 0.2764 - val_loss: 
0.3011 - val_acc: 0.8870 - val_binary_crossentropy: 0.3011
Epoch 4/20
 - 3s - loss: 0.2182 - acc: 0.9295 - binary_crossentropy: 0.2182 - val_loss: 
0.2861 - val_acc: 0.8880 - val_binary_crossentropy: 0.2861
Epoch 5/20
 - 3s - loss: 0.1842 - acc: 0.9400 - binary_crossentropy: 0.1842 - val_loss: 
0.2870 - val_acc: 0.8848 - val_binary_crossentropy: 0.2870
Epoch 6/20
 - 3s - loss: 0.1599 - acc: 0.9477 - binary_crossentropy: 0.1599 - val_loss: 
0.2898 - val_acc: 0.8852 - val_binary_crossentropy: 0.2898
Epoch 7/20
 - 3s - loss: 0.1409 - acc: 0.9563 - binary_crossentropy: 0.1409 - val_loss: 
0.2994 - val_acc: 0.8824 - val_binary_crossentropy: 0.2994
Epoch 8/20
 - 3s - loss: 0.1259 - acc: 0.9618 - binary_crossentropy: 0.1259 - val_loss: 
0.3139 - val_acc: 0.8792 - val_binary_crossentropy: 0.3139
```

```
    Epoch 9/20
     - 3s - loss: 0.1136 - acc: 0.9653 - binary_crossentropy: 0.1136 - val_loss:
0.3287 - val_acc: 0.8755 - val_binary_crossentropy: 0.3287
    Epoch 10/20
     - 3s - loss: 0.1021 - acc: 0.9707 - binary_crossentropy: 0.1021 - val_loss:
0.3449 - val_acc: 0.8727 - val_binary_crossentropy: 0.3449
    Epoch 11/20
     - 3s - loss: 0.0923 - acc: 0.9747 - binary_crossentropy: 0.0923 - val_loss:
0.3632 - val_acc: 0.8712 - val_binary_crossentropy: 0.3632
    Epoch 12/20
     - 3s - loss: 0.0829 - acc: 0.9778 - binary_crossentropy: 0.0829 - val_loss:
0.3812 - val_acc: 0.8693 - val_binary_crossentropy: 0.3812
    Epoch 13/20
     - 3s - loss: 0.0749 - acc: 0.9805 - binary_crossentropy: 0.0749 - val_loss:
0.3996 - val_acc: 0.8670 - val_binary_crossentropy: 0.3996
    Epoch 14/20
     - 3s - loss: 0.0677 - acc: 0.9844 - binary_crossentropy: 0.0677 - val_loss:
0.4238 - val_acc: 0.8648 - val_binary_crossentropy: 0.4238
    Epoch 15/20
     - 3s - loss: 0.0608 - acc: 0.9867 - binary_crossentropy: 0.0608 - val_loss:
0.4385 - val_acc: 0.8646 - val_binary_crossentropy: 0.4385
    Epoch 16/20
     - 3s - loss: 0.0548 - acc: 0.9886 - binary_crossentropy: 0.0548 - val_loss:
0.4602 - val_acc: 0.8630 - val_binary_crossentropy: 0.4602
    Epoch 17/20
     - 3s - loss: 0.0493 - acc: 0.9906 - binary_crossentropy: 0.0493 - val_loss:
0.4828 - val_acc: 0.8601 - val_binary_crossentropy: 0.4828
    Epoch 18/20
     - 3s - loss: 0.0439 - acc: 0.9923 - binary_crossentropy: 0.0439 - val_loss:
0.5161 - val_acc: 0.8589 - val_binary_crossentropy: 0.5161
    Epoch 19/20
     - 3s - loss: 0.0390 - acc: 0.9940 - binary_crossentropy: 0.0390 - val_loss:
0.5295 - val_acc: 0.8580 - val_binary_crossentropy: 0.5295
    Epoch 20/20
     - 3s - loss: 0.0349 - acc: 0.9948 - binary_crossentropy: 0.0349 - val_loss:
0.5514 - val_acc: 0.8565 - val_binary_crossentropy: 0.5514
```

我们可以创建一个更大的模型，看看它多长时间开始过拟合。接下来，向这个基准添加一个容量大得多的网络——远远超出解决问题所需的容量：

```
    bigger_model = keras.models.Sequential([
     keras.layers.Dense(512, activation=tf.nn.relu, input_shape=(NUM_WORDS,)),
     keras.layers.Dense(512, activation=tf.nn.relu),
     keras.layers.Dense(1, activation=tf.nn.sigmoid)
```

```
])

bigger_model.compile(optimizer='adam',
                     loss='binary_crossentropy',
                     metrics=['accuracy','binary_crossentropy'])

bigger_model.summary()
_________________________________________________________________
   Layer (type)                 Output Shape              Param #
   =================================================================
   dense_6 (Dense)              (None, 512)               5120512
   _________________________________________________________________
   dense_7 (Dense)              (None, 512)               262656
   _________________________________________________________________
   dense_8 (Dense)              (None, 1)                 513
   =================================================================
   Total params: 5,383,681
   Trainable params: 5,383,681
   Non-trainable params: 0
   _________________________________________________________________
```

再次使用相同的数据训练该模型：

```
   bigger_history = bigger_model.fit(train_data, train_labels,
                                     epochs=20,
                                     batch_size=512,
                                     validation_data=(test_data, test_labels),
                                     verbose=2)

   Train on 25000 samples, validate on 25000 samples
   Epoch 1/20
    - 6s - loss: 0.3481 - acc: 0.8512 - binary_crossentropy: 0.3481 - val_loss:
0.2956 - val_acc: 0.8800 - val_binary_crossentropy: 0.2956
   Epoch 2/20
    - 6s - loss: 0.1474 - acc: 0.9462 - binary_crossentropy: 0.1474 - val_loss:
0.3600 - val_acc: 0.8643 - val_binary_crossentropy: 0.3600
   Epoch 3/20
    - 6s - loss: 0.0576 - acc: 0.9824 - binary_crossentropy: 0.0576 - val_loss:
0.4228 - val_acc: 0.8669 - val_binary_crossentropy: 0.4228
   Epoch 4/20
    - 6s - loss: 0.0111 - acc: 0.9980 - binary_crossentropy: 0.0111 - val_loss:
0.5609 - val_acc: 0.8688 - val_binary_crossentropy: 0.5609
   Epoch 5/20
    - 6s - loss: 0.0014 - acc: 1.0000 - binary_crossentropy: 0.0014 - val_loss:
```

```
0.6633 - val_acc: 0.8688 - val_binary_crossentropy: 0.6633
  Epoch 6/20
   - 6s - loss: 3.1242e-04 - acc: 1.0000 - binary_crossentropy: 3.1242e-04 -
val_loss: 0.7067 - val_acc: 0.8696 - val_binary_crossentropy: 0.7067
  Epoch 7/20
   - 6s - loss: 1.7861e-04 - acc: 1.0000 - binary_crossentropy: 1.7861e-04 -
val_loss: 0.7352 - val_acc: 0.8702 - val_binary_crossentropy: 0.7352
  Epoch 8/20
   - 6s - loss: 1.2336e-04 - acc: 1.0000 - binary_crossentropy: 1.2336e-04 -
val_loss: 0.7565 - val_acc: 0.8706 - val_binary_crossentropy: 0.7565
  Epoch 9/20
   - 6s - loss: 9.1178e-05 - acc: 1.0000 - binary_crossentropy: 9.1178e-05 -
val_loss: 0.7747 - val_acc: 0.8708 - val_binary_crossentropy: 0.7747
  Epoch 10/20
   - 6s - loss: 7.0124e-05 - acc: 1.0000 - binary_crossentropy: 7.0124e-05 -
val_loss: 0.7901 - val_acc: 0.8708 - val_binary_crossentropy: 0.7901
  Epoch 11/20
   - 6s - loss: 5.5512e-05 - acc: 1.0000 - binary_crossentropy: 5.5512e-05 -
val_loss: 0.8039 - val_acc: 0.8711 - val_binary_crossentropy: 0.8039
  Epoch 12/20
   - 6s - loss: 4.4797e-05 - acc: 1.0000 - binary_crossentropy: 4.4797e-05 -
val_loss: 0.8167 - val_acc: 0.8711 - val_binary_crossentropy: 0.8167
  Epoch 13/20
   - 6s - loss: 3.6816e-05 - acc: 1.0000 - binary_crossentropy: 3.6816e-05 -
val_loss: 0.8278 - val_acc: 0.8713 - val_binary_crossentropy: 0.8278
  Epoch 14/20
   - 6s - loss: 3.0683e-05 - acc: 1.0000 - binary_crossentropy: 3.0683e-05 -
val_loss: 0.8389 - val_acc: 0.8714 - val_binary_crossentropy: 0.8389
  Epoch 15/20
   - 6s - loss: 2.5789e-05 - acc: 1.0000 - binary_crossentropy: 2.5789e-05 -
val_loss: 0.8493 - val_acc: 0.8714 - val_binary_crossentropy: 0.8493
  Epoch 16/20
   - 6s - loss: 2.1778e-05 - acc: 1.0000 - binary_crossentropy: 2.1778e-05 -
val_loss: 0.8598 - val_acc: 0.8716 - val_binary_crossentropy: 0.8598
  Epoch 17/20
   - 6s - loss: 1.8315e-05 - acc: 1.0000 - binary_crossentropy: 1.8315e-05 -
val_loss: 0.8724 - val_acc: 0.8715 - val_binary_crossentropy: 0.8724
  Epoch 18/20
   - 6s - loss: 1.5310e-05 - acc: 1.0000 - binary_crossentropy: 1.5310e-05 -
val_loss: 0.8847 - val_acc: 0.8716 - val_binary_crossentropy: 0.8847
  Epoch 19/20
   - 6s - loss: 1.2654e-05 - acc: 1.0000 - binary_crossentropy: 1.2654e-05 -
val_loss: 0.8981 - val_acc: 0.8715 - val_binary_crossentropy: 0.8981
```

```
  Epoch 20/20
   - 6s - loss: 1.0461e-05 - acc: 1.0000 - binary_crossentropy: 1.0461e-05 -
val_loss: 0.9131 - val_acc: 0.8714 - val_binary_crossentropy: 0.9131
```

图 6.15 直观地展示了训练损失和验证损失，实线表示训练损失，虚线表示验证损失（验证损失越低，表示模型越好）。在此示例中，较小的网络开始过拟合的时间比基准模型晚（前者在 6 个周期之后，后者在 4 个周期之后），并且开始过拟合后，它的效果下降速度也慢得多。

```
  def plot_history(histories, key='binary_crossentropy'):
  plt.figure(figsize=(16,10))

  for name, history in histories:
    val = plt.plot(history.epoch, history.history['val_'+key],
                   '--', label=name.title()+' Val')
    plt.plot(history.epoch, history.history[key], color=val[0].get_color(),
             label=name.title()+' Train')

  plt.xlabel('Epochs')
  plt.ylabel(key.replace('_',' ').title())
  plt.legend()

  plt.xlim([0,max(history.epoch)])

plot_history([('baseline', baseline_history),
              ('smaller', smaller_history),
              ('bigger', bigger_history)])
```

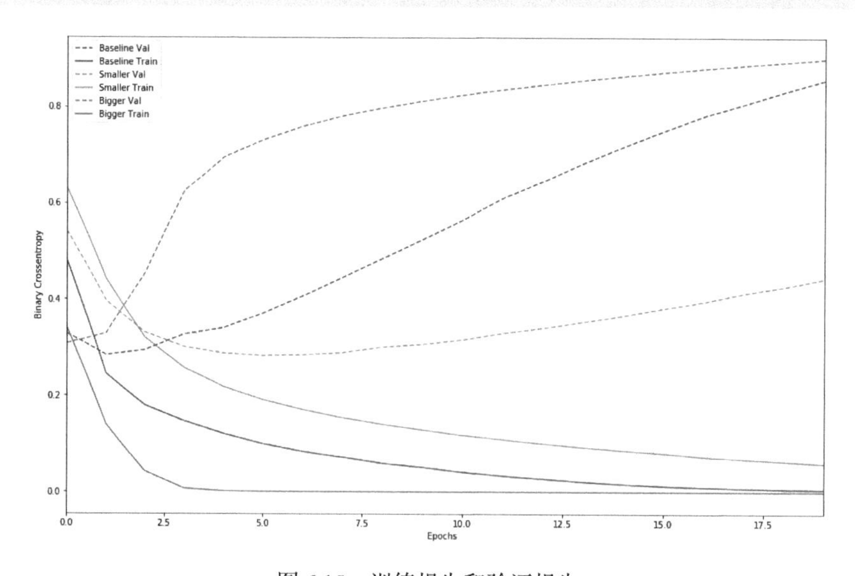

图 6.15　训练损失和验证损失

注意，较大的网络几乎仅仅 1 个周期之后便立即开始过拟合，并且之后会严重得多。网络容量越大，便越快对训练数据进行建模（产生较低的训练损失），但越容易过拟合（导致训练损失与验证损失之间的差异很大）。

6.3.5 过拟合应对策略

1. 添加权重正则化修正项

我们添加正则化修正项的依据是奥卡姆剃刀定律。首先简单介绍一下奥卡姆剃刀定律，奥卡姆剃刀定律的定义是：如果对于同一现象有两种解释，最可能正确的解释是“最简单”的解释，即做出最少量假设的解释。这也适用于神经网络学习的模型：给定一些训练数据和一个网络架构，有多组权重值（多个模型）可以解释数据，而简单模型比复杂模型更不容易过拟合。

在这种情况下，“简单模型”是一种参数值分布的熵较低的模型（或者具有较少参数的模型）。因此，要缓解过拟合，一种常见的方法是限制网络的复杂性，具体方法是强制要求其权重仅采用较小的值，使权重值的分布更“规则”。这称为“权重正则化”，可通过向网络的损失函数添加与权重较大相关的代价来实现。这个代价分为两种类型：

- L1 正则化，其中所添加的代价与权重系数的绝对值（权重“L1 范数”）成正比。
- L2 正则化，其中所添加的代价与权重系数值的平方（权重“L2 范数”）成正比。L2 正则化在神经网络领域也称为权重衰减。不要因为名称不同而感到困惑：从数学角度来讲，权重衰减与 L2 正则化完全相同。

在 tf.keras 中，权重正则化的添加方法为：将权重正则化修正项实例作为关键字参数传递给层。现在，我们来添加 L2 权重正则化。

```
    l2_model = keras.models.Sequential([
    keras.layers.Dense(16, kernel_regularizer=keras.regularizers.l2(0.001),
                       activation=tf.nn.relu, input_shape=(NUM_WORDS,)),
    keras.layers.Dense(16, kernel_regularizer=keras.regularizers.l2(0.001),
                       activation=tf.nn.relu),
    keras.layers.Dense(1, activation=tf.nn.sigmoid)
])

l2_model.compile(optimizer='adam',
             loss='binary_crossentropy',
             metrics=['accuracy', 'binary_crossentropy'])

l2_model_history = l2_model.fit(train_data, train_labels,
                            epochs=20,
                            batch_size=512,
                            validation_data=(test_data, test_labels),
                            verbose=2)

    Train on 25000 samples, validate on 25000 samples
    Epoch 1/20
```

```
 - 3s - loss: 0.5232 - acc: 0.8118 - binary_crossentropy: 0.4838 - val_loss: 
0.3806 - val_acc: 0.8779 - val_binary_crossentropy: 0.3387
Epoch 2/20
 - 2s - loss: 0.3075 - acc: 0.9089 - binary_crossentropy: 0.2609 - val_loss: 
0.3351 - val_acc: 0.8880 - val_binary_crossentropy: 0.2851
Epoch 3/20
 - 2s - loss: 0.2582 - acc: 0.9281 - binary_crossentropy: 0.2059 - val_loss: 
0.3369 - val_acc: 0.8861 - val_binary_crossentropy: 0.2829
Epoch 4/20
 - 2s - loss: 0.2336 - acc: 0.9391 - binary_crossentropy: 0.1781 - val_loss: 
0.3478 - val_acc: 0.8827 - val_binary_crossentropy: 0.2915
Epoch 5/20
 - 2s - loss: 0.2204 - acc: 0.9445 - binary_crossentropy: 0.1625 - val_loss: 
0.3598 - val_acc: 0.8794 - val_binary_crossentropy: 0.3011
Epoch 6/20
 - 2s - loss: 0.2074 - acc: 0.9501 - binary_crossentropy: 0.1482 - val_loss: 
0.3733 - val_acc: 0.8766 - val_binary_crossentropy: 0.3139
Epoch 7/20
 - 2s - loss: 0.2003 - acc: 0.9524 - binary_crossentropy: 0.1399 - val_loss: 
0.3875 - val_acc: 0.8736 - val_binary_crossentropy: 0.3264
Epoch 8/20
 - 2s - loss: 0.1922 - acc: 0.9563 - binary_crossentropy: 0.1304 - val_loss: 
0.3968 - val_acc: 0.8722 - val_binary_crossentropy: 0.3349
Epoch 9/20
 - 2s - loss: 0.1863 - acc: 0.9576 - binary_crossentropy: 0.1239 - val_loss: 
0.4127 - val_acc: 0.8709 - val_binary_crossentropy: 0.3498
Epoch 10/20
 - 2s - loss: 0.1843 - acc: 0.9588 - binary_crossentropy: 0.1208 - val_loss: 
0.4287 - val_acc: 0.8673 - val_binary_crossentropy: 0.3647
Epoch 11/20
 - 2s - loss: 0.1787 - acc: 0.9612 - binary_crossentropy: 0.1142 - val_loss: 
0.4393 - val_acc: 0.8654 - val_binary_crossentropy: 0.3742
Epoch 12/20
 - 2s - loss: 0.1752 - acc: 0.9619 - binary_crossentropy: 0.1101 - val_loss: 
0.4626 - val_acc: 0.8622 - val_binary_crossentropy: 0.3970
Epoch 13/20
 - 2s - loss: 0.1747 - acc: 0.9629 - binary_crossentropy: 0.1083 - val_loss: 
0.4641 - val_acc: 0.8640 - val_binary_crossentropy: 0.3973
Epoch 14/20
 - 2s - loss: 0.1640 - acc: 0.9668 - binary_crossentropy: 0.0972 - val_loss: 
0.4750 - val_acc: 0.8612 - val_binary_crossentropy: 0.4085
Epoch 15/20
 - 2s - loss: 0.1570 - acc: 0.9700 - binary_crossentropy: 0.0907 - val_loss: 
0.4934 - val_acc: 0.8612 - val_binary_crossentropy: 0.4270
Epoch 16/20
 - 2s - loss: 0.1540 - acc: 0.9713 - binary_crossentropy: 0.0871 - val_loss: 
0.5082 - val_acc: 0.8586 - val_binary_crossentropy: 0.4410
Epoch 17/20
 - 2s - loss: 0.1516 - acc: 0.9726 - binary_crossentropy: 0.0844 - val_loss: 
0.5187 - val_acc: 0.8576 - val_binary_crossentropy: 0.4510
Epoch 18/20
```

```
 - 2s - loss: 0.1493 - acc: 0.9724 - binary_crossentropy: 0.0812 - val_loss: 0.5368 - val_acc: 0.8556 - val_binary_crossentropy: 0.4681
 Epoch 19/20
 - 2s - loss: 0.1449 - acc: 0.9753 - binary_crossentropy: 0.0761 - val_loss: 0.5439 - val_acc: 0.8573 - val_binary_crossentropy: 0.4750
 Epoch 20/20
 - 2s - loss: 0.1445 - acc: 0.9752 - binary_crossentropy: 0.0749 - val_loss: 0.5579 - val_acc: 0.8555 - val_binary_crossentropy: 0.4877
```

l2(0.001) 表示层的权重矩阵中的每个系数都会将 0.001 * weight_coefficient_value**2 添加到网络的总损失中。注意，由于此惩罚仅在训练时添加，因此该网络在训练时的损失将远高于测试时。

以下是 L2 正则化惩罚的影响：

```
plot_history([('baseline', baseline_history),
              ('l2', l2_model_history)])
```

从图 6.16 可以看到，L2 正则化模型的过拟合抵抗能力比基准模型强得多，虽然这两个模型的参数数量相同。

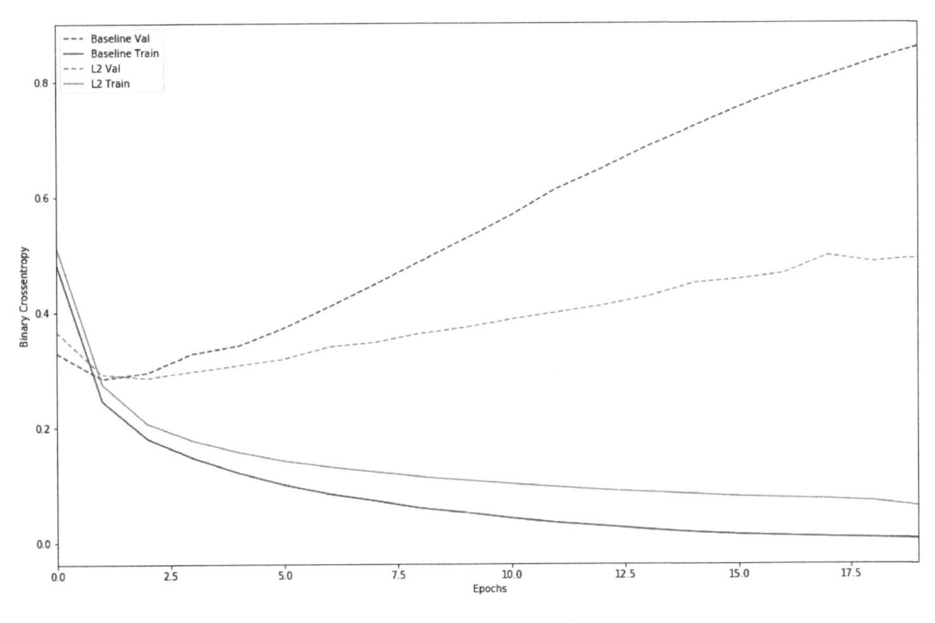

图 6.16　权重正则化结果

2. 添加丢弃层

丢弃是由 Hinton 及其在多伦多大学的学生开发的，是最有效且最常用的神经网络正则化技术之一。丢弃（应用于某个层）是指在训练期间随机“丢弃”（设置为 0）该层的多个输出特征，假设某个指定的层通常会在训练期间针对给定的输入样本返回一个向量[0.2, 0.5, 1.3, 0.8, 1.1]，在应用丢弃后，此向量将随机分布几个 0 条目，例如[0, 0.5, 1.3, 0, 1.1]。“丢弃率”指变为 0 的特征所占的比例，通常设置在 0.2 和 0.5 之间。在测试时，网络不会丢弃任何单元，而是将层的输出值按等同于丢弃率的比例进行缩减，以便平衡以下事实：测试时的活跃单元数大于训练时的活跃单元数。

在 tf.keras 中，可以通过丢弃层将丢弃引入网络中，以便事先将其应用于层的输出。

下面我们在 IMDB 网络中添加两个丢弃层，看看它们在降低过拟合方面表现如何：

```
dpt_model = keras.models.Sequential([
    keras.layers.Dense(16, activation=tf.nn.relu, input_shape=(NUM_WORDS,)),
    keras.layers.Dropout(0.5),
    keras.layers.Dense(16, activation=tf.nn.relu),
    keras.layers.Dropout(0.5),
    keras.layers.Dense(1, activation=tf.nn.sigmoid)
])

dpt_model.compile(optimizer='adam',
                  loss='binary_crossentropy',
                  metrics=['accuracy','binary_crossentropy'])

dpt_model_history = dpt_model.fit(train_data, train_labels,
                                  epochs=20,
                                  batch_size=512,
                                  validation_data=(test_data, test_labels),
                                  verbose=2)

    Train on 25000 samples, validate on 25000 samples
    Epoch 1/20
     - 3s - loss: 0.6315 - acc: 0.6358 - binary_crossentropy: 0.6315 - val_loss:
0.5245 - val_acc: 0.8326 - val_binary_crossentropy: 0.5245
    Epoch 2/20
     - 2s - loss: 0.4959 - acc: 0.8044 - binary_crossentropy: 0.4959 - val_loss:
0.3973 - val_acc: 0.8743 - val_binary_crossentropy: 0.3973
    Epoch 3/20
     - 2s - loss: 0.3848 - acc: 0.8720 - binary_crossentropy: 0.3848 - val_loss:
0.3270 - val_acc: 0.8864 - val_binary_crossentropy: 0.3270
    Epoch 4/20
     - 2s - loss: 0.3106 - acc: 0.9069 - binary_crossentropy: 0.3106 - val_loss:
0.2972 - val_acc: 0.8883 - val_binary_crossentropy: 0.2972
    Epoch 5/20
     - 2s - loss: 0.2692 - acc: 0.9194 - binary_crossentropy: 0.2692 - val_loss:
0.2902 - val_acc: 0.8866 - val_binary_crossentropy: 0.2902
    Epoch 6/20
     - 2s - loss: 0.2264 - acc: 0.9343 - binary_crossentropy: 0.2264 - val_loss:
0.3005 - val_acc: 0.8848 - val_binary_crossentropy: 0.3005
    Epoch 7/20
     - 2s - loss: 0.2026 - acc: 0.9406 - binary_crossentropy: 0.2026 - val_loss:
```

```
0.3158 - val_acc: 0.8846 - val_binary_crossentropy: 0.3158
   Epoch 8/20
    - 2s - loss: 0.1801 - acc: 0.9482 - binary_crossentropy: 0.1801 - val_loss:
0.3287 - val_acc: 0.8830 - val_binary_crossentropy: 0.3287
   Epoch 9/20
    - 2s - loss: 0.1604 - acc: 0.9545 - binary_crossentropy: 0.1604 - val_loss:
0.3329 - val_acc: 0.8806 - val_binary_crossentropy: 0.3329
   Epoch 10/20
    - 2s - loss: 0.1483 - acc: 0.9588 - binary_crossentropy: 0.1483 - val_loss:
0.3490 - val_acc: 0.8786 - val_binary_crossentropy: 0.3490
   Epoch 11/20
    - 2s - loss: 0.1322 - acc: 0.9625 - binary_crossentropy: 0.1322 - val_loss:
0.3758 - val_acc: 0.8774 - val_binary_crossentropy: 0.3758
   Epoch 12/20
    - 2s - loss: 0.1249 - acc: 0.9644 - binary_crossentropy: 0.1249 - val_loss:
0.3953 - val_acc: 0.8764 - val_binary_crossentropy: 0.3953
   Epoch 13/20
    - 2s - loss: 0.1151 - acc: 0.9663 - binary_crossentropy: 0.1151 - val_loss:
0.4445 - val_acc: 0.8766 - val_binary_crossentropy: 0.4445
   Epoch 14/20
    - 2s - loss: 0.1096 - acc: 0.9686 - binary_crossentropy: 0.1096 - val_loss:
0.4400 - val_acc: 0.8753 - val_binary_crossentropy: 0.4400
   Epoch 15/20
    - 2s - loss: 0.0995 - acc: 0.9726 - binary_crossentropy: 0.0995 - val_loss:
0.4778 - val_acc: 0.8760 - val_binary_crossentropy: 0.4778
   Epoch 16/20
    - 2s - loss: 0.0959 - acc: 0.9734 - binary_crossentropy: 0.0959 - val_loss:
0.4899 - val_acc: 0.8759 - val_binary_crossentropy: 0.4899
   Epoch 17/20
    - 2s - loss: 0.0929 - acc: 0.9740 - binary_crossentropy: 0.0929 - val_loss:
0.5084 - val_acc: 0.8754 - val_binary_crossentropy: 0.5084
   Epoch 18/20
    - 2s - loss: 0.0917 - acc: 0.9733 - binary_crossentropy: 0.0917 - val_loss:
0.5460 - val_acc: 0.8745 - val_binary_crossentropy: 0.5460
   Epoch 19/20
    - 2s - loss: 0.0841 - acc: 0.9775 - binary_crossentropy: 0.0841 - val_loss:
0.5420 - val_acc: 0.8754 - val_binary_crossentropy: 0.5420
   Epoch 20/20
    - 2s - loss: 0.0803 - acc: 0.9786 - binary_crossentropy: 0.0803 - val_loss:
0.5750 - val_acc: 0.8744 - val_binary_crossentropy: 0.5750
   plot_history([('baseline', baseline_history),
              ('dropout', dpt_model_history)])
```

添加丢弃层的结果如图 6.17 所示。

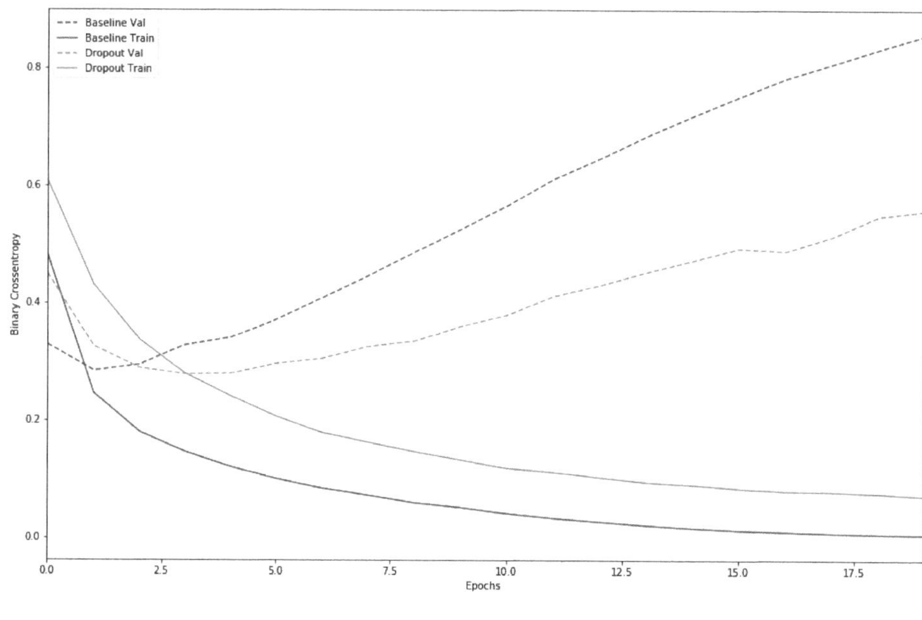

图 6.17 添加丢弃层结果

从图 6.17 中可以清晰地观测到添加丢弃层可明显改善基准模型。

参考文献

[1] Chua L O. CNN: a paradigm for complexity[M]. WORLD SCIENTIFIC, 1998.

[2] Chua L O, Roska T. CNN: a new paradigm of nonlinear dynamics in space[C]// Proceedings of the first world congress on World congress of nonlinear analysts, volume III. 1996.

[3] Zhang K, Zuo W, Chen Y, et al. Beyond a Gaussian Denoiser: Residual Learning of Deep CNN for Image Denoising[J]. IEEE TRANSACTIONS ON IMAGE PROCESSING.

[4] Jin K H, Mccann M T, Froustey E, et al. Deep Convolutional Neural Network for Inverse Problems in Imaging[J]. IEEE Transactions on Image Processing, 2017:1-1.

[5]. Srivastava N, Hinton G, Krizhevsky A, et al. Dropout: A Simple Way to Prevent Neural Networks from Overfitting[J]. Journal of Machine Learning Research, 2014, 15(1):1929-1958.

[6] Ioffe S, Szegedy C. Batch normalization: accelerating deep network training by reducing internal covariate shift[C]// International Conference on International Conference on Machine Learning. JMLR.org, 2015.

第 7 章

循环神经网络

我们周围的许多重要数据都是时序数据，例如自然语言。时序数据是同一统一指标按时间顺序记录的数据列。为了能够使用深度学习更好地处理时序数据，循环神经网络及其改良算法应运而生。循环神经网络是一类功能强大并且应用广泛的对序列数据建模而设计的神经网络体系结构，其中的基本思想是序列中的每个新元素都贡献一些新的信息，这些信息更新了模型的当前状态。

本章主要讲解循环神经网络及其改良算法，并展示如何使用 TensorFlow 中的文本序列完成自然语言处理的一般流程。内容分为三部分，第一部分着重介绍时序数据的特性以及循环神经网络的模型；第二部分着重分析普通的循环神经网络模型的缺点，从而引入并详细分析长短时记忆模型；第三部分使用循环神经网络模型西游记进行仿写，通过这一过程加深对模型的应用和理解。

7.1 循环神经网络概述

7.1.1 时序数据

从前面的介绍我们认识到数据结构的重要性，利用图像的空间结构可以得到先进的模型，并取得很好的效果。模型对训练数据集合（图像的向量空间表示）的数据特征的认知能力是深度学习算法成功的关键。时序数据就是指时间序列数据，在同一数据列中的各个数据必须是同口径的，要求具有可比性。时序数据可以是时期数，也可以是时点数。时间序列数据结构是非常重要和有用的结构类型，从数据科学的角度考虑，这种基本结构出现在所有领域的许多数据集中。在计算机视觉中，视频是随着时间推移而演变的一系列图像内容。在语音识别中需要处理音频信号，在基因组中有基因序列，在医疗保健方面有纵向医疗记录，在股票市场有财务数据，这些都是非常典型的时间序列数据。

自然语言文本数据是一种具有强时间序列关系的特别重要的数据类型。深度学习方法利用文本中的顺序结构：字符、单词、句子、段落和文档构建了自然语言理解（Natural Language Understanding，NLU）系统的前沿课题和解决方案集合，通过效果对比，很多传统方法会显得捉襟见肘。有很多类型的 NLU 任务值得解决，从文档分类到构建强大的语言模型，从自动回答问题到生成人类级别的对话代理。这些任务极其艰巨，吸引了学术界和工业界整个人工智能社区的努力和关注。在本章中，我们重点介绍基本的构建过程和向量映射关系，并展示如何使用 TensorFlow 中的文本序列。我们将深入研究 TensorFlow 中序列模型的核心元素，并从头开始实现其中的一些，以获得全面的理解。我们从最重要和最流行的序列（特别是文本）深度学习模型——循环神经网络开始学习。

7.1.2 循环神经网络模型

循环神经网络（Recurrent Neural Network，RNN）是一类功能强大并且应用广泛的对序列数据建模而设计的神经网络体系结构。循环神经网络模型的基本思想是序列中的每个新元素都贡献一些新的信息，这些信息更新了模型的当前状态。在前一章中，我们探讨了卷积神经网络模型，我们讨论了这些架构是如何受到当前人类大脑处理视觉信息方式的科学认知的启发的。这些科学认知往往与我们日常生活中关于如何处理顺序信息的普通直觉相当接近。当我们收到新信息时，很明显，历史和记忆并没有被抹去，而是更新了。当我们阅读某个文本中的一个句子时，随着每个新单词的出现，当前的信息状态就会更新，它不仅依赖于观察到的新单词，还依赖于它之前的单词。马尔可夫链模型是统计学和概率论中一个基本的数学结构，它常被用作通过机器学习对序列模式建模的基本单元。形象地说，我们可以将数据序列看作数据的链表，链表中的每个节点都以某种方式依赖于前一个节点，这样历史就不会被删除，而是继续下去，保持了历史的完整性。循环神经网模型也基于这种链结构的概念，并且在如何准确地维护和更新信息方面有所不同。顾名思义，递归神经网络应用某种形式的循环。如图 7.1 所示，在 t 时间点上，网络观察输入 x_t（句子中的一个单词），并将其状态向量从之前的向量 s_{t-1} 更新为 s_t。当我们处理新的输入（下一个单词）时，它将以某种依赖于 h_t 的方式完成，从而依赖于序列的历史（前面看到的单词影响了我们对当前单词的理解）。这个循环结构展开后可以简单地看作一个长的信息链，链中的每个节点根据它从前一个节点的输出获得的消息执行相同的处理步骤。

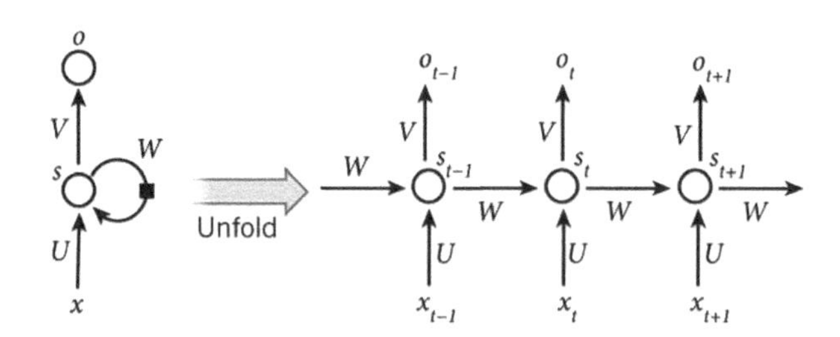

图 7.1 循环神经网络运算过程

在处理语言数据时，通常使用序列，例如单词（字母序列）、句子（单词序列）和文档。前馈网络通过使用向量连接和连续词袋模型（Continuous Bag-of-Words Model，CBOW）在序列上容纳任意的特征函数。特别是，CBOW 表示允许将任意长度序列编码为固定大小的向量。然而，CBOW

表示是非常有限的，并迫使人们忽略特征的顺序。

卷积网络也允许将一个序列编码成一个固定大小的向量。虽然来自卷积网络的表示是对 CBOW 表示的一种改进，因为它们对词序具有一定的敏感性，但它们的顺序敏感性主要限于局部模式，忽略了顺序中相距很远的模式的顺序。循环神经网络允许在固定大小的向量中表示任意大小的顺序输入，同时注意输入的结构特性。

7.2　长短时记忆神经网络架构

循环神经网络，特别是具有门控结构（如 LSTM 和 GRU）的循环神经网络，在捕获顺序输入中的统计规则方面非常强大。循环神经网络及其变体可以说是深度学习对统计自然语言处理工具集的最大贡献。循环神经网络可以设计成一个抽象接口，用于将一系列输入转换为固定大小的输出，然后可以作为更大网络中的组件进行插入。

Hochreiter 和 Schmidhuber 提出的长短时记忆（Long Short-Term Memory，LSTM）结构是为了解决消失梯度问题而设计的，也是第一个引入门机制的深度学习算法。在很多问题上，长短时记忆神经网络都取得了相当巨大的成功，并得到了广泛的使用。长短时记忆神经网络通过完整的体系设计来解决长期依赖问题，长短时记忆体系结构显式地将状态向量分成两半，其中一半作为内存单元，另一半作为工作内存。记忆细胞是用来保存误差梯度的记忆，通过对记忆内容的选择调用使模型跨越时间，并通过可微的闸门控制组件中的平滑数学函数模拟逻辑门。在每个输入状态，门用来决定有多少新的输入应该写入存储单元，以及当前存储单元的内容应该被遗忘。

所有循环神经网络都具有一种重复神经网络模块的链式的形式。在标准的循环神经网络中，这个重复的模块只有一个非常简单的结构，例如一个 tanh 层。长短时记忆神经网络同样是这样的结构，但是重复的模块拥有不同的结构，如图 7.2 所示。不同于单一神经网络层，这里是由 4 种神经网络运算子单元连接而成的，以一种非常特殊的方式进行交互。

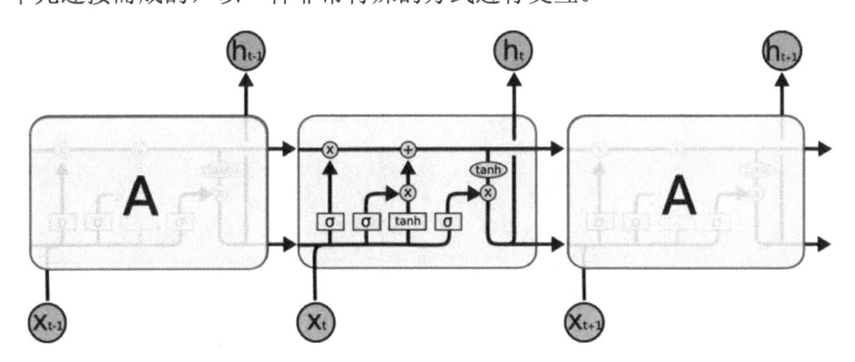

图 7.2　长短时记忆神经网络总体架构

长短时记忆神经网络的关键就是细胞状态，如图 7.3 所示，水平线在图上方贯穿运行。细胞状态类似于传送带，直接在整个链上运行，只有一些少量的线性交互。信息在上面流传保持不变很容易。

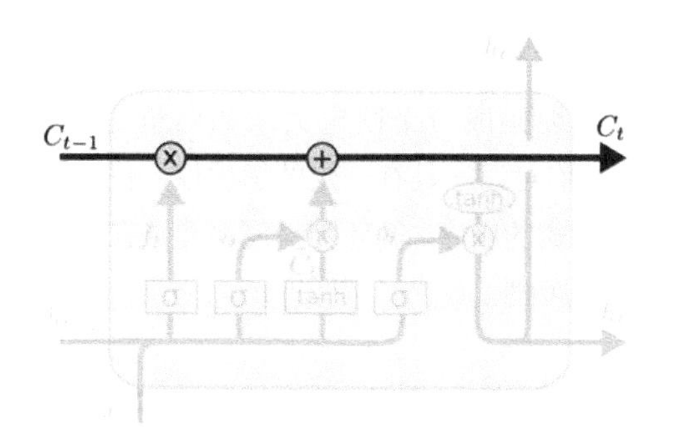

图 7.3　长短时记忆神经网络输入输出节点

长短时记忆神经网络有通过精心设计的称作为“门”的结构来去除或者增加信息到细胞状态的能力，如图 7.4 所示。门是一种让信息选择式通过的方法，它包含一个 Sigmoid 神经网络层和一个 Pointwise 乘法操作。

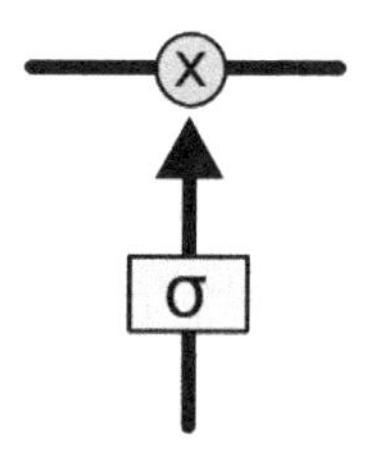

图 7.4　长短时记忆神经网络门结构

Sigmoid 层输出 0 到 1 之间的数值，描述每个部分有多少量可以通过。0 代表不许任何量通过，1 代表允许任意量通过。

长短时记忆神经网络拥有 3 个门，用于保护和控制细胞状态。

在长短时记忆神经网络中的第一步是决定我们会从细胞状态中丢弃什么信息，这个决定通过一个称为忘记门的层完成。该门会读取 h_{t-1} 和 x_t，输出一个 0~1 的数值给每个在细胞状态 C_{t-1} 中的数字，如图 7.5 所示。1 表示完全保留，0 表示完全舍弃。

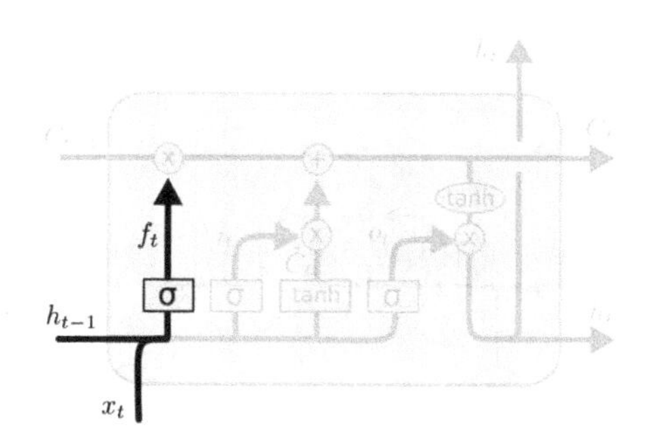

图 7.5　长短时记忆神经网络第一步

让我们回到语言模型的例子中基于已经看到的词来预测下一个词。在这个问题中，细胞状态可能包含当前主语的性别，因此正确的代词可以被选择出来。当看到新的主语时，我们希望忘记旧的主语。

$$f_t = \sigma(W_f \cdot [h_{t-1}, x_t] + b_f)$$

第二步是确定什么样的新信息被存放在细胞状态中，如图 7.6 所示。这里包含两部分：首先，Sigmoid 层（输入门层）决定什么值我们将要更新；然后，一个 tanh 层创建一个新的候选值向量：C_t，会被加入状态中。下一步，我们会用这两个信息来产生对状态的更新。

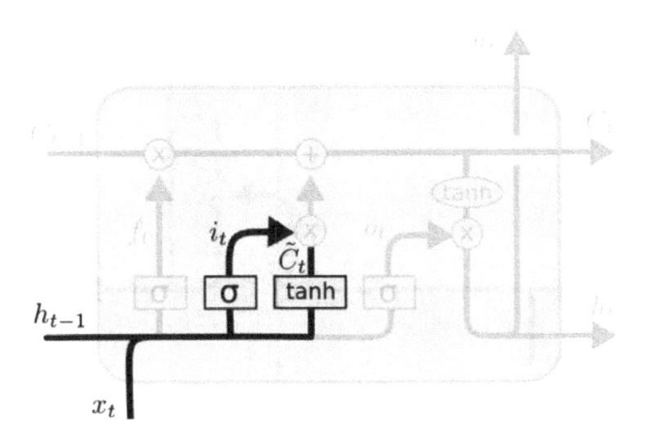

图 7.6　长短时记忆神经网络第二步

在语言模型的例子中，我们希望增加新的主语的性别到细胞状态中，来替代旧的需要忘记的主语。

$$i_t = \sigma(W_i \cdot [h_{t-1}, x_t] + b_i)$$

$$\tilde{C}_t = \tanh(W_C \cdot [h_{t-1}, x_t] + b_C)$$

第三步是更新旧细胞状态，将 C_{t-1} 更新为 C_t，如图 7.7 所示。前面的步骤已经决定了将会做什么，我们现在就是实际去完成。

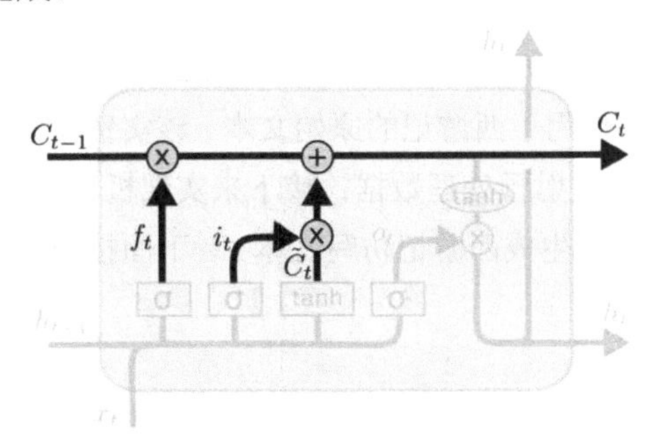

图 7.7　长短时记忆神经网络第三步

我们把旧状态与 f_t 相乘，丢弃掉确定需要丢弃的信息。接着加上 i_t*C_t。这就是新的候选值，根据我们决定更新每个状态的程度进行变化。

在语言模型的例子中，这就是我们实际根据前面确定的目标，丢弃旧代词的性别信息并添加新信息的地方。

$$C_t = f_t * C_{t-1} + i_t * \tilde{C}_t$$

最终，我们需要确定输出什么值，如图 7.8 所示。这个输出将会基于当前的细胞状态，不过是过滤后的版本。首先，我们运行一个 Sigmoid 层来确定细胞状态的哪个部分将输出出去。接着，把细胞状态通过 tanh 进行处理（得到一个在-1 和 1 之间的值），并将它和 Sigmoid 门的输出相乘，最终仅输出我们确定输出的那部分。

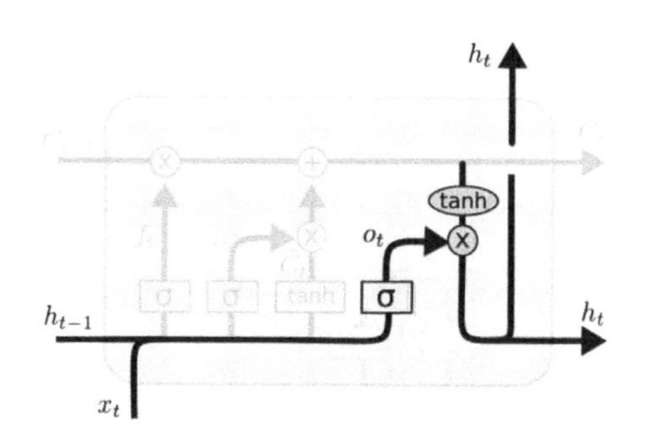

图 7.8　长短时记忆神经网络第四步

$$o_t = \sigma(W_o[h_{t-1}, x_t] + b_o)$$
$$h_t = o_t * \tanh(C_t)$$

比如在语言模型的例子中，模型代码检测到了一个代词，可能需要输出一个与动词相关的信息。例如，输出代词可能有是单数还是复数的变化，这种变化就会在相关的动词信息中包含需要进行的动词词形变化。

7.3　实例：仿写西游记

文本数据是常见的时序数据，本实例通过循环神经网络的推导能力完成一个简单的中文段落仿写应用，考虑到版权问题，我们使用了西游记的原始文本。该实例首先通过简单的字符串处理进行人为的分段操作，通过文本概率映射预处理数据；接下来实现机器学习的一般步骤，即模型构造、训练模型以及应用模型进行预测（生成西游记仿写文本）。下面按照上述的开发流程来逐一研究具体的实现代码。

7.3.1　文本的读取和分段

代码分为两部分，第一部分是文本的读取和分段，具体代码如下：

```
#-------------------------------数据预处理---------------------------#

novel_file ='novel.txt'

# 小说文本转换段落
novels = []
with open(novel_file, "r", encoding='utf-8',) as f:
    for line in f:
```

```
        try:
            content = line.replace(' ','')
            print(line)
            if '_' in content or '(' in content or '（' in content or '《' in
content or '[' in content:
                continue
            if len(content) < 5 or len(content) > 300:
                continue
            content = '[' + content + ']'
            novels.append(content)
        except Exception as e:
            pass
```

首先需要使用 Python 中常规的文件操作 open()函数读取整个文本，open()函数用于打开一个文件，创建一个 file 对象，相关的方法才可以调用它进行读写。这里还使用 with 语句简化异常管理过程，with 语句适用于对资源进行访问的场合，确保无论使用过程中是否发生异常都会执行必要的“清理”操作，释放资源，比如文件使用后自动关闭、线程中锁的自动获取和释放等。

```
with open(novel_file, "r", encoding='utf-8',) as f:
```

有了 f 这个文本对象，我们就可以遍历整个文本对象了：

```
  for line in f:
```

通过组合多行文本形成标准的文本块信息，我们在程序中设定了 300 个汉字加标点符号为文本块的容量上限。

```
if len(content) < 5 or len(content) > 300:
                continue
            content = '[' + content + ']'
```

对生成的结果添加中括号作为分隔符，然后保存到一个定义好的字符串数组中。

novels = []# 声明过程在最前面，和下面的两行代码不在一起，请读者注意。

```
content = '[' + content + ']'
novels.append(content)
```

第二部分的主要功能是将文本信息转换成机器学习模型能够接受的实数矩阵，具体代码如下：

```
# 按西游记小说的字数排序
novels = sorted(novels,key=lambda line: len(line))
print('段落总数: ', len(novels))

# 统计每个字出现的次数
all_words = []
for novel in novels:
    all_words += [word for word in novel]
```

```
    counter = collections.Counter(all_words)
    count_pairs = sorted(counter.items(), key=lambda x: -x[1])
    words, _ = zip(*count_pairs)

    # 取所有不重复的段落文字
    words = words[:len(words)] + (' ',)
    # 每个字映射为一个数字 ID
    word_num_map = dict(zip(words, range(len(words))))
    # 把段落文本转换为向量形式
    to_num = lambda word: word_num_map.get(word, len(words))
    novels_vector = [list(map(to_num, novel)) for novel in novels]
    #[[314, 3199, 367, 1556, 26, 179, 680, 0, 3199, 41, 506, 40, 151, 4, 98, 1],
    #[339, 3, 133, 31, 302, 653, 512, 0, 37, 148, 294, 25, 54, 833, 3, 1, 965, 1315,
377, 1700, 562, 21, 37, 0, 2, 1253, 21, 36, 264, 877, 809, 1]
    #....]

    # 每次取 8 段西游记段落文本进行训练
    batch_size = 8
    n_chunk = len(novels_vector) // batch_size
    x_batches = []
    y_batches = []
    for i in range(n_chunk):
        start_index = i * batch_size
        end_index = start_index + batch_size

        batches = novels_vector[start_index:end_index]
        length = max(map(len,batches))
        xdata = np.full((batch_size,length), word_num_map[' '], np.int32)
        for row in range(batch_size):
            xdata[row,:len(batches[row])] = batches[row]
        ydata = np.copy(xdata)
        ydata[:,:-1] = xdata[:,1:]
        """
        xdata             ydata
        [6,2,4,6,9]       [2,4,6,9,9]
        [1,4,2,8,5]       [4,2,8,5,5]
        """
        x_batches.append(xdata)
        y_batches.append(ydata)
```

首先，我们按照小说字数对数组元素进行重新排序，代码非常简洁。对于 Python 内置函数 sorted()，先拿来跟 list（列表）中的成员函数 list.sort()进行对比。在本质上，list 的排序与内建函数 sorted 的排序是差不多的，参数也基本上是一样的。主要的区别在于，list.sort()是对已经存在的列表进行操作，进而可以改变进行操作的列表。而内建函数 sorted 返回的是一个新的 list，而不是在原来的基础上进行的操作。同时，我们注意到使用 key 参数传入了一个 lambda 函数表达式，将排序要求简单地表述了出来，line 为排序元素，len(line)为排序规则，对应代码如下：

```
novels = sorted(novels,key=lambda line: len(line))
```

接下来，需要统计每个单独的字（比如一）出现的次数，首先利用 for 循环取得所有段落中的字并存入字的数组中，这个数组中的字是有重复的，比如一出现了 10 次，这个数组中就会有 10 个元素记录一。利用 Python 集合中的统计方法 Counter 统计字出现的次数，通过排序和打包得到字和次数关系（使用一个数字 ID 表示）的字典，对应代码如下：

```
all_words = []
for novel in novels:
    all_words += [word for word in novel]
counter = collections.Counter(all_words)
count_pairs = sorted(counter.items(), key=lambda x: -x[1])
words, _ = zip(*count_pairs)
# 取所有不重复的段落文字
words = words[:len(words)] + (' ',)
# 每个字映射为一个数字 ID
word_num_map = dict(zip(words, range(len(words))))
```

接下来，利用上面的文字的字典转换提取的所有文本段落。每个字都对应一个数字，整个段落就转化成了实数数组，这个数据结构已经能够满足循环神经网络的模型训练过程的需要了。

```
to_num = lambda word: word_num_map.get(word, len(words))
novels_vector = [list(map(to_num, novel)) for novel in novels]
```

接下来，我们要整理数据输出方式满足模型训练的批量梯度下降的过程，同时要定义输入数据和标签数据之间的关系，这个关系用语言来表示就是通过前面的一些文字来预测后面紧接的一个文字是什么，因此标签数据会比我们输入的数据向后错位一个字符（重要步骤会使用注释在下面的代码中说明）。

```
batch_size = 8
n_chunk = len(novels_vector) // batch_size
x_batches = []
y_batches = []
for i in range(n_chunk):
    start_index = i * batch_size
    end_index = start_index + batch_size
    batches = novels_vector[start_index:end_index]
    length = max(map(len,batches))
    xdata = np.full((batch_size,length), word_num_map[' '], np.int32)
    for row in range(batch_size):
        xdata[row,:len(batches[row])] = batches[row]
```

```
        ydata = np.copy(xdata)
        ydata[:,:-1] = xdata[:,1:] # 通过错位来形成标签数据
        """
        xdata           ydata
        [6,2,4,6,9]      [2,4,6,9,9]
        [1,4,2,8,5]      [4,2,8,5,5]
        """
        x_batches.append(xdata)
        y_batches.append(ydata)
```

7.3.2 循环神经网络模型定义

完成文本数据预处理和数据集合构造的过程之后，我们进入模型构造的核心环节。循环神经网络模型定义的核心实现代码如下：

```
    #输入数据站位符张量声明
    input_data = tf.placeholder(tf.int32, [batch_size, None])
    #标签数据站位符张量声明
    output_targets = tf.placeholder(tf.int32, [batch_size, None])
    # 定义RNN网络结构
    def neural_network(model='lstm', rnn_size=128, num_layers=2):
        if model == 'rnn': # 利用输入的字符串rnn来选择使用标准RNN神经元
            cell_fun = tf.nn.rnn_cell.BasicRNNCell
        elif model == 'gru': # 利用输入的字符串gru来选择使用GRU神经元
            cell_fun = tf.nn.rnn_cell.GRUCell
        elif model == 'lstm': # 利用输入的字符串lstm来选择使用LSTM神经元
            cell_fun = tf.nn.rnn_cell.BasicLSTMCell
     # 利用选择好的RNN神经元来构造RNN网络
        cell = cell_fun(rnn_size, state_is_tuple=True)
        cell = tf.nn.rnn_cell.MultiRNNCell([cell] * num_layers,
state_is_tuple=True)
     # RNN网络初始化
        initial_state = cell.zero_state(batch_size, tf.float32)

        with tf.variable_scope('rnnlm'):
            softmax_w = tf.get_variable("softmax_w", [rnn_size, len(words)+1])
    # 输出层权重矩阵定义
            softmax_b = tf.get_variable("softmax_b", [len(words)+1]) #输出层偏置矩
阵定义
            with tf.device("/cpu:0"):
                embedding = tf.get_variable("embedding", [len(words)+1,
rnn_size])
                inputs = tf.nn.embedding_lookup(embedding, input_data)
```

```
    outputs, last_state = tf.nn.dynamic_rnn(cell, inputs,
initial_state=initial_state, scope='rnnlm')
    output = tf.reshape(outputs,[-1, rnn_size])
  # 输出层代码设计，输出层就是标准的全连接神经网络构造
    logits = tf.matmul(output, softmax_w) + softmax_b
    probs = tf.nn.softmax(logits)
    return logits, last_state, probs, cell, initial_state
```

其中一些常规操作已经使用注释说明了，我们的关注点主要在循环神经网络的模型设计过程，核心代码为 embedding 和 rnn 两个调用过程，对应代码如下：

```
  inputs = tf.nn.embedding_lookup(embedding, input_data)
  outputs, last_state = tf.nn.dynamic_rnn(cell, inputs,
initial_state=initial_state, scope='rnnlm')
```

实际上，tf.nn.embedding_lookup 的作用就是找到要寻找的 embedding data 中对应的行下的 vector。代码如下：

```
  tf.nn.embedding_lookup(params, ids, partition_strategy='mod', name=None,
validate_indices=True, max_norm=None)
```

tf.nn.dynamic_rnn 输入的参数如下：

```
  tf.nn.dynamic_rnn(
      cell,
      inputs,
      sequence_length=None,
      initial_state=None,
      dtype=None,
      parallel_iterations=None,
      swap_memory=False,
      time_major=False,
      scope=None
  )
```

tf.nn.dynamic_rnn 的返回值有两个：outputs 和 state。

为了描述输出的形状，先介绍几个变量：batch_size 是输入的这批数据的数量；max_time 是这批数据中序列的最长长度，如果输入了 3 个句子，max_time 对应的就是最长句子的单词数量；cell.output_size 就是 rnn cell 中神经元的个数。

outputs：outputs 是一个张量。如果 time_major==True，那么 outputs 的形状为[max_time, batch_size, cell.output_size]（要求 rnn 输入与 rnn 输出形状保持一致）；如果 time_major==False（默认），那么 outputs 的形状为[batch_size, max_time, cell.output_size]。

state：state 也是一个张量。state 是最终的状态，也就是序列中最后一个 cell 输出的状态。一般情况下，state 的形状为[batch_size,cell.output_size]，但当输入的 cell 为 BasicLSTMCell 时，state 的

形状为[2，batch_size,cell.output_size]，其中 2 对应着 LSTM 中的 cell state 和 hidden state。

7.3.3 模型训练和结果分析

当模型建立以后，下面我们遵循机器学习开发流程进行模型的训练。训练代码如下（其中的一些标准化流程笔者使用注释的方式进行了说明）：

```
  # 训练仿写西游记段落循环神经网络模型
  def train_neural_network():
    # 调用定义模型
  logits, last_state, _, _, _ = neural_network()
  # 计算模型的损失
  targets = tf.reshape(output_targets, [-1])
      loss = tf.contrib.legacy_seq2seq.sequence_loss_by_example([logits],
[targets], [tf.ones_like(targets, dtype=tf.float32)], len(words))
  cost = tf.reduce_mean(loss)
  # 定义训练的优化方法及其相关的学习速率等参数
      learning_rate = tf.Variable(0.0, trainable=False)
      tvars = tf.trainable_variables()
      grads, _ = tf.clip_by_global_norm(tf.gradients(cost, tvars), 5)
      optimizer = tf.train.AdamOptimizer(learning_rate)
      train_op = optimizer.apply_gradients(zip(grads, tvars))
   # 实际训练仿写西游记
      with tf.Session() as sess:
          sess.run(tf.initialize_all_variables())

          saver = tf.train.Saver(tf.all_variables())

          for epoch in range(50):
              sess.run(tf.assign(learning_rate, 0.1 * (0.97 ** epoch)))
              n = 0
              for batche in range(n_chunk):
                  train_loss, _ , _ = sess.run([cost, last_state, train_op],
feed_dict={input_data: x_batches[n], output_targets: y_batches[n]})
                  n += 1
                  print(epoch, batche, train_loss)
              if epoch % 7 == 0: # 训练过程中保存训练结果
                  saver.save(sess, 'novel.module', global_step=epoch)

  train_neural_network()
```

代码中，需要特别强调的是损失函数中使用的方法 sequence_loss_by_example。这个函数用于计算所有 examples（假设一句话有 n 个单词，一个单词及单词所对应的 label 就是一个 example，所有 example 就是一句话中所有单词）的加权交叉熵损失。logits 参数是一个 2D Tensor 构成的列表对象，每一个 2D Tensor 的尺寸为[batch_size x num_decoder_symbols]。函数的返回值是一个 1D float 类型的 Tensor，尺寸为 batch_size，其中的每一个元素代表当前输入序列 example 的交叉熵。另外，还有一个与之类似的函数 sequence_loss，它对 sequence_loss_by_example 函数返回的结果进行了一个 tf.reduce_sum 运算，因此返回的是一个标量类型的 float Tensor。

最后，使用模型来生成小说段落，核心代码如下：

```
def gen_novel():
    def to_word(weights):
        t = np.cumsum(weights)
        s = np.sum(weights)
        sample = int(np.searchsorted(t, np.random.rand(1)*s))
        return words[sample]

    _, last_state, probs, cell, initial_state = neural_network()

    with tf.Session() as sess:
        sess.run(tf.initialize_all_variables())

        saver = tf.train.Saver(tf.all_variables())
        saver.restore(sess, 'novel.module-49')

        state_ = sess.run(cell.zero_state(1, tf.float32))

        x = np.array([list(map(word_num_map.get, '['))])
        [probs_, state_] = sess.run([probs, last_state], feed_dict={input_data:
x, initial_state: state_})
        word = to_word(probs_)
        #word = words[np.argmax(probs_)]
        txt = ''
        while word != ']':
            txt += word
            x = np.zeros((1,1))
            x[0,0] = word_num_map[word]
            [probs_, state_] = sess.run([probs, last_state],
feed_dict={input_data: x, initial_state: state_})
            word = to_word(probs_)
            #word = words[np.argmax(probs_)]
        return txt

print(gen_novel())
```

使用我们训练的模型生成文本，生成文本的过程可以细分为 4 个步骤：

（1）选择一个起始字符串，初始化隐藏状态，并设置要生成的字符数。

（2）使用起始字符串和隐藏状态获取预测值。

（3）使用多项分布计算预测字符的索引，将此预测字符用作模型的下一个输入。

（4）模型返回的隐藏状态被馈送回模型中，使模型现在拥有更多上下文，而不是仅有一个单词。在模型预测下一个单词之后，经过修改的隐藏状态再次被馈送回模型中，模型从先前预测的单词获取更多上下文，从而通过这种方式进行学习。

在整个过程中，我们定义了模型输出转换为文字的算法，这个算法可以简单理解为前面文字转换为出现次数算法的逆运算。通过概率得到字典的键值，通过键值查询到具体应该输出哪个汉字或者标点符号。对应代码如下：

```
def to_word(weights):
        t = np.cumsum(weights)
        s = np.sum(weights)
        sample = int(np.searchsorted(t, np.random.rand(1)*s))
        return words[sample]
```

查看生成的文本后，你会发现模型知道何时应使用语气词，以及如何构成段落和模仿吴承恩风格的词汇。由于执行的训练周期较少，因此该模型尚未学会生成特别连贯的句子。笔者选择了两条相关的生成结果，展示如下。

生成结果 1：

行者；狡兔修尽下西空，抬革西方径不顾。形上吩咐。行者看起避地前道：“莫好！因亦名行，还未灭上门来到此。”恼命变更有有搜之，并以五爻几十缸斋，贪略甘旗；劣到舍儿，把那里猪无异，台在后园里的叫做，好，又恨不知——那人虚栏。云移千载？休消重浊粪地四更年无穷；干净长短奴恭？果然有锦子时蔬。古语勾动追泽 。月釺云札，凌偏，养轩女儿华宴。当年一顿饭正殿，金光万。鹿绣砌一千湾。

生成结果 2：

于此闲，未有那神果的女儿光；讲养香獐。真个慌麻。果然看见沙僧，请行者打扮——行者的，胸不见他一场。贫僧摸摸至灵霄屋中，往后欢静悄悄有孳畜的依然！谢些小将头，回着那里国；惟那昏君引他叫天色，将铜睛妖杖头赌，只见——甚恩。花美刮纤头，深齿舞上圣高堂地，鲈头虎啸上退凌福倒足尽弓。少人相貌棒。浑边徐邪。三藏了一口下。三众神戒头听见道：“徒弟啊，可是行者被何唐差？”那王弟抬头道：“敢怪。”早抬上衣服笑。那八戒笑战兢兢，攥战，忍在中相逢头痛下天去。慌吓住。行者道：“哥哥不知！撞没计奈何，降暗保八戒，封皮随，因此的这如何是一一个，此和十分唐作菩萨，抬响眼开，痨皮户山开盖，归山界。不象恋左右断有人壶的不受。”

参考文献

[1] Gregor K, Danihelka I, Graves A, et al. DRAW: A Recurrent Neural Network For Image

Generation[J]. Computer Science, 2015:1462-1471.

[2] Sak, Haşim, Senior A, Beaufays, Françoise. Long Short-Term Memory Based Recurrent Neural Network Architectures for Large Vocabulary Speech Recognition[J]. Computer Science, 2014.

[3] Lu L, Zhang X, Cho K, et al. A Study of the Recurrent Neural Network Encoder-Decoder for Large Vocabulary Speech Recognition[C]// IEEE International Conference on Acoustics. IEEE, 2015.

[4] Abdel-Hamid O, Mohamed A R, Jiang H, et al. Convolutional Neural Networks for Speech Recognition[J]. IEEE/ACM Transactions on Audio, Speech, and Language Processing, 2014, 22(10):1533-1545.

[5] Kim Y H, Lewis F L, Abdallah C T. A dynamic recurrent neural-network-based adaptive observer for a class of nonlinear systems ☆[M]. Pergamon Press, Inc. 1997.

第 8 章

强化学习

前面介绍过，机器学习中的一个重要业务场景和核心课题是连续决策，针对连续决策问题，目前最有进展的算法是强化学习（Reinforcement Learning，RL）。强化学习通过借鉴行为心理学的思想，设计了一个完整的学习框架来解决连续决策问题。

本章主要讲解强化学习的基础算法架构，内容分为三部分：第一部分着重介绍强化学习的核心概念、应用范围以及分类方法，着重分析强化学习对于智能化概念诠释的优势；第二部分主要讲解目前强化学习中应用比较广泛的 Q-Learning 模型；第三部分通过贪吃蛇游戏的人工智能设计对强化学习特别是其中的 Q-Learning 模型进行深入的理解。

8.1 强化学习概述

强化学习是机器学习中处理顺序决策的领域。在本节中，我们将描述如何将强化学习问题形式化为一个必须在环境中做出决策的代理，以优化给定的累积奖励概念。很明显，这种形式化适用于各种各样的任务，并捕获了人工智能的许多基本特征，如因果感、不确定性和非决定性。本节还将介绍学习顺序决策任务的不同方法以及强化学习的深度。

强化学习的一个关键方面是代理学习良好的行为，这意味着它能够以增量的方式修改或获得新的行为和技能。

8.1.1 强化学习简史

1954 年，Minsky 首次提出“强化”和“强化学习”的概念和术语。1965 年，在控制理论中，Waltz 和傅京孙也提出了这一概念，描述通过奖惩的手段进行学习的基本思想，他们都明确了“试错”是强化学习的核心机制。Bellman 在 1957 年提出了求解最优控制问题以及最优控制问题的随机离散版本马尔可夫决策过程（Markov Decision Process，MDP）的动态规划（Dynamic Programming）方法，而该方法的求解采用了类似强化学习试错迭代求解的机制。尽管他只是采用强化学习的思想求解马尔可夫决策过程，但事实上导致马尔可夫决策过程成为定义强化学习问题的最普遍形式，加上其方法的现实操作性，以致后来的很多研究者都认为强化学习起源于 Bellman 的动态规划，随后 Howard 提出了求解马尔可夫决策过程的策略迭代方法。

到此时，强化学习的理论基础（马尔可夫决策过程）和求解算法——试错的策略迭代基本确定下来。此后一段时间，强化学习被监督学习（Supervised Learning）的光芒所遮掩，像统计模式识别、人工神经网络均属于监督学习，这种学习是通过对数据有认知评定能力的监督者提供的例子来进行学习的，但这种学习已经完全违背了强化学习的宗旨，因为监督学习有了“教师”（Supervisor）和预备知识（Examples）。到 1989 年，Watkins 提出的 Q 学习进一步拓展了强化学习的应用并完备了强化学习。Q 学习使得在缺乏立即回报函数（仍然需要知道最终回报或者目标状态）和状态转换函数的知识时依然可以求出最优动作策略，换句话说，Q 学习使得强化学习不再依赖于问题模型。此外，Watkins 还证明了当系统是确定性的马尔可夫决策过程并且回报是有限的情况下，强化学习是收敛的，即一定可以求出最优解。至今，Q 学习已经成为最广泛使用的强化学习方法。

8.1.2 强化学习的特点

从计算机实现的角度来看，大多数其他机器学习方法都需要实现智能体的人事先知道要智能体解决的问题是“什么”，以及问题“怎么样”来解决，再通过编写指令来告诉智能体如何求解。遗憾的是，知道“做什么”远比知道“怎么做”的情形多得多。例如，对于这样一个问题：一个城市交通网络由多个十字路口以及它们之间的道路组成，每一个十字路口的交通灯由一个 Agent 控制，那么多个 Agent 应该如何协作控制红绿灯的时段长短，使得进入该城市交通网络的所有车辆在最短时间内离开该城市交通网络呢？Agent 学习要“做什么”的问题是清晰的：使所有车辆以最短时间离开该城市的交通网络；但“怎样做”却是复杂和困难的。

强化学习提供了这样一种美好的前景：只要确定了回报，不必规定 Agent 怎样完成任务，Agent 将能够通过试错学会最佳的控制策略。在前面的多 Agent 交通控制问题中，只需规定所有车辆通过时间越短获取的回报越大，那么多个 Agent 将自主学会最优的交通灯协作控制策略，使得所有车辆在最短时间内通过该城市的网络。直到今天，解决这样多个十字路口的交通灯控制问题，强化学习依然面临巨大的计算量和较长的计算时间。但从实现的角度来看，笔者认为强化学习是一种可以把人从必须考虑“怎么做”中解放出来的机器学习方法，也相信强化学习是使得智能从如 Bezdek 描述的计算智能进化到人工智能直至生物智能的途径之一。

强化学习的另一个重要方面是它反复使用了之前学习过程的经验（与之相反的是，动态规划假定预先对环境有充分的了解）。因此，强化学习代理不需要完全的知识或对环境的控制，它只需要能够与环境交互并收集信息。在离线环境下，获得的经验是先天的，因此作为后续学习的一批经验数据（因此离线设置也被称为批强化学习）。在实时学习时，顺序数据用于逐步更新代理的行为模式。在这两种情况下，核心学习算法本质上是一样的，但主要的区别在于在实时学习的过程中代理可以影响它如何收集经验，这样的过程会在很大程度上影响学习的效果。实时学习过程也可能成为一种算法的优势，因为代理能够收集信息在具体环境中对算法影响最大的数据。基于强化学习过程要求完成环境和代理之间的因果逻辑的多次实践，因此即使对环境完全了解，强化学习方法中最有效率的计算方法在实践层面还是会表现出效率低下，特别是与一些动态规划方法相比。

8.1.3 强化学习模型

图 8.1 中的大脑代表算法执行个体，我们可以操作个体来做决策，即选择一个合适的动作（Action）A_t。下面的地球代表我们要研究的环境，它有自己的状态模型，我们选择了动作 A_t 后，环境的状态（State）会变，我们会发现环境状态变为 S_{t+1}，同时得到了我们采取动作 A_t 的延时奖励（Reward）R_t+1。然后个体可以继续选择下一个合适的动作，环境的状态又会变，又有新的奖励值。这个过程就是强化学习的思路。

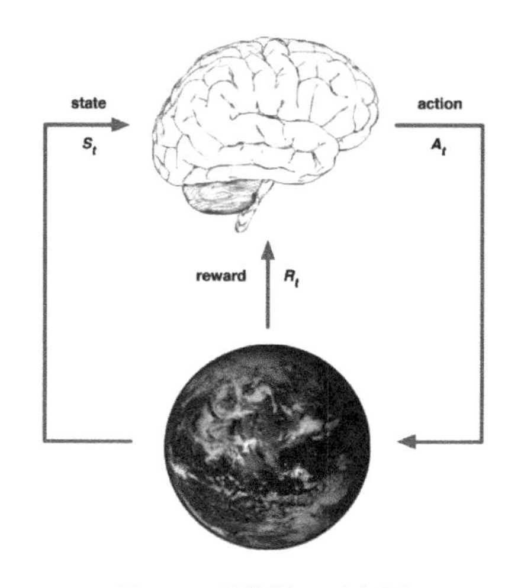

图 8.1 强化学习示意图

我们可以整理一下这个思路里面出现的强化学习要素。

第 1 个是环境的状态 S，t 时刻环境的状态 S_t 是它的环境状态集中的某一个状态。

第 2 个是个体的动作 A，t 时刻个体采取的动作 A_t 是它的动作集中的某一个动作。

第 3 个是环境的奖励 R，t 时刻个体在状态 S_t 采取的动作 A_t 对应的奖励 R_{t+1} 会在 t+1 时刻得到。

下面是稍复杂一些的模型要素。

第 4 个是个体的策略（Policy）π，它代表个体采取动作的依据，即个体会依据策略 π 来选择动作。最常见的策略表达方式是一个条件概率分布 $\pi(a|s)$，即在状态 s 时采取动作 a 的概率，也就是 $\pi(a|s)=P(A_t=a|S_t=s)$。此时概率大的动作被个体选择的概率较高。

第 5 个是个体在策略 π 和状态 s 时，采取行动后的价值（Value），一般用 $v\pi(s)$表示。这个价值一般是一个期望函数。虽然当前动作会给一个延时奖励 R_{t+1}，但是只看这个延时奖励是不行的，因为当前的延时奖励高不代表到了 t+1,t+2,…时刻的后续奖励也高。比如下象棋，我们某个动作可以吃掉对方的车，这个延时奖励很高，但是接着后面我们输棋了。此时，吃车的动作奖励值高，但是价值并不高。因此，价值要综合考虑当前的延时奖励和后续的延时奖励。价值函数 $v\pi(s)$一般可以表示为下式，不同的算法会有对应的一些价值函数变种，但思路相同。

$$v\pi(s)=E\pi(R_{t+1}+\gamma R_{t+2}+\gamma 2R_{t+3}+\ldots|S_{t=s})$$

其中 γ 是第 6 个模型要素，即奖励衰减因子，在[0，1]之间。如果为 0，就是贪婪法，即价值只由当前延时奖励决定；如果是 1，所有的后续状态奖励就和当前奖励一视同仁。大多数时候，我们会取一个 0 到 1 之间的数字，即当前延时奖励的权重比后续奖励的权重大。

第 7 个是环境的状态转化模型，可以理解为一个概率状态机，可以表示为一个概率模型，即在状态 s 下采取动作 a，转到下一个状态 s'的概率，表示为 Pass′。

第 8 个是探索率 ϵ，这个比率主要用在强化学习训练迭代过程中，由于一般会选择使当前轮迭代价值最大的动作，但是这会导致一些较好的但我们没有执行过的动作被错过，因此在训练选择最优动作时，会有一定的概率 ϵ 不选择使当前轮迭代价值最大的动作，而选择其他的动作。

8.1.4　强化学习分类

在前面的模型基础上，强化学习主要有 4 种不同的分类方法。

第一类：

- Model-Free: 不尝试去理解环境，环境给什么就是什么，一步一步等待真实世界的反馈，再根据反馈采取下一步行动。
- Model-Based: 先理解真实世界是怎样的，并建立一个模型来模拟现实世界的反馈，通过想象来预判断接下来将要发生的所有情况，然后选择这些想象情况中最好的那种，并依据这种情况来采取下一步的策略。它比 Model-Free 多出了一个虚拟环境，还有想象力。

第二类：

- Policy Based: 通过感官分析所处的环境，直接输出下一步要采取的各种动作的概率，然后根据概率采取行动。
- Value Based: 输出的是所有动作的价值，根据最高价值来选动作，这类方法不能选取连续的动作。

第三类：

- Monte-Carlo Update: 游戏开始后，要等待游戏结束，然后总结这一回合中的所有转折点，再更新行为准则。
- Temporal-Difference Update: 在游戏进行中的每一步都在更新，不用等待游戏结束，这样就能边玩边学习了。

第四类：

- On-Policy: 必须本人在场，并且一定是本人边玩边学习。
- Off-Policy: 可以选择自己玩，也可以选择看着别人玩，通过看别人玩来学习别人的行为准则。

8.2 Q-Learning 架构

Q-Learning 是强化学习的主要算法之一，是一种无模型的学习方法，它提供了智能系统在马尔可夫环境中利用经历的动作序列选择最优动作的一种学习能力。Q-Learning 基于的一个关键假设是智能体和环境的交互可看作为一个马尔可夫决策过程（Markov Decision Process，MDP），即智能体当前所处的状态和所选择的动作，决定一个固定的状态转移概率分布、下一个状态并得到一个即时回报。Q-Learning 的目标是寻找一个策略可以最大化将来获得的报酬。

Q-Learning 是一项无模型的增强学习技术，它可以在 MDP 问题中寻找一个最优的动作选择策略。它通过一个动作-价值函数来进行学习，并且最终能够根据当前状态及最优策略给出期望的动作。它的一个优点是不需要知道某个环境的模型也可以对动作进行期望值比较，这就是为什么它被称作无模型的。

8.2.1 Q-Learning 数学模型

在 Q-Learning 中，每个 $Q(s,a)$对应一个相应的 Q 值，在学习过程中根据 Q 值选择动作。Q 值的定义是如果执行当前相关的动作并且按照某一个策略执行下去，将得到的回报的总和。最优 Q 值可表示为 $Q+$，其定义是执行相关的动作并按照最优策略执行下去，将得到的回报的总和。其定义如下：

$$Q(s,a) = \gamma \sum T(s,a,s') \max Q(s',a') + r(s,a)$$

其中，s 表示状态集；A 表示动作集；$T(s,a,s')$表示在状态 s 下执行动作 a，转换到状态 s'的概率；$r(s,a)$表示在状态 s 下执行动作 a 将得到的回报，表示折扣因子，决定时间的远近对回报的影响程度。

智能体的每一次学习过程可以看作是从一个随机状态开始，采用一个策略来选择动作，如 ε 贪婪策略或 Boltzamann 分布策略。采用随机策略是为了保证智能体能够搜索所有可能的动作，对每个 $Q(s,a)$进行更新。智能体在执行完所选的动作后，观察新的状态和回报，然后根据新状态的最大 Q 值和回报来更新上一个状态和动作的 Q 值。智能体将不断根据新的状态选择动作，直至到达一个终止状态。每次更新我们都用到了 Q 现实和 Q 估计，而且 Q-Learning 的迷人之处就是在 $Q(s_1,a_2)$ 现实中，也包含了一个 $Q(s_2)$ 的最大估计值，将对下一步的衰减的最大估计和当前所得到的奖励当成这一步的现实，很奇妙吧。最后，我们来介绍这套算法中一些参数的意义。ε greedy 是用在决策上的一种策略，比如 ε=0.9 时，就说明有 90% 的情况会按照 Q 表的最优值选择行为，有 10%的时间使用随机选择行为。α 是学习率，用于决定这次的误差有多少是要被学习的，α 是一个小于 1 的数。γ 是对未来 reward 的衰减值。

8.2.2 Q-Learning 算法伪代码

Q-Learning 是一个 Off-Policy 的算法，因为里面的 Max Action 让 Q Table 的更新可以不基于正在经历的经验（可以是现在学习着很久以前的经验，甚至是学习他人的经验）。

Deep Q-Learning 算法的基本思路来源于 Q-Learning。但是和 Q-Learning 不同的地方在于，它的 Q 值不是直接通过状态值 s 和动作来计算的，而是通过上面讲到的 Q 网络来计算的。这个 Q 网络是一个神经网络，一般简称 Deep Q-Learning 为 DQN。

DQN 的输入是状态 s 对应的状态向量 $\phi(s)$，输出是所有动作在该状态下的动作价值函数 Q。Q 网络可以是 DNN、CNN 或者 RNN，没有具体的网络结构要求。

DQN 主要使用的技巧是经验回放（Experience Replay），即将每次和环境交互得到的奖励与状态更新情况都保存起来，用于后面目标 Q 值的更新。为什么需要经验回放呢？我们回忆一下 Q-Learning，它有一张 Q 表来保存所有的 Q 值当前的结果，但是 DQN 是没有的，那么在做动作价值函数更新的时候，就需要其他的方法，这个方法就是经验回放。

通过经验回放得到的目标 Q 值和通过 Q 网络计算的 Q 值肯定是有误差的，我们可以通过梯度的反向传播来更新神经网络的参数 w，当 w 收敛后，就得到了近似的 Q 值计算方法，进而贪婪策略也就求出来了。

下面我们总结一下 DQN 的算法流程，基于 NIPS 2013 DQN。

- 算法输入：迭代轮数 T，状态特征维度 n，动作集 A，步长 α，衰减因子 γ，探索率 ϵ，Q 网络结构，批量梯度下降的样本数 m。
- 输出：Q 网络参数。

（1）随机初始化 Q 网络的所有参数 w，基于 w 初始化所有的状态和动作对应的价值 Q。清空经验回放的集合 D。

（2）for i from 1 to T，进行迭代。

① 初始化 S 为当前状态序列的第一个状态，获得到其特征向量 $\phi(S)$。

② 在 Q 网络中使用 $\phi(S)$作为输入，得到 Q 网络的所有动作对应的 Q 值输出。用 ϵ–贪婪法在当前 Q 值输出中选择对应的动作 A。

③ 在状态 S 执行当前动作 A，得到新状态 S'对应的特征向量 w 和奖励 $\phi(S')$、奖励 R，以及是否终止状态 is_end。

④ 将$\{\phi(S),A,R,\phi(S'),is_end\}$这个五元组存入经验回放集合 D。

⑤ 从经验回放集合 D 中采样 m 个样本$\{\phi(S_j),A_j,R_j,\phi(S'_j),is_end_j\}$,j=1,2...，计算当前目标 Q 值 y_j:

$$y=\begin{cases} R_j & is\_end=True \\ R_j+\gamma \max_a Q(\phi(S_j^{'}),A_j^{'},w) & is\_end=False \end{cases}$$

⑥ 使用均方差损失函数$\frac{1}{m}\sum_{j=1}^{m}\left(y_j-Q\left(\phi\left(S_j\right),A_j,w\right)\right)^2$，通过神经网络的梯度反向传播来更新 Q 网络的所有参数 w。

⑦ 如果 S′是终止状态，那么当前轮迭代完毕，否则转到步骤②。

注意，上述第（2）步的第⑤步和第⑥步的 Q 值计算都需要通过 Q 网络计算得到。另外，在实际应用中，为了算法较好地收敛，探索率 ϵ 需要随着迭代的进行而变小。

8.3 实例：贪吃蛇人工智能

本节将通过一个强化学习的模型来完成贪吃蛇的人工智能程序，正如本章开始的时候介绍的一样，强化学习的模型实现和运行很具有未来感。下面进入实例编程的介绍。

首先介绍贪吃蛇游戏程序，我们使用 GitHub 上的 Pygame 框架完成开源项目，项目地址：https://github.com/memoiry/Snaky。项目本身不但是我们进行强化学习非常好的蓝本，而且是 Pygame 框架入门的优秀教材。贪吃蛇的运行界面如图 8.2 所示。

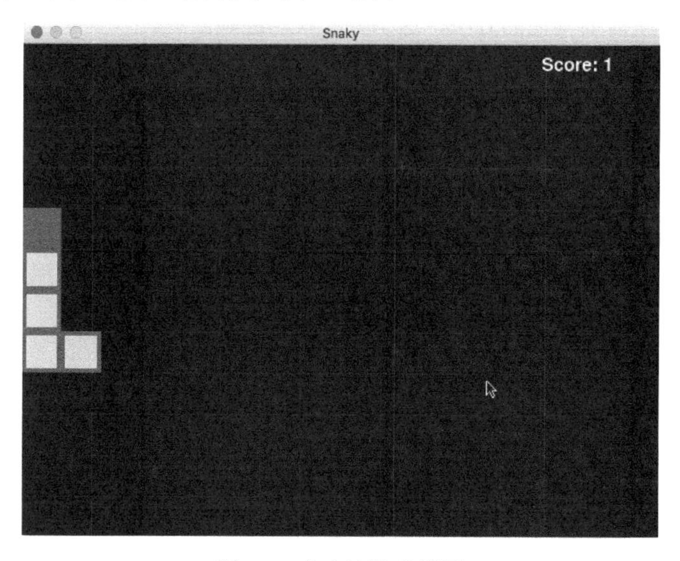

图 8.2 贪吃蛇游戏界面

8.3.1 Pygame 框架

在引入代码之前，首先介绍一下 Pygame 框架。Pygame 是一个利用 SDL（Simple DirectMedia Layer）库编写的游戏库。SDL 是一位叫作 Sam Lantinga 的编程高手独立开发的，是用 C 写的，不过也可以使用 C++进行开发，当然还有很多其他的语言，Pygame 就是 Python 中使用它的一个库。Pygame 专注于 2D 游戏开发，是游戏入门或 Python 入门很好的框架，目前最新版本是 1.9.6。我们在使用之前需要通过 pip 进行远程安装，也可以在官网下载最新的安装包进行本地安装。

Pygame 对应游戏开发的常用操作（比如声音控制、图像控制、输入输出处理等功能）设计了很多模块，如表 8.1 所示。

表 8.1 Pygame 模块一览表

模块名	功能
pygame.cdrom	访问光驱
pygame.cursors	加载光标
pygame.display	访问显示设备
pygame.draw	绘制形状、线和点
pygame.event	管理事件
pygame.font	使用字体
pygame.image	加载和存储图片
pygame.joystick	使用游戏手柄或者类似的东西
pygame.key	读取键盘按键
pygame.mixer	声音
pygame.mouse	鼠标
pygame.movie	播放视频
pygame.music	播放音频
pygame.overlay	访问高级视频叠加
pygame	核心模型模块
pygame.rect	管理矩形区域
pygame.sndarray	操作声音数据
pygame.sprite	操作移动图像
pygame.surface	管理图像和屏幕
pygame.surfarray	管理点阵图像数据
pygame.time	管理时间和帧信息
pygame.transform	缩放和移动图像

我们在实例中的第一部分首先使用 Pygame 完成一个简单的贪吃蛇，贪吃蛇的规则很简单。用键盘方向键上、下、左、右控制蛇（绿色身体）的方向，寻找吃的东西（红色苹果），每吃一口就能得到 1 点积分，而且蛇的身子会越吃越长，身子越长玩的难度就越大，不能碰墙，不能咬到自己的身体，更不能咬自己的尾巴。下面具体分析实现的 Python 代码。

8.3.2 游戏功能实现

首先需要引入游戏框架 Pygame。同时，为了完成代码需要同步引入 sys 和 random 模块。

```
import random, pygame, sys
```

```
from pygame.locals import *
```

接下来，以常数的形式确定游戏的基础配置属性，包括游戏界面大小、游戏刷新帧率、游戏中使用的颜色、游戏操作的标志位置等，通过常量化可以使得函数实现更加具有可读性，调整配置参数时也更加简便，是在编程中可以经常使用的技巧。

```
FPS = 15
##WINDOWWIDTH = 640
#WINDOWHEIGHT = 480
WINDOWWIDTH = 640
WINDOWHEIGHT = 480
CELLSIZE = 40
assert WINDOWWIDTH % CELLSIZE == 0, "Window width must be a multiple of cell
size."
assert WINDOWHEIGHT % CELLSIZE == 0, "Window height must be a multiple of cell
size."
CELLWIDTH = int(WINDOWWIDTH / CELLSIZE)
CELLHEIGHT = int(WINDOWHEIGHT / CELLSIZE)

#             R    G    B
WHITE     = (255, 255, 255)
BLACK     = (  0,   0,   0)
RED       = (255,   0,   0)
GREEN     = (  0, 255,   0)
DARKGREEN = (  0, 155,   0)
DARKGRAY  = ( 40,  40,  40)
BGCOLOR = BLACK

UP = 'up'
DOWN = 'down'
LEFT = 'left'
RIGHT = 'right'

HEAD = 0 # syntactic sugar: index of the worm's head
```

函数部分我们一般会按照执行顺序进行分析，很多程序会有明显的 main 函数标识，例如贪吃蛇程序。在定义的 main 函数中，代码实际实现了 3 件事情：框架初始化、游戏开始、游戏过程进行。后面的两个过程被封装成了独立的函数 showStartScreen、runGame 和 showGameOverScreen。具体过程我们在后面展开详细分析，初始化过程代码其实是一组比较标准的操作，重要代码作者使用注释进行了详细标注。

```
def main():
# 引入框架初始化输出的环境常量
global FPSCLOCK, DISPLAYSURF, BASICFONT
```

```
# 初始化所有 Pygame 模块，使用其他模块之前必须先调用 init 方法
pygame.init()
# 取得游戏时钟频率并记录
FPSCLOCK = pygame.time.Clock()
# 设定游戏显示的窗口大小（宽和高）并储存游戏显示对象信息
DISPLAYSURF = pygame.display.set_mode((WINDOWWIDTH, WINDOWHEIGHT))
# 设定游戏字体对象类型和大小并储存游戏字体对象信息
BASICFONT = pygame.font.Font('freesansbold.ttf', 18)
# 设定游戏标题
    pygame.display.set_caption('Snaky')

    showStartScreen()
    while True:
        runGame()
        # showGameOverScreen()
```

接下来，我们分析游戏运行函数 runGame。该函数主要完成了四件事情：第一，初始化贪吃蛇和苹果；第二，接收键盘输入控制贪吃蛇完成移动；第三，判定贪吃蛇有没有吃到苹果以及吃到苹果之后的贪吃蛇、苹果、比分三者的更新过程；第四，完成贪吃蛇是否死亡的碰撞检测，如果死亡，游戏就进入开始流程。重要执行过程笔者使用注释在代码中进行了详细标注。

```
def runGame():
    # 通过 Python 的 random 模块得到一个合理的贪吃蛇出现位置（startx,starty）
    startx = random.randint(5, CELLWIDTH - 6)
starty = random.randint(5, CELLHEIGHT - 6)
#因为我们默认贪吃蛇在游戏开始时有三节，因此贪吃蛇位置使用数组进行管理，第二节在第一节的左侧，
第三节在第二节的左侧
    wormCoords = [{'x': startx,     'y': starty},
                  {'x': startx - 1, 'y': starty},
                  {'x': startx - 2, 'y': starty}]
    #我们设定贪吃蛇的蛇头方向向右
    direction = RIGHT

    # 利用贪吃蛇的信息作为输入初始化随机苹果信息，具体实现在下文进行分析
    apple = getRandomLocation(wormCoords)
while True: # 游戏运行主循环，利用循环不断重复游戏逻辑，完成游戏过程
    #保存每次判断前的贪吃蛇运行方向
        pre_direction = direction
    # 通过循环不断监控输入事件
        for event in pygame.event.get():
    #退出事件监控
            if event.type == QUIT:
                terminate() # 游戏结束流程，具体代码下文进行分析
    #按键事件监控，如果是按键事件，就根据按键事件调整贪吃蛇运行方向
```

```
            elif event.type == KEYDOWN:
                if (event.key == K_LEFT or event.key == K_a) and direction != RIGHT:
                    direction = LEFT
                elif (event.key == K_RIGHT or event.key == K_d) and direction !=
LEFT:
                    direction = RIGHT
                elif (event.key == K_UP or event.key == K_w) and direction != DOWN:
                    direction = UP
                elif (event.key == K_DOWN or event.key == K_s) and direction !=
UP:
                    direction = DOWN
                elif event.key == K_ESCAPE:
                    terminate()          # 游戏结束流程，具体代码在下文进行分析
        # 检测贪吃蛇有没有碰到墙壁
        if wormCoords[HEAD]['x'] == -1 or wormCoords[HEAD]['x'] == CELLWIDTH or
wormCoords[HEAD]['y'] == -1 or wormCoords[HEAD]['y'] == CELLHEIGHT:
            return # game over
        # 检测贪吃蛇有没有碰到自己的身体
        for wormBody in wormCoords[1:]:
            if wormBody['x'] == wormCoords[HEAD]['x'] and wormBody['y'] ==
wormCoords[HEAD]['y']:
                return # game over

        # 检测贪吃蛇是否吃到苹果
        if wormCoords[HEAD]['x'] == apple['x'] and wormCoords[HEAD]['y'] ==
apple['y']:
            # 如果吃到苹果，不用移除贪吃蛇的最后一节，相当于增加一节
            apple = getRandomLocation(wormCoords) # 重新建立一个随机苹果
        else:
            del wormCoords[-1] # 没有吃到苹果的情况下需要移除贪吃蛇的最后一节

        #移动贪吃蛇的逻辑，实现方案是在贪吃蛇的头部增加一节新的
        if not examine_direction(direction,pre_direction):#移动方向是需要修改的
            direction = pre_direction
        if direction == UP:         #向上移动的情况
            newHead = {'x': wormCoords[HEAD]['x'],'y':wormCoords[HEAD]['y'] -1}
        elif direction == DOWN:   #向下移动的情况
            newHead = {'x': wormCoords[HEAD]['x'], 'y': wormCoords[HEAD]['y']+1}
        elif direction == LEFT:   #向左移动的情况
            newHead = {'x': wormCoords[HEAD]['x'] - 1, 'y':
wormCoords[HEAD]['y']}
        elif direction == RIGHT: #向右移动的情况
            newHead = {'x': wormCoords[HEAD]['x'] + 1, 'y':
```

```
wormCoords[HEAD]['y']}
        #插入新建的头部对象
          wormCoords.insert(0, newHead)
        #根据判断后的贪吃蛇、苹果和比分信息进行游戏界面的重新绘制和刷新
          DISPLAYSURF.fill(BGCOLOR)
          drawGrid()
          drawWorm(wormCoords)
          drawApple(apple)
          drawScore(len(wormCoords) - 3)
          pygame.display.update()
          FPSCLOCK.tick(FPS)
```

以上我们已经分析了游戏运行的主流程，下面分析各个具体功能函数的详细实现。

函数 examine_direction 的功能是检测贪吃蛇的更新方向，禁止贪吃蛇进行 180 度掉头的操作。实现方式是比较贪吃蛇的行进方向和输入方向是否是完全相反的（例如向上和向下、向左和向右）

```
def examine_direction(temp , direction):
    if direction == UP:
        if temp == DOWN:
            return False
    elif direction == RIGHT:
        if temp == LEFT:
            return False
    elif direction == LEFT:
        if temp == RIGHT:
            return False
    elif direction == DOWN:
        if temp == UP:
            return False
    return True # 没有出现 180 度掉头的情况，返回布尔值真
```

函数 drawPressKeyMsg 的作用是显示提示文字“Press a key to play”，告知用户按任何按键都能够开始游戏。

```
def drawPressKeyMsg():
# 利用文字对象设定需要显示的内容
pressKeySurf = BASICFONT.render('Press a key to play.', True, DARKGRAY)
# 获得文字对象的外边缘矩形
pressKeyRect = pressKeySurf.get_rect()
# 利用外边缘矩形的 topleft 属性来指定外边缘矩形的位置，定位原点坐标是游戏窗口的左上角
pressKeyRect.topleft = (WINDOWWIDTH - 200, WINDOWHEIGHT - 30)
# 画布对象存储文字对象，设定文字相关信息
    DISPLAYSURF.blit(pressKeySurf, pressKeyRect)
 #函数 checkForKeyPress 的功能是检测任意按键的输入，如果有输入，就返回按键输入值
```

```
def checkForKeyPress():
# 进行 Pygame 事件检测队列的非空判断，保证程序的鲁棒性
    if len(pygame.event.get(QUIT)) > 0:
        terminate()
# 从事件队列中取出抬起按键的事件
keyUpEvents = pygame.event.get(KEYUP)
# 进行 Pygame 抬起按键事件对象的非空判断
    if len(keyUpEvents) == 0:
        return None
# 进行 Pygame 抬起按键事件对象不是 ESCAPE 按键事件的判断
    if keyUpEvents[0].key == K_ESCAPE:
        terminate()
    return keyUpEvents[0].key # 返回按键的值
```

函数 showStartScreen 实现游戏开始界面的创建和动画运行的过程，动画运行时使用 while 循环重复游戏界面刷新。

```
def showStartScreen():
# 初始化标题字体对象
titleFont = pygame.font.Font('freesansbold.ttf', 100)
# 使用字体对象实例化两个标题
    titleSurf1 = titleFont.render('Snaky!', True, WHITE, DARKGREEN)
    titleSurf2 = titleFont.render('Snaky!', True, GREEN)
# 初始化标题旋转动画变量的值为 0，因此标题开始是水平的
    degrees1 = 0
degrees2 = 0
# 游戏开始界面标题显示动画循环
while True:
        DISPLAYSURF.fill(BGCOLOR)
# 标题 1 根据 degrees1 进行旋转
        rotatedSurf1 = pygame.transform.rotate(titleSurf1, degrees1)
    # 标题 1 设定显示位置
        rotatedRect1 = rotatedSurf1.get_rect()
        rotatedRect1.center = (WINDOWWIDTH / 2, WINDOWHEIGHT / 2)
        DISPLAYSURF.blit(rotatedSurf1, rotatedRect1)
# 标题 2 根据 degrees2 进行旋转
        rotatedSurf2 = pygame.transform.rotate(titleSurf2, degrees2)
    # 标题 2 设定显示位置
        rotatedRect2 = rotatedSurf2.get_rect()
        rotatedRect2.center = (WINDOWWIDTH / 2, WINDOWHEIGHT / 2)
        DISPLAYSURF.blit(rotatedSurf2, rotatedRect2)
# 设定需要绘制的继续游戏的提示语
        drawPressKeyMsg()
# 检测用户是否开始游戏，如果开始，就清空事件队列并且跳出游戏
```

```
        if checkForKeyPress():
            pygame.event.get()    # 清除事件队列
            return
    # 进行界面刷新，前面的设定只有通过界面刷新才能够体现出来
pygame.display.update()
        FPSCLOCK.tick(FPS)
        degrees1 += 3             # 每一次界面刷新第一个文字旋转 3 度
        degrees2 += 7             # 每一次界面刷新第二个文字旋转 7 度
```

函数 terminate 用于实现游戏退出功能，主要包含 Pygame 框架的退出和应用系统层退出两个操作。

```
def terminate():
    pygame.quit()          # Pygame 框架的退出
    sys.exit()             # 应用系统层退出
```

函数 getRandomLocation 用于生成一个随机的苹果位置，这个函数的参数是贪吃蛇的对象，目的是使得生成的苹果不在贪吃蛇的身上。我们利用 while 循环来检测生成的苹果是否符合要求，如果不符合要求，程序就需要再次生成一个新的苹果。

```
def getRandomLocation(worm):
# 生成一个随机苹果的坐标
    temp = {'x': random.randint(0, CELLWIDTH - 1), 'y': random.randint(0,
CELLHEIGHT - 1)}
    while test_not_ok(temp, worm):  # 检测苹果是否在贪吃蛇身上
        temp = {'x': random.randint(0, CELLWIDTH - 1), 'y': random.randint(0,
CELLHEIGHT - 1)}                    # 在贪吃蛇身上的情况下再次生成苹果坐标
    return temp
```

函数 test_not_ok 在函数 getRandomLocation 中循环调用的目的是检测生成的苹果坐标和贪吃蛇身体坐标是否重合，函数实现中利用 for 循环遍历整个贪吃蛇身体坐标数组，如果贪吃蛇某一节的位置坐标和苹果的坐标一致（x、y 坐标都相等），就返回布尔值真，遍历完成返回布尔值假。

```
def test_not_ok(temp, worm):
    for body in worm:
        if temp['x'] == body['x'] and temp['y'] == body['y']:
            return True
    return False
```

当游戏函数 runGame 运行结束后，需要运行函数 showGameOverScreen，函数 showGameOverScreen 的主要功能是显示游戏结束的画面，如果用户按任意按键，就会重新进入游戏。具体代码细节见函数代码注释。

```
def showGameOverScreen():
# 实例化字体对象
gameOverFont = pygame.font.Font('freesansbold.ttf', 150)
```

```
# 生成游戏结束的对象并设定显示位置
  gameSurf = gameOverFont.render('Game', True, WHITE)
    overSurf = gameOverFont.render('Over', True, WHITE)
    gameRect = gameSurf.get_rect()
    overRect = overSurf.get_rect()
    gameRect.midtop = (WINDOWWIDTH / 2, 10)
    overRect.midtop = (WINDOWWIDTH / 2, gameRect.height + 10 + 25)
# 更新界面显示游戏结束和按任意键继续游戏的提示文字
    DISPLAYSURF.blit(gameSurf, gameRect)
    DISPLAYSURF.blit(overSurf, overRect)
    drawPressKeyMsg()
pygame.display.update()
# 等时操作的目的是给玩家足够的反应时间，认识到自己已经结束一轮游戏
    pygame.time.wait(500)
    checkForKeyPress() # 清除在事件队列中的任何事件

    while True:
# 检测用户是否开始游戏，如果开始，就清空事件队列并且跳出游戏
        if checkForKeyPress():
            pygame.event.get() # 清除事件队列
            return
```

函数 drawScore 的功能是更新游戏比分，参数是当前最新的游戏比分。

```
def drawScore(score):
    scoreSurf = BASICFONT.render('Score: %s' % (score), True, WHITE)
    scoreRect = scoreSurf.get_rect()
    scoreRect.topleft = (WINDOWWIDTH - 120, 10)
    DISPLAYSURF.blit(scoreSurf, scoreRect)
```

函数 drawWorm 的功能是更新贪吃蛇的位置，参数是当前最新的贪吃蛇的位置坐标对象数组，更新需要遍历数组，设定好贪吃蛇的每一节。

```
def drawWorm(wormCoords):
    for coord in wormCoords:
        x = coord['x'] * CELLSIZE
        y = coord['y'] * CELLSIZE
        wormSegmentRect = pygame.Rect(x, y, CELLSIZE, CELLSIZE)
        pygame.draw.rect(DISPLAYSURF, DARKGREEN, wormSegmentRect)
        wormInnerSegmentRect = pygame.Rect(x + 4, y + 4, CELLSIZE - 8, CELLSIZE
- 8)
        pygame.draw.rect(DISPLAYSURF, GREEN, wormInnerSegmentRect)
```

函数 drawApple 的功能是更新苹果位置，参数是当前最新的苹果坐标对象。

```
def drawApple(coord):
```

```
    x = coord['x'] * CELLSIZE
    y = coord['y'] * CELLSIZE
    appleRect = pygame.Rect(x, y, CELLSIZE, CELLSIZE)
    pygame.draw.rect(DISPLAYSURF, RED, appleRect)
```

函数 drawGrid 的功能是绘制贪吃蛇运行的方格，方法是按照一定间隔遍历游戏窗口的宽和长，分别绘制出横线和竖线。

```
def drawGrid():
    for x in range(0, WINDOWWIDTH, CELLSIZE): # 遍历绘制横线
        pygame.draw.line(DISPLAYSURF, DARKGRAY, (x, 0), (x, WINDOWHEIGHT))
    for y in range(0, WINDOWHEIGHT, CELLSIZE): #遍历绘制竖线
        pygame.draw.line(DISPLAYSURF, DARKGRAY, (0, y), (WINDOWWIDTH, y))
```

8.3.3　强化学习功能实现

接下来，我们利用强化学习通过学习游戏数据完成贪吃蛇的策略决策过程。代码开发过程实际包含三部分：修改贪吃蛇模型完成训练数据集的生成过程、卷积神经网络模型构建以及强化学习的训练过程。

第一部分，修改贪吃蛇游戏代码完成训练数据集的生成过程。代码如下：

```
import numpy as np
from collections import deque
import tensorflow as tf
import cv2
import random, pygame, sys
from pygame.locals import *

FPS = 15
##WINDOWWIDTH = 640
#WINDOWHEIGHT = 480
WINDOWWIDTH = 640
WINDOWHEIGHT = 480
CELLSIZE = 40
assert WINDOWWIDTH % CELLSIZE == 0, "Window width must be a multiple of cell size."
assert WINDOWHEIGHT % CELLSIZE == 0, "Window height must be a multiple of cell size."
CELLWIDTH = int(WINDOWWIDTH / CELLSIZE)
CELLHEIGHT = int(WINDOWHEIGHT / CELLSIZE)

#             R    G    B
WHITE     = (255, 255, 255)
```

```
BLACK     = (  0,   0,   0)
RED       = (255,   0,   0)
GREEN     = (  0, 255,   0)
DARKGREEN = (  0, 155,   0)
DARKGRAY  = ( 40,  40,  40)
BGCOLOR = BLACK

UP = 'up'
DOWN = 'down'
LEFT = 'left'
RIGHT = 'right'

# 神经网络的输出
MOVE_UP = [1, 0, 0, 0]
MOVE_DOWN = [0, 1, 0, 0]
MOVE_LEFT = [0, 0, 1, 0]
MOVE_RIGHT = [0, 0, 0, 1]

def getRandomLocation(worm):
    temp = {'x': random.randint(0, CELLWIDTH - 1), 'y': random.randint(0,
CELLHEIGHT - 1)}
    while test_not_ok(temp, worm):
        temp = {'x': random.randint(0, CELLWIDTH - 1), 'y': random.randint(0,
CELLHEIGHT - 1)}
    return temp

def test_not_ok(temp, worm):
    for body in worm:
        if temp['x'] == body['x'] and temp['y'] == body['y']:
            return True
    return False

HEAD = 0 # syntactic sugar: index of the worm's head
 # Set a random start point.
startx = random.randint(5, CELLWIDTH - 6)
starty = random.randint(5, CELLHEIGHT - 6)
wormCoords = [{'x': startx,     'y': starty},
              {'x': startx - 1, 'y': starty},
              {'x': startx - 2, 'y': starty}]
direction = RIGHT

# Start the apple in a random place.
apple = getRandomLocation(wormCoords)
```

```
def game():
    global FPSCLOCK, DISPLAYSURF, BASICFONT

    pygame.init()
    FPSCLOCK = pygame.time.Clock()
    DISPLAYSURF = pygame.display.set_mode((WINDOWWIDTH, WINDOWHEIGHT))
    BASICFONT = pygame.font.Font('freesansbold.ttf', 18)
    pygame.display.set_caption('Snaky')

    showStartScreen()
    # while True:
    #     runGame()
    #     # showGameOverScreen()

def runGame(action = MOVE_UP):
    global direction, wormCoords, apple
    # while True: # main game loop
    pre_direction = direction
    if action == MOVE_LEFT and direction != RIGHT:
        direction = LEFT
    elif action == MOVE_RIGHT and direction != LEFT:
        direction = RIGHT
    elif action == MOVE_UP and direction != DOWN:
        direction = UP
    elif action == MOVE_DOWN and direction != UP:
        direction = DOWN
    reward = 0
    # check if the worm has hit itself or the edge
    if wormCoords[HEAD]['x'] == -1 or wormCoords[HEAD]['x'] == CELLWIDTH or
wormCoords[HEAD]['y'] == -1 or wormCoords[HEAD]['y'] == CELLHEIGHT:
        reward = -1
        startx = random.randint(5, CELLWIDTH - 6)
        starty = random.randint(5, CELLHEIGHT - 6)
        wormCoords = [{'x': startx,     'y': starty},
                      {'x': startx - 1, 'y': starty},
                      {'x': startx - 2, 'y': starty}]
        direction = RIGHT

        # Start the apple in a random place.
        apple = getRandomLocation(wormCoords)
        screen_image = pygame.surfarray.array3d(pygame.display.get_surface())
        pygame.display.update()
```

```
        FPSCLOCK.tick(FPS)
        return reward, screen_image
    for wormBody in wormCoords[1:]:
        if wormBody['x'] == wormCoords[HEAD]['x'] and wormBody['y'] ==
wormCoords[HEAD]['y']:
            reward = -1
            startx = random.randint(5, CELLWIDTH - 6)
            starty = random.randint(5, CELLHEIGHT - 6)
            wormCoords = [{'x': startx,     'y': starty},
                    {'x': startx - 1, 'y': starty},
                    {'x': startx - 2, 'y': starty}]
            direction = RIGHT

            # Start the apple in a random place.
            apple = getRandomLocation(wormCoords)
            apple = getRandomLocation(wormCoords)
            screen_image =
pygame.surfarray.array3d(pygame.display.get_surface())
            pygame.display.update()
            FPSCLOCK.tick(FPS)
            return reward, screen_image

    # check if worm has eaten an apply
    if wormCoords[HEAD]['x'] == apple['x'] and wormCoords[HEAD]['y'] ==
apple['y']:
        # don't remove worm's tail segment
        apple = getRandomLocation(wormCoords) # set a new apple somewhere
        reward = 1    # 击中奖励
    else:
        del wormCoords[-1] # remove worm's tail segment

     # move the worm by adding a segment in the direction it is moving
    if not examine_direction(direction, pre_direction):
        direction = pre_direction
    if direction == UP:
        newHead = {'x': wormCoords[HEAD]['x'], 'y': wormCoords[HEAD]['y'] - 1}
    elif direction == DOWN:
        newHead = {'x': wormCoords[HEAD]['x'], 'y': wormCoords[HEAD]['y'] + 1}
    elif direction == LEFT:
        newHead = {'x': wormCoords[HEAD]['x'] - 1, 'y': wormCoords[HEAD]['y']}
    elif direction == RIGHT:
        newHead = {'x': wormCoords[HEAD]['x'] + 1, 'y': wormCoords[HEAD]['y']}
    wormCoords.insert(0, newHead)
```

```
        DISPLAYSURF.fill(BGCOLOR)
        drawGrid()
        drawWorm(wormCoords)
        drawApple(apple)
        drawScore(len(wormCoords) - 3)
        # 获得游戏界面像素
        screen_image = pygame.surfarray.array3d(pygame.display.get_surface())
        pygame.display.update()
        FPSCLOCK.tick(FPS)

        # 返回游戏界面像素和对应的奖励
        return reward, screen_image
    # def getRandomLocation(worm):
    #     temp = {'x': random.randint(0, CELLWIDTH - 1), 'y': random.randint(0,
CELLHEIGHT - 1)}
    #     while test_not_ok(temp, worm):
    #         temp = {'x': random.randint(0, CELLWIDTH - 1), 'y': random.randint(0,
CELLHEIGHT - 1)}
    #     return temp

    # def test_not_ok(temp, worm):
    #     for body in worm:
    #         if temp['x'] == body['x'] and temp['y'] == body['y']:
    #             return True
    #     return False
```

修改游戏代码的目的非常明确，就是让构建的强化学习算法来接管玩游戏的过程。玩游戏的过程其实包括两方面：一方面是能够获得游戏的反馈，在贪吃蛇游戏中就是指游戏画面；另一方面，就是发出玩游戏的指令，在贪吃蛇游戏中其实只有4个指令，对应着使贪吃蛇向上、向下、向左和向右运动。因此，我们可以清楚地确定算法输出应该也是4个离散数据。

```
# 强化学习结果的输出
MOVE_UP = [1, 0, 0, 0]
MOVE_DOWN = [0, 1, 0, 0]
MOVE_LEFT = [0, 0, 1, 0]
MOVE_RIGHT = [0, 0, 0, 1]
```

其实，我们希望在过程中游戏可以一直运行，这样我们的算法就能够不断地获取游戏数据。因此，我们对游戏结束的逻辑进行了调整，去掉了重新开始的玩家输入过程：

```
    # while True:
    #     runGame()
    #     # showGameOverScreen()
```

我们定义了奖励值，吃苹果动作完成的时候，奖励值会加1，在贪吃蛇碰到墙壁或者与自身相撞的时候奖励值减 1。更重要的是，我们会拿到游戏运行主循环一个周期后的屏幕图像。后面在构建机器学习算法时，我们会使用这个图像作为输入值。

```
screen_image = pygame.surfarray.array3d(pygame.display.get_surface())
    pygame.display.update()
```

第二部分，卷积神经网络模型构建。由于使用了游戏图像作为输入值，按照第 7 章的学习实践得到的经验，我们需要构造一个卷积神经网络模型来完成图像特征提取的过程。由于构建过程是标准化的，并且很大程度上借鉴了第 7 章实例 1 的模型架构，因此大家可以结合下面的实现代码进行理解分析。

```
# 定义 CNN
def convolutional_neural_network(input_image):
    weights = {'w_conv1':tf.Variable(tf.zeros([8, 8, 4, 32])),
              'w_conv2':tf.Variable(tf.zeros([4, 4, 32, 64])),
              'w_conv3':tf.Variable(tf.zeros([3, 3, 64, 64])),
              'w_fc4':tf.Variable(tf.zeros([128, 64])),
              'w_out':tf.Variable(tf.zeros([64, output]))}

    biases = {'b_conv1':tf.Variable(tf.zeros([32])),
             'b_conv2':tf.Variable(tf.zeros([64])),
             'b_conv3':tf.Variable(tf.zeros([64])),
             'b_fc4':tf.Variable(tf.zeros([64])),
             'b_out':tf.Variable(tf.zeros([output]))}

    conv1 = tf.nn.relu(tf.nn.conv2d(input_image, weights['w_conv1'], strides
= [1, 4, 4, 1], padding = "VALID") + biases['b_conv1'])
    conv1 = tf.nn.max_pool(conv1, ksize=[1, 2, 2, 1], strides=[1, 2, 2, 1],
padding='SAME')
    conv2 = tf.nn.relu(tf.nn.conv2d(conv1, weights['w_conv2'], strides = [1,
2, 2, 1], padding = "VALID") + biases['b_conv2'])
    conv2 = tf.nn.max_pool(conv2, ksize=[1, 2, 2, 1], strides=[1, 2, 2, 1],
padding='SAME')
    conv3 = tf.nn.relu(tf.nn.conv2d(conv2, weights['w_conv3'], strides = [1,
1, 1, 1], padding = "VALID") + biases['b_conv3'])
    conv3 = tf.nn.max_pool(conv3, ksize=[1, 2, 2, 1], strides=[1, 2, 2, 1],
padding='SAME')
    conv3_flat = tf.reshape(conv3, [-1, 128])
    fc4 = tf.nn.relu(tf.matmul(conv3_flat, weights['w_fc4']) +
biases['b_fc4'])

    output_layer = tf.matmul(fc4, weights['w_out']) + biases['b_out']
    return output_layer
```

第三部分，强化学习的训练过程，也是本实例的重点，通过运行经验和实际运行过程中模型的奖励变化来调节整个学习过程。正如本章介绍的那样，由于过程是动态的，因此对比单纯的深度学习过程，效率是有所下降的。首先是强化学习参数设定的过程，对具体参数的意义使用注释进行了说明。

```
#ML
# learning_rate
LEARNING_RATE = 0.99
# 更新梯度
INITIAL_EPSILON = 1.0
FINAL_EPSILON = 0.05
# 测试观测次数
EXPLORE = 500000
OBSERVE = 100
# 存储过往经验大小
REPLAY_MEMORY = 1024

BATCH =16
```

接下来是强化学习算法构建的过程，笔者依然采用注释和核心执行过程重点分析的方法进行说明。

```
def train (input_image):
    # 调用卷积神经网络
predict_action = convolutional_neural_network(input_image)
   # 计算损失函数
    argmax = tf.placeholder("float", [None, output])
    gt = tf.placeholder("float", [None])

    action = tf.reduce_sum(tf.multiply(predict_action, argmax),
reduction_indices = 1)
cost = tf.reduce_mean(tf.square(action - gt))
# 定义机器学习过程
    optimizer = tf.train.AdamOptimizer(1e-2).minimize(cost)

    game()
    D = deque()

    _, image = runGame()
    # 转换为灰度值
    image = cv2.cvtColor(cv2.resize(image, (120, 160)), cv2.COLOR_BGR2GRAY)
    # 转换为二值
    ret, image = cv2.threshold(image, 1, 255, cv2.THRESH_BINARY)
    input_image_data = np.stack((image, image, image, image), axis = 2)
    # 实际机器学习模型的训练过程
```

```
    with tf.Session() as sess:
        sess.run(tf.initialize_all_variables())

        saver = tf.train.Saver()

        n = 0
        epsilon = INITIAL_EPSILON
        while True:
        # 通过概率选择动作
            action_t = predict_action.eval(feed_dict = {input_image :
[input_image_data]})[0]

            argmax_t = np.zeros([output], dtype=np.int)

            if(random.random() <= epsilon):
                maxIndex = random.randrange(output)
            else:
                maxIndex = np.argmax(action_t)
            argmax_t[maxIndex] = 1
            if epsilon > FINAL_EPSILON:
                epsilon -= (INITIAL_EPSILON - FINAL_EPSILON) / EXPLORE

            #for event in pygame.event.get():  macOS 需要事件循环，否则白屏
            #   if event.type == QUIT:
            #       pygame.quit()
            #       sys.exit()
            # 通过选定的动作进行游戏并获得游戏结果
            reward, image = runGame(list(argmax_t))
            image = cv2.cvtColor(cv2.resize(image, (120, 160)),
cv2.COLOR_BGR2GRAY)
            ret, image = cv2.threshold(image, 1, 255, cv2.THRESH_BINARY)
            image = np.reshape(image, (160, 120, 1))
            input_image_data1 = np.append(image, input_image_data[:, :, 0:3],
axis = 2)
            # 训练样本储存
            D.append((input_image_data, argmax_t, reward, input_image_data1))

            if len(D) > REPLAY_MEMORY:
                D.popleft()

            if n > OBSERVE:
                minibatch = random.sample(D, BATCH)
                input_image_data_batch = [d[0] for d in minibatch]
```

```
                argmax_batch = [d[1] for d in minibatch]
                reward_batch = [d[2] for d in minibatch]
                input_image_data1_batch = [d[3] for d in minibatch]

                gt_batch = []

                out_batch = predict_action.eval(feed_dict = {input_image :
input_image_data1_batch})

                for i in range(0, len(minibatch)):
                    gt_batch.append(reward_batch[i] + LEARNING_RATE *
np.max(out_batch[i]))

                optimizer.run(feed_dict = {gt : gt_batch, argmax : argmax_batch,
input_image : input_image_data_batch})

            input_image_data = input_image_data1
            n = n+1

            # if n % 10000 == 0:
            #   saver.save(sess, 'game.cpk', global_step = n)  # 保存模型

            print(n, "epsilon:", epsilon, " " ,"action:", maxIndex, "
" ,"reward:", reward)

    train (input_image)
```

经验池 D 采用了队列的数据结构，是 TensorFlow 中基础的数据结构，可以通过 dequeue()和 enqueue([y])方法进行取出和压入数据。经验池 D 用来存储实验过程中的数据，后面的训练过程会从中随机取出一定量的 batch 进行训练。

```
    D = deque()
```

在实验一段时间后，经验池 D 中已经保存了一些样本数据后，可以从这些样本数据中随机抽样，进行模型训练。

```
     if len(D) > REPLAY_MEMORY:
                D.popleft()
    if n > OBSERVE:
```

input_image_data_batch、argmax_batch、reward_batch、input_image_data1_batch 是从经验池 D 中提取的马尔科夫序列，gt_batch 为标签值，若游戏结束，则不存在下一步中状态对应的 Q 值（回忆 Q 值更新过程），直接添加 reward_batch，若未结束，则用折合因子（0.99）和下一步中状态的最大 Q 值的乘积添加至 gt_batch。执行梯度下降训练，train_step 的入参是 input_image_data_batch、 argmax_batch 和 gt_batch。

```
    minibatch = random.sample(D, BATCH)
                input_image_data_batch = [d[0] for d in minibatch]
                argmax_batch = [d[1] for d in minibatch]
                reward_batch = [d[2] for d in minibatch]
                input_image_data1_batch = [d[3] for d in minibatch]

                gt_batch = []

                out_batch = predict_action.eval(feed_dict = {input_image :
input_image_data1_batch})

                for i in range(0, len(minibatch)):
                    gt_batch.append(reward_batch[i] + LEARNING_RATE *
np.max(out_batch[i]))

                optimizer.run(feed_dict = {gt : gt_batch, argmax : argmax_batch,
input_image : input_image_data_batch})
```

参考文献

[1] Barto A G, Sutton R S, Anderson C W. Neuronlike adaptive elements that can solve difficult learning control problems[J]. IEEE Transactions on Systems, Man, and Cybernetics, 2012, SMC-13(5):834-846.

[2] Sutton, R. S. and A. G. Barto. 2017. Reinforcement Learning: An Introduction (2nd Edition, in progress). MIT Press.

[3] O'Donoghue B, Munos R, Kavukcuoglu K, et al. Combining policy gradient and Q-learning[J]. 2016.

[4] Bertsekas D P. Dynamic Programming and Optimal Control[M]// Dynamic programming and optimal control. 2000.

[5] Owen D B. Applied Dynamic Programming[J]. Journal of the Operational Research Society, 1964, 15(2):155-156.

[6] Human-level control through deep reinforcement learning[J]. Nature, 2015, 518(7540):529-533.

第三篇　无人驾驶

第 9 章　无人驾驶系统

无人驾驶是指通过软件算法代替人类操作过程使交通工具能够自行完成行驶过程的整套流程。无人驾驶智能汽车是利用车载传感器来感知车辆周围的环境，并根据感知获得道路、车辆位置和障碍物信息，控制车辆的转向和速度，从而使车辆能够安全、可靠地在道路上行驶，具体包括感知、定位、决策、控制等多个关联子系统。

本章主要讲解无人驾驶系统的软硬件构成，以百度 Apollo 系统为例说明子系统的组织关系。内容分为三部分：第一部分着重介绍无人驾驶系统的软硬件构成，分析无人驾驶各个子系统的核心算法和基础业务流程；第二部分主要分析百度 Apollo 无人驾驶系统的架构和子系统之间的数据流动关系；第三部分细致讲解利用 Docker 技术在 Ubuntu 系统中完成百度 Apollo 无人驾驶系统的开发环境搭建。

9.1　无人驾驶系统概述

自动驾驶汽车意味着将先进技术应用到轿车、卡车和巴士上。其中包括自动车辆引导和制动、变道系统、使用摄像头和感应器避免碰撞、使用人工智能实时分析信息、安装高性能计算和深度学习系统，通过高清 3D 地图适应新环境。光探测和测距系统（又称激光雷达）和人工智能是导航和防碰撞的关键。前者是安装在车辆顶部的光和雷达设备，通过雷达和光束实现 360 度成像，测算周围物体的速度和距离。因为车辆的前后左右都装有感应器，这一设备可以获取必要的信息，保证快速移动的车辆在车道内行驶，避开其他车辆，在必要时瞬间启动制动和转向，从而避免事故发生。高精度地图对无人驾驶至关重要。百度在中国绘制的高精度地图达到了厘米级精度。高精度地图比GPS 更精确，后者的准确度只有 5~10 米（约 16~32 英寸）。百度使用约 250 辆测绘车收集道路信息用以制作准确度在 5~10 米的传统导航地图以及高精度地图。另外，所有测绘车可以实现快速升级，支持高精度地图数据收集。针对自动驾驶汽车，百度使用的厘米级高精度地图包含交通信号灯、车道标记（如白线、黄线、双车道或单车道、实线、虚线）、路缘石、障碍物、电线杆、立交桥、地下通道等详细信息。上述信息都有地理编码，因此导航系统可以准确定位地形、物体和道路轮廓，从而引导车辆行驶。百度已经绘制了约 670 万公里（400 万英里）的中国普通公路和高速公路，用以制作传统导航地图。百度的道路导航系统可以实现 95%以上的路标和车道标记准确度。因为建设和其他变动，高速公路地图需要定期更新。传统地图需要每 3 个月重新绘制一次，但是无人驾驶车辆地图需要不断更新，以便掌握路况变化。数字图像处理技术的精确度已经非常高，以人脸识别为例，人的错误率为 8‰，而安装了图像识别软件的计算机错误率只有 2.3‰。就可见度（安全视力距离）而言，人只能看见道路前方 50 米（55 码），而装了激光雷达和摄像头的自动驾驶车辆可以达到 200 米（219 码）。摄像头和感应器收集大量信息，只有瞬间进行处理才能避开旁边车道的车辆。为了适应新情况，无人驾驶汽车需要具备高性能的计算能力、先进的算法和深度学习系统，这意味着软件才是关键，而非实体车辆本身。先进的软件可以让汽车学习道路上其他车辆的经验，并根据天气、行驶和路况的改变调整导航系统。车载系统也可以通过机器间的交流了解路面上其他车辆的情况。如果没有先进的人工智能模型和高精度的地图进行信息分析，以及了解周边环境变化的能力，无人驾驶汽车将很难安全行驶，也无法处理世界各地道路和高速公路的复杂情况。自动驾驶系统包含自动驾驶中的各个子系统以及子系统中的数据流和控制流的流程关系，具体来说，重要的核心子系统包括环境感知、车辆定位、路径规划、车辆控制 4 部分。下面我们分别介绍这 4 个子系统的业务流程、硬件基础和核心软件算法。

9.1.1 环境感知概述

环境感知通过各类传感器采集周边和自身的信息，实时发送给处理器，形成对周边环境的认知模型，而且需要保证在行驶过程中对环境信息进行获取并处理过程的连续性和实时性。从目前的大多数技术方案来看，激光雷达对周围环境的三维空间感知完成了 60%~75% 的环境信息获取，其次是相机获取的图像信息，再次是毫米波雷达获取的定向目标距离信息，以及 GPS 定位及惯性导航获取的无人车位置及自身姿态信息，最后是其他超声波传感器、红外线传感器等其他光电传感器获取的各种信息。

激光雷达是环境感知的重要渠道，因为激光的发散度极小，可以远距离传播，在无人车上，激光雷达一般被用于远程扫描，发现需要分析的区域。激光雷达按照光源数量的不同，分成单线激光雷达和多线激光雷达。其中，多线激光雷达水平 360°扫描，在垂直方向具备一个俯仰角，因此不但可以测量物体的距离，还可以扫描障碍物的三维尺寸。

另一种重要的环境感知硬件是相机，相机主要分为单目相机、双目相机、全景相机 3 大类。把相机安置在无人车上，对周围环境进行拍照以获得环境的图像信息，这就是应用在无人车上的机器视觉技术。为了适应无人车上的天气变化、车体震动等造成的干扰，相机需要经过车规级的设计和测试。其中，单目相机造价低廉，应用较广，是机器视觉行业翘楚以色列 Mobileye 公司的主打产品。双目相机由于在运动的无人车上成像精度不高，并没有被特别广泛地应用。全景相机是利用多台相机同时拍照，再经过计算机图像处理，合成再现环境图景的，是应用在无人车上进行目标识别的先进武器。但也有一个小问题——计算量比较大。

毫米波雷达工作在毫米波段。通常毫米波的频段为 30GHz～300GHz（波长为 1mm～10mm）。毫米波的波长介于厘米波和光波之间，因此毫米波兼有微波制导和光电制导的优点。同厘米波导引头相比，毫米波导引头具有体积小、质量轻和空间分辨率高的特点。与红外、激光、电视等光学导引头相比，毫米波导引头穿透雾、烟、灰尘的能力强，具有全天候（大雨天除外）全天时的特点。另外，毫米波导引头的抗干扰能力优于其他微波导引头。

9.1.2　车辆定位概述

车辆定位是利用车辆的硬件测量结果，结合地图软件输入比对得到车辆位置的过程。

在自动驾驶车辆定位中，GPS 是一种标志性技术。GPS 定位的基本原理是，测量出已知位置的卫星到地面 GPS 接收器之间的距离，然后接收器通过与至少 4 颗卫星通信，计算与这些卫星间的距离，就能确定其在地球上的具体位置。为了提高定位精度，可以使用载波相位差分技术（Real Time Kinematic，RTK）。载波相位差分技术能够实时地提供测站点在指定坐标系中的三维定位结果，并达到厘米级精度。在载波相位差分技术作业模式下，基站采集卫星数据并通过数据链将其观测值和站点坐标信息一起传送给移动站，而移动站通过对所采集到的卫星数据和接收到的数据链进行实时载波相位差分处理（历时不足一秒）得出厘米级的定位结果。

惯性测量单元（Inertial Measurement Unit，IMU）是测量物体三轴姿态角（或角速率）以及加速度的装置。陀螺仪及加速度计是惯性测量单元的主要元件，其精度直接影响惯性系统的精度。在实际工作中，由于各种干扰因素不可避免，因此导致陀螺仪及加速度计产生误差。从初始对准开始，其导航误差就随时间而增长，尤其是位置误差，这是惯导系统的主要缺点。所以需要利用外部信息进行辅助，实现组合导航，使其有效地减小误差随时间积累的问题。为了提高可靠性，还可以为每个轴配备更多的传感器。一般而言，惯性测量单元要安装在被测物体的重心上。

高精细地图是指高精度、精细化定义的地图，其精度需要达到分米级才能够区分各个车道。随着定位技术的发展，高精度的定位已经成为可能。而精细化定义则是需要格式化存储交通场景中的各种交通要素，包括传统地图的道路网数据、车道网络数据、车道线以及交通标志等数据。

与传统电子地图不同，高精度电子地图的主要服务对象是无人驾驶车，或者机器驾驶员。和人类驾驶员不同，机器驾驶员缺乏与生俱来的视觉识别、逻辑分析能力。比如，人可以很轻松、准确地利用图像、GPS 定位自己，鉴别障碍物、人、交通信号灯等，但这对当前的机器人来说是非常困难的任务。因此，高精度电子地图是当前无人驾驶车技术中必不可少的一个组成部分。高精度电子地图包含大量行车辅助信息，其中最重要的是对路网精确的三维表征（厘米级精度），比如路面的几何结构、道路标示线的位置、周边道路环境的点云模型等。有了这些高精度的三维表征，车载机器人就可以通过比对车载 GPS、IMU、LiDAR 或摄像头数据来精确地确认自己当前的位置。此外，高精地图还包含丰富的语义信息，比如交通信号灯的位置及类型、道路标示线的类型、识别哪些路面可以行驶等。这些能极大地提高无人驾驶车鉴别周围环境的能力。此外，高精度地图还能帮助无人驾驶车识别车辆、行人及未知障碍物。这是因为高精地图一般会过滤掉车辆、行人等活动障碍物。如果无人驾驶车在行驶过程中发现当前高精地图中没有的物体，便有很大概率是车辆、行人或障碍物。因此，高精地图可以提高无人驾驶车发现并鉴别障碍物的速度和精度。

9.1.3 路径规划概述

路径规划是解决无人车从起点到终点，走怎样的路径的问题。规划的总体要求是不要撞到障碍物，保证自身的安全和可能相遇的车辆和行人的安全。在此基础上，再去依次追求下列目标：车体平稳、乘坐舒适、寻求路径最短等。

路径规划问题可以分成两类：总体路径规划和局部路径规划。前者是指给无人车设定目的地，从出发地到目的地走哪条路最好。后者指在行进过程中，遇到障碍、行人、车辆甚至小动物等，怎样获得理想的行进路径，利用卫星定位和自身保存的离线地图规划出理想路径，过程与我们用手机导航去一个地方非常类似。这里的结构化道路指的是边缘比较清晰规则、路面平坦、有明显的车道线及其他人工标记的行车道路。

在路径规划过程中有几种典型算法。Dijkstra's 算法，在起点周围不会遇到障碍的所有可能点中寻找最短路径，规划结果比较优越，但在没有足够约束条件的情况下，计算量巨大；随机采样算法，是在 Dijkstra's 算法的基础上改良的；为了减少计算量，加入了启发式算法，配合随机采样，只计算样本中的最短路径，解决了计算量的问题，但路径可能不连续；基于差值曲线的路径规划，降低了计算量，同时解决了路径不连续的问题，是比较有优势的一种算法；基于数值最优，把无人车姿态和环境约束条件都加入模型的一种算法，可以得到较好的规划结果，但对计算能力依赖性强。

9.1.4 车辆控制概述

我们希望控制的对象（无人车）能够按照希望（规划好）的路径行驶，我们会将环境当前的反馈（当前的位置）和参考线进行比较，得到当前偏离参考线的距离（误差），基于这个误差，设计一定的算法来产生输出信号，使得这个误差不断变小，这样的过程就是反馈控制的一般过程。那么，如何基于这个误差来产生控制指令呢？最直观的感觉就是让误差在我们的控制下逐渐变小直到为 0。PID 控制是目前利用最为广泛的控制理论。下面简单介绍一下 PID 控制理论的原理。

PID 就是指比例（Proportion）、积分（Integral）、导数（Derivative），这 3 项表示我们如何使用误差来产生控制指令，整个流程如图 9.1 所示。

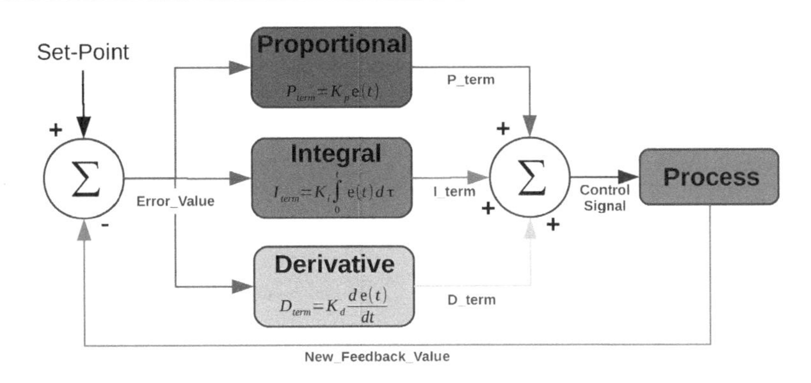

图 9.1　PID 控制流程

首先根据反馈和参考值求出误差，这里的误差根据具体的情况可以是各种度量，比如控制车辆按照指定的路径形式，就是车辆当前位置和参考线的距离，控制车辆的速度达到设定的值，就是当前速度和设定速度的差值。求出误差以后，再根据误差求比例、积分和微分 3 项，其中 K_p、K_i、和 K_d 是 3 项的系数，它们决定着这 3 项对最后输出的影响的比重。将 P、I、D 三项求和作为最后的输出信号。

另一种常用的控制算法是 MPC 控制。MPC 控制算法的作用机理是在每一个采样时刻，根据获得的当前测量信息在线求解一个有限时间开环优化问题，并将得到的控制序列的第一个元素作用于被控对象。在下一个采样时刻重复上述过程：用新的测量值作为此时预测系统未来动态的初始条件，刷新优化问题并重新求解。

9.2　Apollo 简介

汽车软件系统是模块化的，这意味着它们彼此之间大多数时候是独立运行的，也就很少进行信息交换。然而，更为严峻的问题是，今天的汽车依然从宏观意义上缺少适合的人工智能。只有人类驾驶员掌握方向盘时，现代的车载软件才能正常运作，一旦脱离人类的驾驶，这些汽车就几近瘫痪。当今这些依赖人类驾驶的汽车缺少的是一套强大的机器人操作系统，这个系统应该包含必备的人工智能使其能够灵活应对新情况，并能从过往经历中“学习”提升。

想要把汽车驾驶得同人类一样好或者比人类更好，无人驾驶汽车的软件系统必须足够聪明——能够知道自己在哪里，知道周边有什么，能够预计将会发生什么并做出应对计划。除了做出正确的判断（在哪里转向、何时停下、何时刹车以及何时变道等）外，自动化驾驶汽车的操作系统还必须能监控汽车底层的运动状态，例如告知汽车的人造“肌肉”（制动器）踩刹车，或者对方向进行微调。百度 Apollo 的架构就尝试串联整个无人驾驶的工作流程，提出了包括系统底层、硬件使用、数据学习方面的综合解决方案。

9.2.1　Apollo 架构概述

Apollo 平台架构如图 9.2 所示。

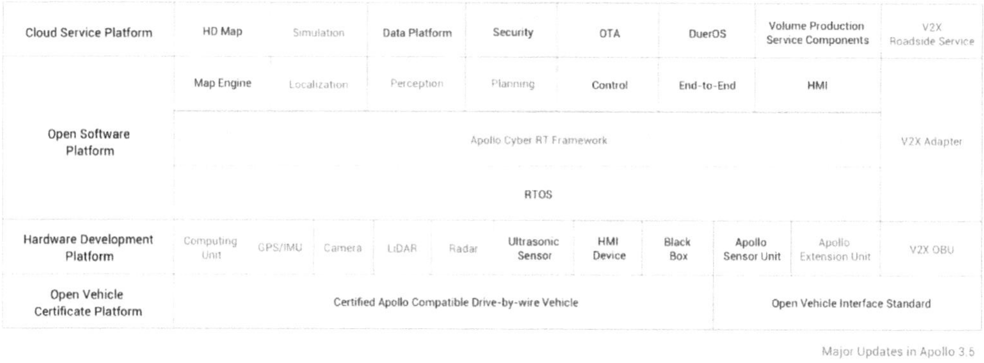

Cloud Service Platform	HD Map	Simulation	Data Platform	Security	OTA	DuerOS	Volume Production Service Components	V2X Roadside Service			
Open Software Platform	Map Engine	Localization	Perception	Planning	Control	End-to-End	HMI	V2X Adapter			
	Apollo Cyber RT Framework										
	RTOS										
Hardware Development Platform	Computing Unit	GPS/IMU	Camera	LiDAR	Radar	Ultrasonic Sensor	HMI Device	Black Box	Apollo Sensor Unit	Apollo Extension Unit	V2X OBU
Open Vehicle Certificate Platform	Certified Apollo Compatible Drive-by-wire Vehicle								Open Vehicle Interface Standard		

Major Updates in Apollo 3.5

图 9.2 Apollo 平台架构

Apollo 平台架构包括:

- 云服务平台（Cloud Service Platform）：包括 HD 地图、仿真、数据平台、安全、云更新、指令控制等。
- 软件平台（Open Software Platform）：包括地图工程、定位、感知、预测、规划、安全、控制、自主规划、人机交互等。
- 硬件平台（Hardware Development Platform）：包括计算单元、GPS、相机、激光雷达、毫米波雷达、超声波雷达、人机接口设备、黑盒（数据记录）等。
- 车辆平台（Open Vehicle Certificate Platform）：进行线控车辆。

面向自动驾驶的 Apollo 平台 3.5 版本主要包括以下软件模块:

- 感知：感知模块用于识别自动驾驶车辆的周围环境。该模块包含两个重要的子模块：障碍物检测和交通灯检测。
- 预测：预测模块用于预测感知障碍物的未来运动轨迹。
- 定位：定位模块利用 GPS、LiDAR 和 IMU 等各种信息来估计自动驾驶车辆本身的位置。
- 路由：路由模块告诉自动驾驶车辆如何经过一系列车道或道路到达目的地。
- 规划：规划模块规划自动驾驶车辆所采取的时空轨迹。
- 控制：控制模块通过生成控制命令（如加速、刹车和转向）来执行规划模块提供的时空轨迹。
- CANBus：Can 总线是传递控制命令到车辆硬件的接口。它负责将硬件系统信息传递给软件系统。
- 高精地图：该模块类似于一个库。它更多作为查询引擎提供有关道路的临时结构化信息，而不是发布和订阅消息。
- 人机接口：Apollo 平台的人机接口或者 DreamView 是一个查看自动驾驶车辆状态的模块，同时也可用来测试其他模块和实时控制车辆。
- 监测：车辆中包括硬件在内的所有模块的监控系统。

Apollo 系统运行流程如图 9.3 所示。

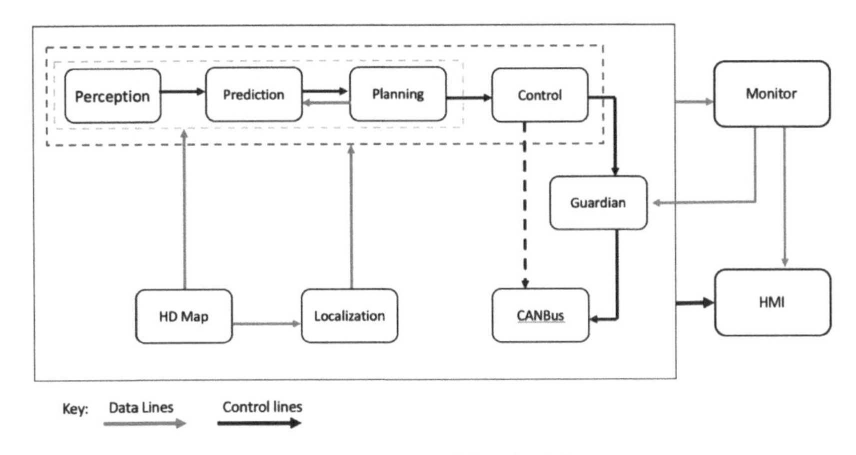

图 9.3　Apollo 系统运行流程

9.2.2　Apollo 子系统交互关系

根据上一小节的图 9.3，本小节我们对 Apollo 系统的各组成部分（子系统）进行详细介绍。

1. 感知

感知依赖 LiDAR 点云数据和相机原始数据。除了这些传感器数据输入之外，还有交通灯检测依赖定位以及 HD-Map。由于实时 ad-hoc 交通灯检测在计算上是不可行的，因此交通灯检测需要依赖定位确定何时何地开始通过相机捕获的图像检测交通灯。

- 全线支持：粗线支持，可实现远程精确度。相机有高低两种不同的安装方式。
- 异步传感器融合：因为不同传感器的帧速率差异（雷达为 10ms，相机为 33ms，LiDAR 为 100ms），所以异步融合 LiDAR、雷达和相机数据，获取所有信息并得到数据点的功能非常重要。
- 在线姿态估计：在出现颠簸或斜坡时确定与估算角度变化，以确保传感器随汽车移动及角度/姿态相应地变化。
- 视觉定位：基于相机的视觉定位方案正在测试中。
- 超声波传感器：作为安全保障传感器，与 Guardian 一起用于自动紧急制动和停车。

2. 预测

预测模块负责预测所有感知障碍物的未来运动轨迹。输出预测消息封装了感知信息。预测订阅定位和感知障碍物消息，代码如下：

```
//Set perception obstacle callback function
AdapterManager::AddPerceptionObstaclesCallback(&Prediction::RunOnce, this)
//Set localization callback function
AdapterManager::AddLocalizationCallback(&Prediction::OnLocalization, this)
```

当接收到定位更新时，预测模块更新其内部状态。当感知模块发布感知障碍物消息时，触发预测实际执行。

3. 定位

定位模块聚合各种数据以定位自动驾驶车辆。有两种类型的定位模式：OnTimer 和多传感器融合。

第一种基于 RTK 的定位方法，通过计时器的回调函数 OnTimer 实现，代码如下：

```
timer_ = AdapterManager::CreateTimer(cyber::Duration(dutation),
&RTKLocalization:;OnTimer, this)
```

另一种定位方法是多传感器融合（MSF）方法，其中注册了一些事件触发的回调函数，代码如下：

```
void RTKLocalization::ImuCallback(
    const std::shared_ptr<localization::CorrectedImu> &imu_msg) {
  std::unique_lock<std::mutex> lock(imu_list_mutex_);
  if (imu_list_.size() < imu_list_max_size_) {
    imu_list_.push_back(*imu_msg);
  } else {
    imu_list_.pop_front();
    imu_list_.push_back(*imu_msg);
  }
  return;
}
```

4. 路由

为了计算可通行的车道和道路，路由模块需要知道起点和终点。通常，路由起点是自动驾驶车辆的位置。重要的数据接口是一个名为 OnRoutingRequest 的事件触发函数，其中 RoutingResponse 的计算和发布如下：

```
AdapterManager::AddRoutingRequestCallback(&Routing::OnRoutingRequest, this)
```

5. 规划

Apollo 需要使用多个信息源来规划安全无碰撞的行驶轨迹，因此规划模块几乎与其他所有模块进行交互。

首先，规划模块获得预测模块的输出。预测输出封装了原始感知障碍物，规划模块订阅交通灯检测输出而不是感知障碍物输出。然后，规划模块获取路由输出。在某些情况下，若当前路由的结果不可执行，则规划模块还可以通过发送路由请求来触发新的路由计算。

最后，规划模块需要知道定位信息（定位：我在哪里）以及当前的自动驾驶车辆信息（例如底盘的状态是什么）。

6. 控制

如规划模块中所述，控制将规划轨迹作为输入，并生成控制命令传递给 CanBus。Apollo 控制模块主要考虑了车辆的运动学模型和动力学模型。注意提供了 PID 车辆控制算法和 MPC 车辆控制算法。PID 车辆控制算法集中在 LatController 和 LonController 对象中，MPC 车辆控制算法集中在 MPCController 对象中。

7. CANBus

CANBus 有两个数据接口。相关回调函数如下：

```
AdapterManager::AddControlCommandCallback(&Canbus::OnControlCommand, this)
```

第一个数据接口是基于计时器的发布者，回调函数为 OnTimer。如果启用，此数据接口就会定期发布底盘信息。第二个数据接口是一个基于事件的发布者，回调函数为 OnControlCommand，当 CanBus 模块接收到控制命令时会触发该函数。

8. HMI

Apollo 中的 HMI 或 DreamView 是一个 Web 应用程序，功能是：可视化自动驾驶模块的输出，例如规划轨迹、汽车定位、底盘状态等；为用户提供人机交互界面，以查看硬件状态、打开/关闭模块以及启动自动驾驶汽车；提供调试工具，如 PNC Monitor，以有效跟踪模块问题。

9. 监控

包括硬件在内的车辆中所有模块的监控系统。监控模块从其他模块接收数据并传递给 HMI，以便司机查看并确保所有模块都正常工作。如果模块或硬件发生故障，监控就会向 Guardian（新的操作中心模块）发送警报，然后决定需要采取哪些操作来防止系统崩溃。

10. Guardian

这个新模块根据 Monitor 发送的数据做出相应决定。Guardian 有两个主要功能：一是所有模块都正常工作，Guardian 允许控制模块正常工作，此时控制信号被发送到 CANBus，就像 Guardian 不存在一样；二是监控检测到模块崩溃，Guardian 将阻止控制信号到达 CANBus 并使汽车停止。Guardian 有 3 种方式决定如何停车并会依赖最终的 Gatekeeper：如果超声波传感器运行正常而未检测到障碍物，Guardian 就会使汽车缓慢停止；如果超声波传感器没有响应，Guardian 就会硬制动，使汽车马上停止；还有一种特殊情况，如果 HMI 通知驾驶员即将发生碰撞并且驾驶员在 10 秒内没有干预，Guardian 就会使用硬制动使汽车立即停止。

9.3　Apollo 开发环境搭建

正如 9.2 节介绍的那样，Apollo 是一个完整的开源自动驾驶系统，Apollo 系统致力于使开发者能够非常容易地构建、使用和标定使用 Apollo 系统运行的自动驾驶设备。本节将详细介绍 Apollo 软件开发环境搭建的步骤和注意事项。

9.3.1 软件系统代码本地安装

1. git-lfs 安装

Apollo 中的大文件开发版本管理使用 git-lfs（Git Large File Storage，LFS），其主要针对大文件（比如视频、大数据数据库备份、图片集合等）使用 Git 进行管理的业务场景。git-lfs 的实际处理方法是把存储和存储记录区分开来，使用 Git 来管理存储记录，当我们需要下载同步大文件时，通过存储记录到 GitHub 的文件存储服务器集群中获得对应的文件。因此，首先要安装 git-lfs 插件，安装方式有两种：一种是利用 git-lfs 的安装脚本；另一种是使用 Ubuntu 自带的库管理工具 apt-get。具体如下：

```
# 方法 1： Required for Ubuntu 14.04 / 16.04.
curl -s
https://packagecloud.io/install/repositories/github/git-lfs/script.deb.sh |
sudo bash

# 方法 2：Ubuntu 14.04 / 16.04 / 18.04.
sudo apt-get install -y git-lfs
```

建议大家检查一下 Git 版本，因为在最新的 Git 版本中，git-lfs 的版本同步功能已经和 Git 完全整合在了一起，在调用 git clone 或者 git pull 命令时，git-lfs 会在后台自动运行完成大文件的同步工作。如果 Git 的版本是 1.X，这种同步就无法完成，我们还需要显式地调用 git lfs clone 或者 git lfs pull 命令。

2. 下载 Apollo 源代码

接下来是下载 Apollo 源代码的过程。首先需要使用 git 进行下载，然后切换到我们需要的正确分支。

```
git clone git@github.com:ApolloAuto/apollo.git
cd apollo
git checkout [release_branch_name]
```

3. Apollo 系统安装

首先需要指定系统的环境变量 APOLLO_HOME：

```
echo "export APOLLO_HOME=$(pwd)" >> ~/.bashrc && source ~/.bashrc
```

然后调用 Apollo 系统中的脚本进行安装：

```
source ~/.bashrc
```

脚本实际安装过程中使用 Docker 进行 Apollo 系统的安装，Docker 容器是设置 Apollo 构建环境的简单方法，具体步骤如下：

（1）在 Ubuntu 安装 docker-ce 指南并配置相应参数。

① 登录系统，用你的账号使用 sudo 权限。

② 更新 APT 包索引：sudo apt-get update。

③ 安装 Docker：sudo apt-get install docker-engine。

④ 开启 Docker 后台进程：sudo service docker start。

⑤ 校验 Docker 是否安装成功：sudo docker run hello-world。

（2）安装完成后，注销并重新登录系统以启用 Docker。

（可选）如果你已经安装了 Docker（在安装 Apollo 内核之前），就要在/etc/default/docker 中添加以下脚本内容：

```
DOCKER_OPTS = "-s overlay"
```

（3）通过运行以下命令下载并启动 Apollo 发布的 Docker 映像：

```
cd $APOLLO_HOME
bash docker/scripts/release_start.sh
```

（4）如果你需要定制化 Docker 映像，那么可以通过运行以下命令登录已下载的 Docker 映像：

```
bash docker/scripts/release_into.sh
```

（5）按照以下步骤保存本地环境，如果没有修改过本地的 Docker Release Container 里的配置，可跳过此步：

```
# EXIT OUT OF DOCKER ENV
# commit your docker local changes to local docker image.
exit # exit from docker environment
cd $APOLLO_HOME
bash docker/scripts/release_commit.sh
```

9.3.2 开发环境搭建中的注意事项

安装 Docker 之前需要进行 apt-get 源的更新。Docker 的 APT 仓库包含 1.7.1 以及更高的版本。通过设置 APT 使用来自 Docker 仓库的包。

（1）登录机器，用户必须使用 sudo 或者 root 权限。

（2）打开终端。

（3）更新包信息，确保 APT 能使用 HTTPS 方式工作，并且 CA 证书已经安装了，如图 9.4 所示。

```
        sudo apt-get update
        sudo apt-get install apt-transport-https ca-certificates
```

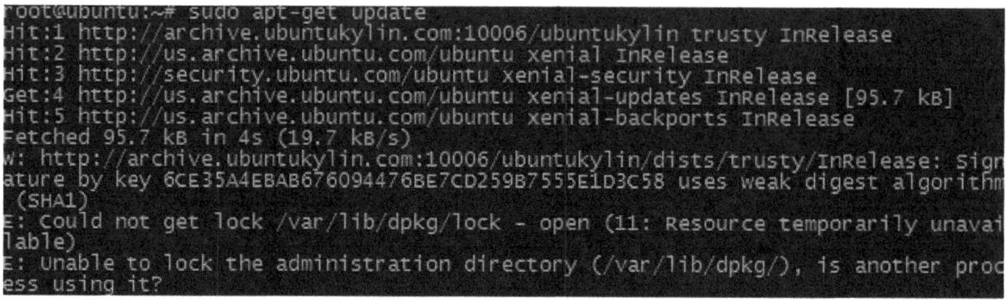

```
root@ubuntu:~# sudo apt-get update
Hit:1 http://archive.ubuntukylin.com:10006/ubuntukylin trusty InRelease
Hit:2 http://us.archive.ubuntu.com/ubuntu xenial InRelease
Hit:3 http://security.ubuntu.com/ubuntu xenial-security InRelease
Get:4 http://us.archive.ubuntu.com/ubuntu xenial-updates InRelease [95.7 kB]
Hit:5 http://us.archive.ubuntu.com/ubuntu xenial-backports InRelease
Fetched 95.7 kB in 4s (19.7 kB/s)
W: http://archive.ubuntukylin.com:10006/ubuntukylin/dists/trusty/InRelease: Sign
ature by key 6CE35A4EBAB676094476BE7CD259B7555E1D3C58 uses weak digest algorithm
 (SHA1)
E: Could not get lock /var/lib/dpkg/lock - open (11: Resource temporarily unavai
lable)
E: Unable to lock the administration directory (/var/lib/dpkg/), is another proc
ess using it?
```

图 9.4 更新包信息

出现这个问题可能是有另一个程序正在运行，导致资源被锁不可用。而导致资源被锁的原因可能是上次运行安装或更新没有正常完成，解决办法就是删除。

```
sudo rm /var/cache/apt/archives/lock
sudo rm /var/lib/dpkg/lock
```

（4）添加一个新的 GPG 密钥：

```
sudo apt-key adv --keyserver hkp://p80.pool.sks-keyservers.net:80
--recv-keys 58118E89F3A912897C070ADBF76221572C52609D
```

（5）找到适合你的 Ubuntu 操作系统的键，这个键决定 APT 将搜索哪个包，可能的键有：

```
Ubuntu version      Repository
Precise 12.04       deb
https://apt.dockerproject.org/repoubuntu-precise main
Trusty 14.04        deb
https://apt.dockerproject.org/repoubuntu-trusty main
Xenial 16.04        deb
https://apt.dockerproject.org/repoubuntu-xenial main
```

（6）运行下面的命令，用占位符<REPO> 为你的操作系统替换键。

```
echo "<REPO>" | sudo tee /etc/apt/sources.list.d/docker.list
```

比如你使用的是 Ubuntu16.04，需要将上面命令的<REPO>部分替换成 deb https://apt.dockerproject.org/repoubuntu-xenial main，执行这条命令，就在文件夹下创建了一个 docker.list 文件，里面的内容是：

```
deb https://apt.dockerproject.org/repoubuntu-xenial main
```

（7）更新 APT 包索引：

```
sudo apt-get update
```

（8）校验 APT 是从一个正确的仓库拉取安装包。

当运行下面的命令的时候，这个键会返回目前可以安装的 Docker 版本，每个键都包括 URL：https://apt.dockerproject.org/repo/。图 9.5 是截取的部分输出内容。

```
命令：apt-cache policy docker-engine
```

```
root@ubuntu:~# apt-cache policy docker-engine
docker-engine:
  Installed: (none)
  Candidate: 1.12.3-0~xenial
  Version table:
     1.12.3-0~xenial 500
        500 https://apt.dockerproject.org/repo ubuntu-xenial/main amd64 Packages
     1.12.2-0~xenial 500
        500 https://apt.dockerproject.org/repo ubuntu-xenial/main amd64 Packages
     1.12.1-0~xenial 500
        500 https://apt.dockerproject.org/repo ubuntu-xenial/main amd64 Packages
     1.12.0-0~xenial 500
        500 https://apt.dockerproject.org/repo ubuntu-xenial/main amd64 Packages
     1.11.2-0~xenial 500
        500 https://apt.dockerproject.org/repo ubuntu-xenial/main amd64 Packages
     1.11.1-0~xenial 500
        500 https://apt.dockerproject.org/repo ubuntu-xenial/main amd64 Packages
     1.11.0-0~xenial 500
        500 https://apt.dockerproject.org/repo ubuntu-xenial/main amd64 Packages
```

图 9.5　部分输出内容

现在当你运行 apt-get upgrade 的时候，APT 就会从新的仓库拉安装包。

官方推荐的系统是 Ubuntu 14.04，但是这个 Ubuntu 版本有一个比较明显的缺点，就是 Ubuntu 官方已经不再提供技术支持，目前 Ubuntu 长期支持的版本是 Ubuntu 16，在 Apollo 的官方开发计划中，也把 Ubuntu 的 Docker 镜像开发作为未来的开发计划。因此，笔者结合 Apollo 官方的 Ubuntu 16 兼容性安装说明，介绍在 Ubuntu 16 上进行 Apollo 安装，保证 the perception_lowcost_vis 正常使用的注意事项。

首先我们需要前置安装 GLFW（Graphics Library Framework，图形库框架），要求是 3.2+版本。通过安装 GLFW 能够很好地支持 Apollo 系统中各种图像相关的操作。在 GLFW 安装完成后，还需要在 glfw_fusion_viewer.cc 文件中的 GLFWFusionViewer::window_init() 方法中的 glfwCreateWindow()函数调用之前运行以下代码：

```
glfwWindowHint(GLFW_CONTEXT_CREATION_API, GLFW_EGL_CONTEXT_API);
```

这两步操作保证了，the perception_lowcost_vis 功能就能够在 Ubuntu 16 上正常运行。

参考文献

[1] Bojarski M, Del Testa D, Dworakowski D, et al. End to End Learning for Self-Driving Cars[J]. 2016.

[2] Rajamani R. Vehicle Dynamics and Control[M]// VEHICLE DYNAMICS AND CONTROL. 2006.

[3] Anderson, Kalra J M, Stanley N, et al. Autonomous Vehicle Technology[J]. Rand, 2014.

[4] Anderson J M, Kalra N, Stanley K D, et al. Autonomous Vehicle Technology: A Guide for Policymakers[M]. RAND Corporation, 2014.

[5] Anderson J M, Kalra N, Stanley K D, et al. Autonomous Vehicle Technology How to Best Realize Its Social Benefits[J]. 2014.

第10章 Cyber 基础

Apollo Cyber RT 是一个专门为自动驾驶场景开发的开源运行时框架，是 Apollo 无人驾驶系统的基石。Cyber RT 定义了明确的数据融合任务接口，建立了完整的开发工具链，以及广泛的传感器驱动支持，从而加速了自动驾驶系统的开发速度；同时，还设计了高效且灵活的消息通信，以及可配置的用户层资源优化调度实现，通过组件化设计降低了应用子系统之间的耦合性，使用户层开发部署过程进一步优化；最后，完全开源而且一致的生态、插件化的用户模块开发方式能够使用户更加容易地定制开发自动驾驶模块。更重要的是，Cyber RT 基于集中计算模型，在性能、延迟和数据吞吐量方面都进行了深度优化，完成针对车辆自动驾驶场景的高性能、高并发、低延迟和高吞吐量运行时环境。

本章主要讲解 Apollo Cyber RT 框架架构，通过实例详细分析 Cyber RT API 的用法，通过启动源码分析深入了解 Apollo 中模块化的核心设计思路。内容分为三部分：第一部分着重介绍 Apollo Cyber RT 框架架构，通过对比 ROS 系统说明 Cyber RT 的主要设计思路；第二部分通过实例详细分析 Cyber RT API 的用法，特别是基于节点的模型设计和基于组件的子系统设计规范的详细分析；第三部分通过启动源码分析深入了解 Apollo 中模块化的核心设计思路。

10.1 Cyber 简介（包括和 ROS 的对比）

10.1.1 什么是 Apollo Cyber RT

Apollo 已经从开发阶段进入产品化阶段，随着大量的实地部署，高可靠性和高性能的需求越来越突出。为此，Apollo 团队花费数年时间来构建 Apollo Cyber RT 用于满足自动驾驶领域的需求。Apollo Cyber RT 是一个专门为自动驾驶场景开发的开源运行时框架，基于集中计算模型，在性能、延迟和数据吞吐量方面都进行了深度优化，完成针对车辆自动驾驶场景的高性能、高并发、低延迟和高吞吐量运行时环境。

Cyber RT 定义了明确的数据融合任务接口，建立了完整的开发工具链，以及广泛的传感器驱动支持，从而加速了自动驾驶系统的开发速度；同时，还设计了高效且灵活的消息通信，以及可配置的用户层资源优化调度实现，通过组件化设计降低了应用子系统之间的耦合性，使用户层开发部署过程进一步优化；最后，完全开源而且一致的生态、插件化的用户模块开发方式能够使用户更加容易地定制开发自动驾驶模块。

10.1.2 ROS 系统

Apollo 最初使用的中间件是开源机器人操作系统（Robot Operating System，ROS）。下面简单总结一下 ROS 系统的优点和缺点。首先介绍优点，概括来说，ROS 系统主要包含三方面的优点：

第一，通信系统架构合理，松耦合设计使应用层开发更加便捷。ROS 是一个分布式的松耦合系统，算法模块以独立进程形式存在，也就是我们常说的 Node。ROS 基于 Socket 实现了 Pub/Sub 的通信方式，不同的算法节点（Node）之间通过 Pub/Sub 发送/接收消息。

第二，框架健全完整，开发、标定、测试工具链齐备。开发者可基于 ROS 提供的 Client Library 和通信层设计应用业务逻辑代码。一般情况下，开发者只需要关注消息处理相关的算法，至于算法何时被调用，全部由框架来处理。

第三，生态系统健全完整，经过十多年的发展，ROS 已建立起强大的生态系统，在机器人社区广受欢迎，在社区内，开发者可以很方便地寻找到很多现成的传感器驱动和算法实现等进行参考。

虽然 ROS 系统拥有上述优点，但是使用 ROS 架构自动驾驶平台产品在其开发实践中遇到了很多挑战。挑战主要集中在执行效率、协调效率和资源控制等方面。首先，ROS 中的算法模块是以独立的进程形式存在的。实际上，ROS 依赖于 Linux，没有经过优化，本身是一个通用系统，内核中的调度器对上面的算法业务逻辑并不清楚，它只是在尽量满足公平的情况下让大家都得到调度。所以，ROS Node 的运行顺序并无任何逻辑。但本质上自动驾驶是一个专用系统，任务应按照一定的业务逻辑执行。之前的改进是在 ROS 层加一个 Node，由其来同步各个算法任务的运行，这就是执行效率的瓶颈所在。其次，ROS 是一个分布式的系统，既然是分布式系统，就要有通信的开销——即使在同一个物理节点上，依然存在着通信的开销，所以 Apollo 前期曾经使用共享内存去降低 ROS 原生的基于 Socket 通信的开销。ROS 也在使用 DDS 解决通信方面的实时性，其支持 Nodelet 模式，这可以去掉进程间通信的开销，但是调度的挑战依然存在。再次，除了调度的不确定性外，ROS 系统中还存在其他很多不确定的资源控制的瓶颈，比如内存的动态申请。

Cyber 设计的主要特性从 Cyber RT 的运行流程上可以比较清楚地看出来，如图 10.1 所示。算法模块通过有向无环图（Directed Acyclic Graph，DAG）配置任务间的逻辑关系。对于每个算法，也有其优先级、运行时间、使用资源等方面的配置。系统启动时，结合 DAG、调度配置等创建相应的任务，从框架内部来讲，就是协程，调度器把任务放到各个 Processor 的队列中，然后由 Sensor 输入的数据驱动整个系统运转。

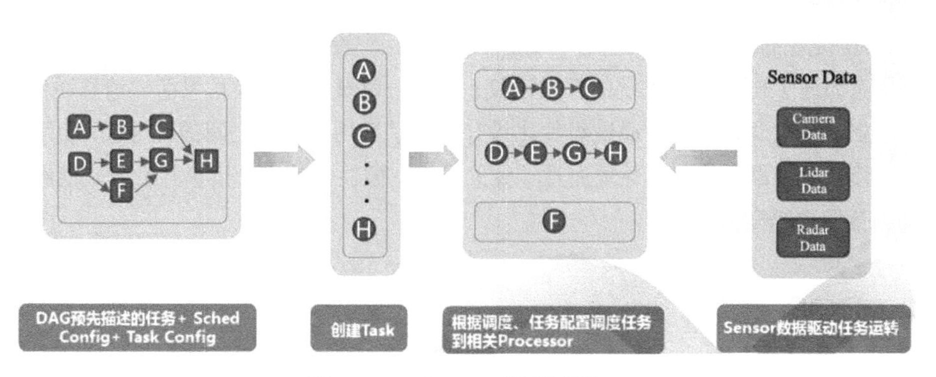

图 10.1 Cyber RT 运行流程

10.1.3 Cyber RT 的架构及核心软件模块分析

Cyber RT 的软件模块架构采用分层架构，如图 10.2 所示。分层架构是软件框架里面常见的架构，最低层是基础库，为了高效 Cyber RT 实现了自己的基础库。

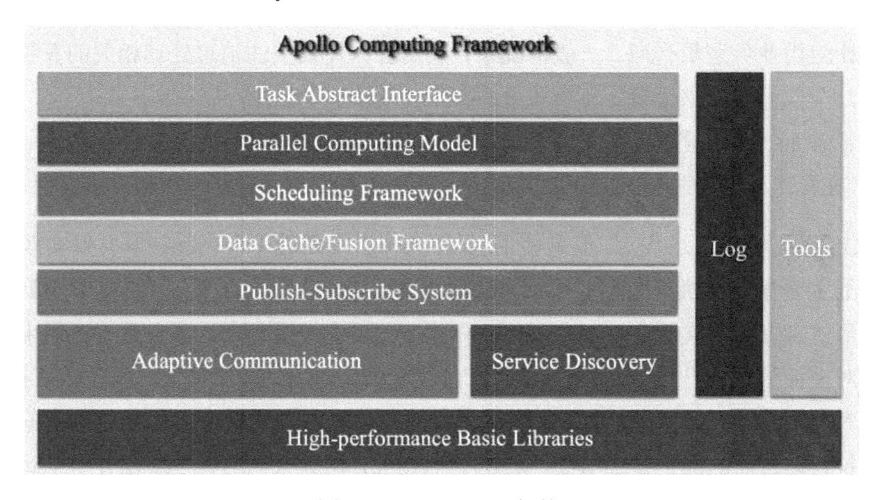

图 10.2 Cyber RT 架构

比如 Lock-Free 的对象池，实现了 Lock-Free 的队列。除了效率原因以外，也希望 Cyber RT 减少依赖。上层是通信相关的，包括服务发现，还有 Publish-Subscribe 通信机制。Cyber RT 也支持跨进程、跨机通信，上层业务逻辑无须关心，通信层会根据算法模块的部署自动选择相应的通信机制。通信层之上是数据缓存/融合层。多路传感器之间数据需要融合，而且算法可能需要缓存一定的数据。比如典型的仿真应用，不同算法模块之间需要有一个数据桥梁，数据层起到了这个模块间通信的桥梁的作用。再往上是计算模型，计算模型包括前面提到的调度和任务，计算模型之上是为开发者提供的 API 接口，比如 Cyber RT 为开发者提供了 Component 类，开发者的算法业务模块只需要继承该类，实现其中的 Proc 接口即可。该接口类似于 ROS 中的 Callback，消息通过参数的方式传递，用户只要在 Proc 中实现算法、消息处理相关的逻辑。Cyber RT 基于协程为开发者提供了并行计算相关的接口。Cyber RT 也为开发者提供了开发调试、录制回放等工具，未来还会开放性能调试工具。核心API机器实例是10.2节的主要内容。

Cyber RT的设计优势之一是将调度、任务从内核空间搬到用户空间。因为ROS的主要挑战之一是没有调度。为了解决ROS遇到的问题，Cyber RT的核心设计将调度、任务从内核空间搬到了用户空间。这一设计使得调度可以和算法业务逻辑紧密结合。从Cyber RT的角度，操作系统的Native Thread相当于物理CPU。在操作系统中，内核中的调度器负责调度任务（进程、线程等）到物理CPU上运行。而在Cyber RT中，是Cyber RT中的调度器调度协程（Coroutine）在Native Thread上有序运行。

Cyber RT的设计优势之二是出色的任务编排策略。每个Processor（Native Thread）一个任务队列，由调度器编排队列中的任务。任务在哪个CPU上运行，任务之间是否需要相邻运行，哪些先运行，哪些后运行，这些都由调度器统一调度，任务基于协程实现。在任务阻塞时，快速让出CPU。每个物理CPU上除运行一个Normal级别的Thread外，另外还运行着一个以上的高优先级的Thread。基于此，实现用户空间的高优先级的任务抢先运行。比如，之前在GPU运行的算法，在GPU上完成运行返回后，这个算法的后续工作应该尽快得到运行。这种调度策略很好地结合了业务逻辑、数据共享和算力的平衡，并且任务不会在不同CPU上随机地调度来调度去，具有非常好的Cache友好性。任务编排策略不仅需要对业务逻辑的深度理解，还需要结合计算机的算力等综合考虑。因此，Cyber RT提供类似经典线程池模式的调度算法，这种模式几乎不存在配置的代价。对此，Cyber RT做了一些改进，比如为了减小锁的瓶颈，任务是多队列的。任务队列也支持优先级，后续还会支持分组，通过组控制算法对资源的使用。

Cyber RT使用协程作为算法任务的载体。我们可以理解协程与线程的关系，就类似于线程之于物理CPU，由Cyber RT中的调度器负责在各个线程上周而复始，切换调度协程。为了算法模块在其他协处理器执行计算时可以让出Processor（Native Thread），并在完成之后回来时可以再次运行，Cyber RT采用了有状态的协程。协程除了切换非常快之外，调度的确定性是Cyber RT采用协程重要的原因。举一个典型的例子，假设用Native Thread去执行一个任务，当任务因为去GPU等加速器运算时，或者因为资源原因被阻塞时，在Thread就绪时，什么时候调度上来其实是一个非确定的过程，完全依赖于操作系统以及其上任务的情况。

除此之外，Cyber RT支持跨进程、跨机通信。在实际部署中，存在着比如某个工具需要运行在独立的进程，安全系统部署在另外的节点。因此，Cyber RT支持跨进程、跨机通信。上层业务逻辑无须关心，通信层会根据算法模块的部署自动选择相应的通信机制。

10.2 Cyber API和API Demo

本节将会介绍Cyber系统的核心API，核心API包括建立Cyber各类对象的基本操作（节点、服务等），以及各个对象之间数据通信的实现方式，Apollo 3.5利用Cyber系统API构建了全新的功能组件（Component），例如定位组件、规划组件、控制组件等。Cyber的系统核心API给各个自动驾驶子组件规范了业务标准，保持与系统的低耦合性的同时又保证了信息通信的时效性和正确率。下面我们分类介绍Cyber的系统核心API，并通过各个针对性的实例来演示使用方法。

10.2.1 Talker-Listener（简单对话系统）

本节介绍的API实例是利用Cyber RT中的Node（最小基本工作单元）、Reader（在Cyber RT系统中负责读取系统消息）、Writer（在Cyber RT系统中负责发送系统消息）三个核心概念相关联的API构建两个客户端程序，分别命名为Talker和Listener，Talker负责发送一条信息，Listener负责接收Talker发送的信息并展示。

1. 创建Node

在Cyber RT框架中，节点是最小的基础实体单元，我们可以类比生物学中的细胞，所有的基础功能都需要与已经存在的细胞相关联起来。在Cyber RT框架中，我们可以使用核心API创建一些具有特殊功能的对象，比如接下来讲到的Writer和Reader，但是能够使你的对象成功创建的前提是必须首先创建出Node的实例，而且在其他对象创建的过程中，需要明确它们与Node实例的关联关系。下面分析一下Node实例创建的API函数。

```
    std::unique_ptr<Node> apollo::cyber::CreateNode(const std::string& node_name,
const std::string& name_space = "");
    Parameters:
    node_name: name of the node, globally unique identifier
    name_space: name of the space where the node is located
    name_space is empty by default. It is the name of the space concatenated with
node_name. The format is /namespace/node_name
    Return value - An exclusive smart pointer to Node
    Error Conditions - when cyber::Init() has not called, the system is in an
uninitialized state, unable to create a node, return nullptr
```

2. 创建Writer

Writer在Cyber RT框架中的功能是发送消息。当Writer创建成功之后，每一个Writer都会在内部持有一个通信通道，通道会绑定通信消息的数据格式。实际创建Writer时，系统需要调用Node类中的CreateWriter方法，这种调用关系在成功创建Writer的同时也绑定了Node和Writer的关联关系。下面分析一下创建Writer的CreateWriter函数。

```
    template <typename MessageT>
      auto CreateWriter(const std::string& channel_name)
          -> std::shared_ptr<Writer<MessageT>>;
    template <typename MessageT>
      auto CreateWriter(const proto::RoleAttributes& role_attr)
          -> std::shared_ptr<Writer<MessageT>>;
    Parameters:
    channel_name: the name of the channel to write to
    MessageT: The type of message to be written out
    Return value - Shared pointer to the Writer object
```

3. 创建 Reader

Reader 在 Cyber RT 框架中的功能是接收对应的系统消息。创建 Reader 的时候实际上建立了其对应的回调功能实现，这个回调会监听指定的通信通道。当新的消息通过通信通道发送过来以后，这个回调会及时响应得到对应的消息数据。实际创建 Reader 时，系统需要调用 Node 类中的 CreateReader 方法，这种调用关系在成功创建 Reader 的同时也绑定了 Node 和 Reader 的关联关系。下面分析一下创建 Reader 的 CreateReader 函数。

```
    template <typename MessageT>
    auto CreateReader(const std::string& channel_name, const
std::function<void(const std::shared_ptr<MessageT>&)>& reader_func)
        -> std::shared_ptr<Reader<MessageT>>;

    template <typename MessageT>
    auto CreateReader(const ReaderConfig& config,
                      const CallbackFunc<MessageT>& reader_func = nullptr)
        -> std::shared_ptr<cyber::Reader<MessageT>>;

    template <typename MessageT>
    auto CreateReader(const proto::RoleAttributes& role_attr,
                      const CallbackFunc<MessageT>& reader_func = nullptr)
    -> std::shared_ptr<cyber::Reader<MessageT>>;
  Parameters:
  MessageT: The type of message to read
  channel_name: the name of the channel to receive from
  reader_func: callback function to process the messages
  Return value - Shared pointer to the Reader object
```

4. 实例代码

```
Talker (cyber/examples/talker.cc)
#include "cyber/cyber.h"
#include "cyber/proto/chatter.pb.h"
#include "cyber/time/rate.h"
#include "cyber/time/time.h"
using apollo::cyber::Rate;
using apollo::cyber::Time;
using apollo::cyber::proto::Chatter;
int main(int argc, char *argv[]) {
  // 初始化 Cyber RT 框架
  apollo::cyber::Init(argv[0]);
  // 创建 talker 节点
  std::shared_ptr<apollo::cyber::Node> talker_node(
      apollo::cyber::CreateNode("talker"));
  // 利用 talker 节点创建 writer 对象
```

```
  auto talker = talker_node->CreateWriter<Chatter>("channel/chatter");
  Rate rate(1.0);
  while (apollo::cyber::OK()) {// 利用循环发送消息
// 系统消息创建
    static uint64_t seq = 0;
    auto msg = std::make_shared<apollo::cyber::proto::Chatter>();
    msg->set_timestamp(Time::Now().ToNanosecond());
    msg->set_lidar_timestamp(Time::Now().ToNanosecond());
    msg->set_seq(seq++);
msg->set_content("Hello, apollo!");
// 系统消息发送
    talker->Write(msg);
    AINFO << "talker sent a message!";
    rate.Sleep();
  }
  return 0;
}
Listener (cyber/examples/listener.cc)
#include "cyber/cyber.h"
#include "cyber/proto/chatter.pb.h"
// 定义回调接口，接口接收消息数据并打印
void MessageCallback(
    const std::shared_ptr<apollo::cyber::proto::Chatter>& msg) {
  AINFO << "Received message seq-> " << msg->seq();
  AINFO << "msgcontent->" << msg->content();
}
int main(int argc, char *argv[]) {
  // 初始化 Cyber RT 框架
  apollo::cyber::Init(argv[0]);
  // 创建 listener 节点
  auto listener_node = apollo::cyber::CreateNode("listener");
  // 利用 listener 节点创建 reader 对象，MessageCallback 回调接口需要作为参数传入
CreateReader 函数中
  auto listener =
      listener_node->CreateReader<apollo::cyber::proto::Chatter>(
          "channel/chatter", MessageCallback);
  apollo::cyber::WaitForShutdown();
  return 0;
}
Bazel BUILD file(cyber/samples/BUILD)
cc_binary(
    name = "talker",
```

```
    srcs = [ "talker.cc", ],
    deps = [
        "//cyber",
        "//cyber/examples/proto:examples_cc_proto",
    ],
)

cc_binary(
    name = "listener",
    srcs = [ "listener.cc", ],
    deps = [
        "//cyber",
        "//cyber/examples/proto:examples_cc_proto",
    ],
)
Build and Run
Build（编译过程）: bazel build cyber/examples/…
```

分别在不同终端中运行 Talker 客户端和 Listener 客户端，运行命令如下：

```
./bazel-bin/cyber/examples/talker
./bazel-bin/cyber/examples/listener
```

结果检验：我们可以看到在 Listener 运行终端中不断有消息打印出来，如果关闭 Talker，打印就将停止服务的创建和输出。

10.2.2 Cyber 服务

在自动驾驶整套系统中，收发消息的需求有可能会进一步复杂化，换句话说，在很多场景下，我们的两个子系统（节点、组件等）的信息存在交互的过程，在子系统中的通信模块不止需要单一的发送和接收，而是同步多次的进行信息交流工作。因此，我们需要一种新的、有别于之前介绍的读写终端（Reader/Writer）的通信模型，这种模型在 Cyber RT 中称为服务。由于读写通信模型依赖于通信通道（Channel），因此通信是单向的。服务实现了双向通信的模型，例如一个节点通过发送请求来获得特定的应答，完成主动式的业务交互过程。下面通过实例分析 Service 模块相关的 API 函数的具体使用方法。

1. 服务实例

我们首先需要进行用户故事（User Story）的定义。在服务的实例中，我们准备创建一个能够应答驾驶信息（Driver.proto）的服务端。当然，为了使功能更完整，需要构建对应的客户端程序。当整套程序运行的时候，能够实现一种机制，当客户端发送请求到服务端以后，服务端完成请求消息的解析和计算处理，并且及时地把部分结果回传。下面分为几个部分来详细介绍这个服务实例。

第一步，定义请求和应答消息体格式。

在 Cyber RT 框架中，所有通信消息都遵循 Protobuf 的规范。Protobuf 是谷歌提出的混合语言标准，主要特点是传输速度快，对 Cyber RT 有很好的支持，在 Apollo 之前的版本中，ROS 搭配的也是 Protobuf，证明 Protobuf 十分适合自动驾驶实时系统的子系统数据调度工作。任何一个符合 Protobuf 定义的消息都能够非常容易地序列化和反序列化，因此在服务的请求和应答的消息体定义中都会遵循这一规范。我们在 examples.proto 中定义了消息名为 Driver 的消息体数据格式，具体的数据定义如下：

```
// 文件名：examples.proto
syntax = "proto2";
package apollo.cyber.examples.proto;
message Driver {
    optional string content = 1;
    optional uint64 msg_id = 2;
    optional uint64 timestamp = 3;
};
```

第二步，创建服务端和客户端。

```
// 文件名：cyber/examples/service.cc
#include "cyber/cyber.h"
#include "cyber/examples/proto/examples.pb.h"
// 指定使用定义的 Driver Protobuf 数据格式定义
using apollo::cyber::examples::proto::Driver;

int main(int argc, char* argv[]) {
// 初始化 Cyber RT 框架
  apollo::cyber::Init(argv[0]);
// 创建例子中的起始节点
  std::shared_ptr<apollo::cyber::Node> node(
    apollo::cyber::CreateNode("start_node"));
// 通过起始节点创建服务端，服务端创建调用 CreateService 方法指定名称和消息处理方法
  auto server = node->CreateService<Driver, Driver>(
    "test_server", [](const std::shared_ptr<Driver>& request,
                      std::shared_ptr<Driver>& response) {// 消息处理方法
      AINFO << "server: I am driver server";
      static uint64_t id = 0;
      ++id;
      response->set_msg_id(id);
      response->set_timestamp(0);
    });
// 通过起始节点创建客户端，调用方法是 CreateClient
  auto client = node->CreateClient<Driver, Driver>("test_server");
// 构建客户端发送的信息
  auto driver_msg = std::make_shared<Driver>();
```

```
    driver_msg->set_msg_id(0);
    driver_msg->set_timestamp(0);
  while (apollo::cyber::OK()) {// 使用 while 循环发送
  // 发送并接收服务端的返回结果
      auto res = client->SendRequest(driver_msg);
      if (res != nullptr) {
        AINFO << "client: responese: " << res->ShortDebugString();
      } else {
        AINFO << "client: service may not ready.";
      }
      sleep(1);
    }
  // 如果对话程序结束，就关闭 Cyber RT 框架
    apollo::cyber::WaitForShutdown();
    return 0;
  }
  Bazel build file
  cc_binary(
      name = "service",
      srcs = [ "service.cc", ],
      deps = [
          "//cyber",
          "//cyber/examples/proto:examples_cc_proto",
      ],
  )
  Build and run
  Build service/client（程序编译）: bazel build cyber/examples/…
  Run（程序运行）: ./bazel-bin/cyber/examples/service
```

第三步，Examining Result（结果检测）：运行成功后，我们在 Apollo 的日志文件 apollo/data/log/service.INFO 中能够看到相应的输出，可以使用 tail 命令进行持续监控。

```
  I1124 16:36:44.568845 14965 service.cc:30] [service] server: i am driver server
  I1124 16:36:44.569031 14949 service.cc:43] [service] client: responese: msg_id:
1 timestamp: 0
  I1124 16:36:45.569514 14966 service.cc:30] [service] server: i am driver server
  I1124 16:36:45.569932 14949 service.cc:43] [service] client: responese: msg_id:
2 timestamp: 0
  I1124 16:36:46.570627 14967 service.cc:30] [service] server: i am driver server
  I1124 16:36:46.571024 14949 service.cc:43] [service] client: responese: msg_id:
3 timestamp: 0
  I1124 16:36:47.571566 14968 service.cc:30] [service] server: i am driver server
  I1124 16:36:47.571962 14949 service.cc:43] [service] client: responese: msg_id:
4 timestamp: 0
```

```
I1124 16:36:48.572634 14969 service.cc:30] [service] server: i am driver server
I1124 16:36:48.573030 14949 service.cc:43] [service] client: responese: msg_id: 5 timestamp: 0
```

服务程序开发过程中需要注意：

（1）当我们使用程序注册服务的时候，服务的表示名称在系统中应该具有唯一性。

（2）当我们进行服务端和客户端系统注册时，它们依赖的节点名称也需要具有唯一性。

参数服务在 Cyber RT 框架中的作用是在 jied 之间同步数据。参数服务 API 封装了数据共享类的基本操作（set、get 和 list）。参数服务被命名为服务，它实际上是在 Cyber RT 服务功能上的封装，使用它和前面的服务一致需要创建服务端和客户端代码。

2. 参数对象

（1）支持的数据类型

所有 Cyber RT 中的参数框架都定义为 apollo::cyber::Parameter 对象，表 10.1 列出了 Cyber RT 支持的 6 种参数类型。

表 10.1　Cyber RT 支持的 6 种参数类型

Cyber RT 参数类型	对应的 C++ 数据类型	对应的 Protobuf 数据类型
apollo::cyber::proto::ParamType::INT	int64_t	int64
apollo::cyber::proto::ParamType::DOUBLE	double	double
apollo::cyber::proto::ParamType::BOOL	bool	bool
apollo::cyber::proto::ParamType::STRING	std::string	string
apollo::cyber::proto::ParamType::PROTOBUF	std::string	string
apollo::cyber::proto::ParamType::NOT_SET	-	-

除了上述 6 种参数类型外，Cyber RT 参数系统还支持使用 Protobuf 对象定义的参数。Protobuf 参数在系统程序运行时会进行序列化的过程，最终将参数对象转换成字符串类型进行传输。

（2）创建参数对象

支持的构造函数如下：

```
Parameter();  // 参数对象名称为空，类型是 NOT_SET
explicit Parameter(const Parameter& parameter);
explicit Parameter(const std::string& name);  // 类型是 NOT_SET
Parameter(const std::string& name, const bool bool_value);
Parameter(const std::string& name, const int int_value);
Parameter(const std::string& name, const int64_t int_value);
Parameter(const std::string& name, const float double_value);
Parameter(const std::string& name, const double double_value);
Parameter(const std::string& name, const std::string& string_value);
```

```
    Parameter(const std::string& name, const char* string_value);
    Parameter(const std::string& name, const std::string& msg_str,
              const std::string& full_name, const std::string& proto_desc);
    Parameter(const std::string& name, const google::protobuf::Message& msg);
```

使用参数对象的实例代码如下：

```
  Parameter a("int", 10);
  Parameter b("bool", true);
  Parameter c("double", 0.1);
  Parameter d("string", "cyber");
  Parameter e("string", std::string("cyber"));
  // Protobuf 类型的参数转换实例
  Chatter chatter;
  Parameter f("chatter", chatter);
  std::string msg_str("");
  chatter.SerializeToString(&msg_str);
  std::string msg_desc("");
  ProtobufFactory::GetDescriptorString(chatter, &msg_desc);
  Parameter g("chatter", msg_str, Chatter::descriptor()->full_name(),
msg_desc);
```

（3）接口和数据读取

接口列表如下：

```
    inline ParamType type() const;
    inline std::string TypeName() const;
    inline std::string Descriptor() const;
    inline const std::string Name() const;
    inline bool AsBool() const;
    inline int64_t AsInt64() const;
    inline double AsDouble() const;
    inline const std::string AsString() const;
    std::string DebugString() const;
    template <typename Type>
    typename std::enable_if<std::is_base_of<google::protobuf::Message,
Type>::value, Type>::type
    value() const;
    template <typename Type>
    typename std::enable_if<std::is_integral<Type>::value && !std::is_same<Type,
bool>::value, Type>::type
    value() const;
    template <typename Type>
    typename std::enable_if<std::is_floating_point<Type>::value, Type>::type
    value() const;
```

```
    template <typename Type>
    typename std::enable_if<std::is_convertible<Type, std::string>::value,
const std::string&>::type
    value() const;
    template <typename Type>
    typename std::enable_if<std::is_same<Type, bool>::value, bool>::type
    value() const;
```

参数接口使用实例如下：

```
    Parameter a("int", 10);
    a.Name();  // return int
    a.Type();  // return apollo::cyber::proto::ParamType::INT
    a.TypeName();  // return string: INT
    a.DebugString();  // return string: {name: "int", type: "INT", value: 10}
    int x = a.AsInt64();  // x = 10
    x = a.value<int64_t>();  // x = 10
    x = a.AsString();  // 没有定义的行为，系统报错
    f.TypeName();  // return string: chatter
    auto chatter = f.value<Chatter>();
```

3. 参数服务

如果某一个节点需要提供参数服务的目的是向其他节点传递 Cyber RT 参数，就需要使用 ParameterService 函数接口创建 Parameter Service，代码如下：

```
  /**
    * @brief Construct a new ParameterService object
    *
    * @param node shared_ptr of the node handler
    */
   explicit ParameterService(const std::shared_ptr<Node>& node);
```

由于所有创建的参数都存储在我们创建的参数服务对象中，因此参数可以通过参数服务 API 之间进行操作，而不需要像前面介绍的一般 Cyber RT 服务需要服务请求的介入。

下面是参数设定的一个简单代码示例。

```
   /**
   * @brief Set the Parameter object
   *
   * @param parameter parameter to be set
   */
   void SetParameter(const Parameter& parameter);
  Getting parameters:
   /**
    * @brief Get the Parameter object
```

```
   *
   * @param param_name
   * @param parameter the pointer to store
   * @return true
   * @return false call service fail or timeout
   */
  bool GetParameter(const std::string& param_name, Parameter* parameter);
 Getting the list of parameters:
  /**
   * @brief Get all the Parameter objects
   *
   * @param parameters pointer of vector to store all the parameters
   * @return true
   * @return false call service fail or timeout
   */
  bool ListParameters(std::vector<Parameter>* parameters);
```

4. 参数客户端

如果某一个节点获取参数服务的目的是与其他节点交互 Cyber RT 参数，就需要使用 ParameterClient 函数接口创建 Parameter Client，代码如下：

```
  /**
   * @brief Construct a new ParameterClient object
   *
   * @param node shared_ptr of the node handler
   * @param service_node_name node name which provide a param services
   */
  ParameterClient(const std::shared_ptr<Node>& node, const std::string&
service_node_name);
```

参数客户端包含参数的基础操作 SetParameter、GetParameter 和 ListParameters，这些操作的使用方式和参数服务中是完全一致的。

参数客户端使用示例如下：

```
#include "cyber/cyber.h"
#include "cyber/parameter/parameter_client.h"
#include "cyber/parameter/parameter_server.h"


using apollo::cyber::Parameter;
using apollo::cyber::ParameterServer;
using apollo::cyber::ParameterClient;


int main(int argc, char** argv) {
```

```
// 初始化 Cyber RT 框架
  apollo::cyber::Init(*argv);
  // 创建例子中的参数节点
  std::shared_ptr<apollo::cyber::Node> node =
      apollo::cyber::CreateNode("parameter")
// 创建参数服务端
  auto param_server = std::make_shared<ParameterServer>(node);
// 创建参数客户端
  auto param_client = std::make_shared<ParameterClient>(node, "parameter");
  // 使用服务端进行参数传递
  param_server->SetParameter(Parameter("int", 1));
  Parameter parameter;
  param_server->GetParameter("int", &parameter);
  AINFO << "int: " << parameter.AsInt64();
// 使用客户端进行参数传递
  param_client->SetParameter(Parameter("string", "test"));
  param_client->GetParameter("string", &parameter);
  AINFO << "string: " << parameter.AsString();
  param_client->GetParameter("int", &parameter);
  AINFO << "int: " << parameter.AsInt64();
  return 0;
}
Build and run
```

- Build（编译），使用如下命令：

```
bazel build cyber/examples/…
```

- Run（运行），使用如下命令：

```
./bazel-bin/cyber/examples/paramserver
```

最后，进行 Result Examine（结果检测），在 log 系统中能够看到对应的参数打印。

10.2.3 日志类库

Cyber RT 框架的日志类库是在 Glog 日志类库上的再封装，Glog 是 Google 的开源日志系统，相比较 Log 4 系列的日志系统，它更加轻巧灵活，而且功能也比较完善。如果要使用 Glog 日志系统需要引入相应的日志依赖库和 Cyber 的基础依赖库。

```
#include "cyber/common/log.h"
#include "cyber/init.h"
```

1. 日志使用配置

默认全局配置路径：cyber/setup.bash。

日志配置能够通过命令行终端进行修改，相关命令如下：

```
export GLOG_log_dir=/apollo/data/log
export GLOG_alsologtostderr=0
export GLOG_colorlogtostderr=1
export GLOG_minloglevel=0
```

2. Cyber RT 日志系统初始化

初始化方法十分简洁，只需要在程序入口的位置调用一次 Cyber RT 的系统初始化方法 Init，Log 系统的初始化操作已经友好地整合在了系统初始化代码中：

```
apollo::cyber::cyber::Init(argv[0]) is initialized.
If no macro definition is made in the previous component, the corresponding log
is printed to the binary log.
```

3. 日志输出宏定义

Cyber RT 系统的日志模块中定义了一组宏命令来描述不同级别的输出，通过宏定义能够更加简洁地进行日志输出控制，相关的宏命令如下：

```
ADEBUG << "hello cyber.";
AINFO  << "hello cyber.";
AWARN  << "hello cyber.";
AERROR << "hello cyber.";
AFATAL << "hello cyber.";
```

4. 日志格式

Cyber RT 系统日志格式：

```
<MODULE_NAME>.log.<LOG_LEVEL>.<datetime>.<process_id>
```

5. 日志文件

目前封装后的 Cyber RT 日志模块和原声的 Glog 模块在日志文件保存方面有一点主要的不同：Cyber RT 日志系统产生的各种级别的日志会按照模块保存到一个模块日志文件中，而 Glog 默认分成不同级别的 Log 文件。

10.2.4　Cyber 模组

使用 Component 创建一个子模组是 Cyber RT 中最重要的应用开发方式。下面来领略一下 Cyber RT 的模组开发风采。首先需要学习 Cyber RT 模组中的核心要点。

1. Component 概述

Component 是 Cyber RT 系统中用来创建实现各种应用组件的基石（基础类）。每一个应用组件都需要继承 Component 类，然后定义实现组件自己的 Init 和 Proc 方法，当完成上述代码开发后，这个应用组件就成为一个合格的 Cyber RT 组件（Apollo 组件）了，我们能够通过配置和框架 API 进行组件的调用。

2. 基于二进制的 Cyber RT 应用和基于 Component 的 Cyber RT 应用的区别

Cyber RT 系统支持两种应用格式：第一类是基于二进制的，前面开发的 Reader/Writer、服务等几个实例使用的就是这类应用格式，这类应用最终编译结果是二进制文件，应用之间的通信需要实现当前应用自有的 Reader 和 Writer，而且需要在调用的地方显示调用应用功能；另一类是前面介绍的基于 Component 的应用，这类应用的编译结果是产生一个共享库，通过继承 Component 类并且写入相关联的 Dag 描述文件，Cyber RT 框架能够动态载入应用模块并进行运行。

在二进制类应用代码层面，我们能够清楚地看到节点建立、关系绑定等基础框架流程代码，在组件代码层，这些基本操作已被固化和封装，我们只需要实现固定接口函数和确定配置文件就能够达到完成应用开发的目的。

3. 主要的组件接口

在 Component 中，Init 接口方法是组件的初始化入口，一般我们在 Init 接口方法中定义组件初始化相关联的各个节点及模块的初始化工作。

在 Component 中，Proc 接口方法可以理解为模块的监听程序，当有系统消息到达时，组件会调用 Proc 接口方法进行处理和应答。

4. 组件开发的优势

组件开发的优势如下：

- 通过配置启动文件，组件能够灵活地组合于不同的系统执行流程中，并且发布过程管理十分灵活。
- 可以通过改变 Dag 文件中相应的信息通道名称来改变组件监听的信息类型，重要的是这个过程不需要再次编译组件程序。
- 组件支持接收不同类型的数据。
- 组件支持提供多种融合应用策略。

5. Dag 文件类型

Dag 文件实例如下：

```
# Define all coms in DAG streaming.
  module_config {
  module_library : "lib/libperception_component.so"
  components {
      class_name : "PerceptionComponent"
      config {
          name : "perception"
          readers {
              channel: "perception/channel_name"
```

```
        }
      }
    }
    timer_components {
      class_name : "DriverComponent"
      config {
        name : "driver"
        interval : 100
      }
    }
    }
```

Dag 文件的配置项如下。

- module_library：此项配置指定了组件编译共享文件的位置，相对于 Cyber RT 系统的工作根目录（工作根目录是指 setup.bash 文件所在的目录）。
- components & timer_component：指定需要载入组件的基础类型。
- class_name：载入组件的组件名称。
- name：系统中 Cyber RT 应用的名称。
- readers：指定从其他组件进行数据接收的相关配置，支持最多 3 个不同的数据通道。

6. 组件实例代码

common_component_example（cyber/examples/common_component_example/*），头文件定义（common_component_example.h）如下：

```
#include <memory>

#include "cyber/class_loader/class_loader.h"
#include "cyber/component/component.h"
#include "cyber/examples/proto/examples.pb.h"

using apollo::cyber::examples::proto::Driver;
using apollo::cyber::Component;
using apollo::cyber::ComponentBase;

class Commontestcomponent : public Component<Driver, Driver> {
 public: //标准接口函数声明
  bool Init() override;
  bool Proc(const std::shared_ptr<Driver>& msg0,
          const std::shared_ptr<Driver>& msg1) override;
};
```

通过宏注册组件 CYBER_REGISTER_COMPONENT（Commontestcomponent），Cpp 文件实现（common_component_example.cc）如下：

```
  #include
"cyber/examples/common_component_smaple/common_component_example.h"

  #include "cyber/class_loader/class_loader.h"
  #include "cyber/component/component.h"
  //初始化逻辑
  bool Commontestcomponent::Init() {
    AINFO << "Commontest component init";
    return true;
  }
  //应用通信回调逻辑
  bool Commontestcomponent::Proc(const std::shared_ptr<Driver>& msg0,
                                 const std::shared_ptr<Driver>& msg1) {
    AINFO << "Start commontest component Proc [" << msg0->msg_id() << "] ["
          << msg1->msg_id() << "]";
    return true;
  }
```

timer_component_example（cyber/examples/timer_component_example/*），头文件定义（timer_component_example.h）如下：

```
  #include <memory>

  #include "cyber/class_loader/class_loader.h"
  #include "cyber/component/component.h"
  #include "cyber/component/timer_component.h"
  #include "cyber/examples/proto/examples.pb.h"

  using apollo::cyber::examples::proto::Driver;
  using apollo::cyber::Component;
  using apollo::cyber::ComponentBase;
  using apollo::cyber::TimerComponent;
  using apollo::cyber::Writer;

  class TimertestComponent : public TimerComponent {
   public: //标准接口函数声明
    bool Init() override;
    bool Proc() override;

   private:
    std::shared_ptr<Writer<Driver>> driver_writer_ = nullptr;
  };
```

通过宏注册组件 CYBER_REGISTER_COMPONENT（TimertestComponent），Cpp 文件实现

（timer_component_example.cc）如下：

```
#include "cyber/examples/timer_component_example/timer_component_example.h"

#include "cyber/class_loader/class_loader.h"
#include "cyber/component/component.h"
#include "cyber/examples/proto/examples.pb.h"
//初始化逻辑
bool TimertestComponent::Init() {
//组件对应 writer 创建
  driver_writer_ = node_->CreateWriter<Driver>("/carstatus/channel");
  return true;
}
//应用通信回调逻辑
bool TimertestComponent::Proc() {
  //应答消息创建
  static int i = 0;
  auto out_msg = std::make_shared<Driver>();
  out_msg->set_msg_id(i++);
//发送应答消息
  driver_writer_->Write(out_msg);
  AINFO << "timertestcomponent: Write drivermsg->"
        << out_msg->ShortDebugString();
  return true;
}
```

最后运行 timertestcomponent 进行演示。

- Build（编译）：命令为 bazel build cyber/examples/timer_component_smaple/…。
- Run（运行）：命令为 mainboard -d cyber/examples/timer_component_smaple/timer.dag。

7. 应用开发注意点

组件载入类库需要在程序中显式注册，注册的宏定义如下：

```
CYBER_REGISTER_CLASS(DriverComponent)
```

如果在注册代码中使用命名空间，那么这个命名空间需要在 Dag 配置文件中被同步声明。

Component 应用组件和 TimerComponent 应用组件的 Dag 配置文件并不一致，上面的例子已经清晰地展示了具体区别，开发时需要注意。

8. 启动模块

cyber_launch 是 Cyber RT 框架中的启动器模块。启动过程可以简单地总结为首先根据启动配置文件（Launch）开启启动过程，这个过程是比较宏观的，具体各个模块的启动会根据启动配置文件中指定的 Dag 文件来执行各个模块的具体启动细节，这样的好处是可以通过配置启动产生不同的主应用程序，而且各个应用程序中的具体模块也可以通过配置来进行区分。cyber_launch 支持两种启动场景动态载入 Component 组件和在应用的子进程中直接启动二进制应用。

启动文件实例如下：

```
<cyber>
    <module>
        <name>driver</name>
        <dag_conf>driver.dag</dag_conf>
        <process_name></process_name>
        <exception_handler>exit</exception_handler>
    </module>
    <module>
        <name>perception</name>
        <dag_conf>perception.dag</dag_conf>
        <process_name></process_name>
        <exception_handler>respawn</exception_handler>
    </module>
    <module>
        <name>planning</name>
        <dag_conf>planning.dag</dag_conf>
        <process_name></process_name>
    </module>
</cyber>
```

- Module：每一个载入的组件或二进制文件称为一个模组。
- name：指的是模组名称。
- dag_conf：组件启动需要的 dag 文件的名称信息。
- process_name：应用启动后的模组在进程中的名字。
- exception_handler：当前模组在启动过程中遇到异常所指定采用的处理策略。这一配置项有两个可以选择的值：exit 和 respawn。exit 意味着当异常出现时系统会终止这个组件进程；respawn 指的是系统会在退出后重启该模组进程，但是重启过程可能会由于系统无法定位模组信息而跳过。用户也可以根据进程启动的具体情况进行控制。

10.2.5 Timer 计时器

计时器模块的主要功能是创建定时任务，定时任务一般有两种：定时运行一次和按照时间周期循环运行。

1. Timer 定时器接口

```
/**
  * @brief Construct a new Timer object
  *
```

```
  * @param period The period of the timer, unit is ms
  * @param callback The tasks that the timer needs to perform
  * @param oneshot True: perform the callback only after the first timing cycle
  *               False: perform the callback every timed period
  */
 Timer(uint32_t period, std::function<void()> callback, bool oneshot);
Or you could encapsulate the parameters into a timer option as follows:
struct TimerOption {
 uint32_t period;                  // The period of the timer, unit is ms
 std::function<void()> callback;  // The tasks that the timer needs to perform
 bool oneshot;  // True: perform the callback only after the first timing cycle
                // False: perform the callback every timed period
};
/**
  * @brief Construct a new Timer object
  *
  * @param opt Timer option
  */
 explicit Timer(TimerOption opt);
```

2. 启动定时器

在定时器实例创建后，需要显式调用 Timer::Start()来启动定时器。

3. 终止定时器

我们启动的定时器是可以通过代码手动终止的，方法是显式调用 Timer::Stop()接口函数。

4. Timer 定时器实例

```
#include <iostream>
#include "cyber/cyber.h"
int main(int argc, char** argv) {
    cyber::Init(argv[0]);
    // 以 10Hz 的频率打印时间
    cyber::Timer timer(100, [](){
        std::cout << cyber::Time::Now() << std::endl;
    }, false);
    timer.Start()
    sleep(1);
    timer.Stop();
}
```

10.2.6 时间（Time）类

Cyber RT 框架中的时间管理 API 需要使用 Time 类，Time 类的功能包括时间显示、时间运算、不同时间显示方式直接进行时间转换等。下面列出常用的 API 接口函数：

```
// 不同的时间实例构造方法
Time(uint64_t nanoseconds); //uint64_t 格式，nanoseconds 单位
Time(int nanoseconds); // int 格式，nanoseconds 单位
Time(double seconds); // double 格式，seconds 单位
Time(uint32_t seconds, uint32_t nanoseconds); // 同时指定秒和纳秒数据来构造实例
Static Time Now(); // 得到当前时间
Double ToSecond() const;//时间转换成秒来表示
Uint64_t ToNanosecond()const;//时间转换成纳秒来表示
Std::string ToString()const;//时间转换成字符串来表示，例如"2018-07-10 20:21:51.123456789"
Bool IsZero() const; // 判断时间是否为零
```

时间类实例如下：

```
#include <iostream>
#include "cyber/cyber.h"
#include "cyber/duration.h"
int main(int argc, char** argv) {
    cyber::Init(argv[0]);
    Time t1(1531225311123456789UL);
    std::cout << t1.ToString() std::endl; // 2018-07-10 20:21:51.123456789
    // 设置时间间隔
    Time t1(100);
    Duration d(200);
    Time t2(300);
    assert(d == (t1-t2)); // True
}
```

10.2.7 Apollo 记录文件的读写操作

RecordReader 是 Cyber RT 框架中用于对记录文件进行读操作的核心 API 类。每一个 RecordReader 都能够通过 open 方法打开对应的一个在系统中存在的记录文件，并且同时建立一个线程进行同步读操作。用户只需要执行 ReadMessage 就能够很容易地读出记录文件的最后信息。读出的信息 Cyber RT 进行了对象化封装，我们需要进一步调用方法 GetCurrentMessageChannelName、GetCurrentRawMessage、GetCurrentMessageTime 来得到信息的名称、内容和时间戳。

RecordWriter 是 Cyber RT 框架中用于对记录文件进行写操作的核心 API 类。每一个 RecordWriter 都能够通过 open 方法创建一个指定名称的记录文件。用户只需要执行 WriteMessage 和 WriteChannel 方法就能够同步写入信息内容和通信通道信息。

记录文件的读写操作实例（cyber/examples/record.cc）如下。本例首先通过 test_write 方法写 100 条信息到 TEST_FILE 文件中，然后通过 test_read 方法逐一读出。

```
#include <string>
```

```
#include "cyber/cyber.h"
#include "cyber/message/raw_message.h"
#include "cyber/proto/record.pb.h"
#include "cyber/record/record_message.h"
#include "cyber/record/record_reader.h"
#include "cyber/record/record_writer.h"

using ::apollo::cyber::record::RecordReader;
using ::apollo::cyber::record::RecordWriter;
using ::apollo::cyber::record::RecordMessage;
using apollo::cyber::message::RawMessage;
//记录文件，消息文件常量信息
const char CHANNEL_NAME_1[] = "/test/channel1";
const char CHANNEL_NAME_2[] = "/test/channel2";
const char MESSAGE_TYPE_1[] = "apollo.cyber.proto.Test";
const char MESSAGE_TYPE_2[] = "apollo.cyber.proto.Channel";
const char PROTO_DESC[] = "1234567890";
const char STR_10B[] = "1234567890";
const char TEST_FILE[] = "test.record";

void test_write(const std::string &writefile) {
  RecordWriter writer;
  writer.SetSizeOfFileSegmentation(0);
  writer.SetIntervalOfFileSegmentation(0);
  //创建记录文件
  writer.Open(writefile);
  //创建写通道信息
  writer.WriteChannel(CHANNEL_NAME_1, MESSAGE_TYPE_1, PROTO_DESC);
  //循环写入100条信息
  for (uint32_t i = 0; i < 100; ++i) {
    auto msg = std::make_shared<RawMessage>("abc" + std::to_string(i));
    writer.WriteMessage(CHANNEL_NAME_1, msg, 888 + i);
  }
 //关闭记录文件写对象
  writer.Close();
}

void test_read(const std::string &readfile) {
  RecordReader reader(readfile);
  RecordMessage message;
  uint64_t msg_count = reader.GetMessageNumber(CHANNEL_NAME_1);
  AINFO << "MSGTYPE: " << reader.GetMessageType(CHANNEL_NAME_1);
```

```
    AINFO << "MSGDESC: " << reader.GetProtoDesc(CHANNEL_NAME_1);

    // 读出所有记录文件上的消息
    uint64_t i = 0;
    uint64_t valid = 0;
    for (i = 0; i < msg_count; ++i) {
      if (reader.ReadMessage(&message)) {
        AINFO << "msg[" << i << "]-> "
              << "channel name: " << message.channel_name
              << "; content: " << message.content
              << "; msg time: " << message.time;
        valid++;
      } else {
        AERROR << "read msg[" << i << "] failed";
      }
    }
    AINFO << "static msg=================";
    AINFO << "MSG validmsg:totalcount: " << valid << ":" << msg_count;
  }
  //实例入口方法
  int main(int argc, char *argv[]) {
    //Cyber 初始化
    apollo::cyber::Init(argv[0]);
  //读操作
    test_write(TEST_FILE);
  //系统等时操作保证读操作完成
    sleep(1);
  //写操作
    test_read(TEST_FILE);
    return 0;
  }
  Build and run
  Build（编译）: bazel build cyber/examples/…
  Run（运行）: ./bazel-bin/cyber/examples/record
  Examining result（结果检测）:
  I1124 16:56:27.248200 15118 record.cc:64] [record] msg[0]-> channel name:
/test/channel1; content: abc0; msg time: 888
  I1124 16:56:27.248227 15118 record.cc:64] [record] msg[1]-> channel name:
/test/channel1; content: abc1; msg time: 889
  I1124 16:56:27.248239 15118 record.cc:64] [record] msg[2]-> channel name:
/test/channel1; content: abc2; msg time: 890
  I1124 16:56:27.248252 15118 record.cc:64] [record] msg[3]-> channel name:
/test/channel1; content: abc3; msg time: 891
```

```
I1124 16:56:27.248297 15118 record.cc:64] [record] msg[4]-> channel name:
/test/channel1; content: abc4; msg time: 892
I1124 16:56:27.248378 15118 record.cc:64] [record] msg[5]-> channel name:
/test/channel1; content: abc5; msg time: 893
...
I1124 16:56:27.250422 15118 record.cc:73] [record] static msg=================
I1124 16:56:27.250434 15118 record.cc:74] [record] MSG validmsg:totalcount:
100:100
```

10.3 Apollo 模块启动源码分析

10.3.1 Apollo 模块启动流程

Apollo 启动过程很好地体现了高内聚松耦合的系统设计，我们先从启动脚本文件 scripts/bootstrap.sh 开始剖析。服务启动命令 bash scripts/bootstrap.sh start 实际上执行了 scripts/bootstrap.sh 脚本中的 start 函数，如下所示：

```
function start() {
    ./scripts/monitor.sh start
    ./scripts/dreamview.sh start
    if [ $? -eq 0 ]; then
        sleep 2  # wait for some time before starting to check
        http_status="$(curl -o /dev/null -I -L -s -w '%{http_code}'
${DREAMVIEW_URL})"
        if [ $http_status -eq 200 ]; then
            echo "Dreamview is running at" $DREAMVIEW_URL
        else
            echo "Failed to start Dreamview. Please check /apollo/data/log or
/apollo/data/core for more information"
        fi
    fi
}
```

start 函数内部分别调用脚本文件 scripts/monitor.sh 与 scripts/dreamview.sh 内部的 start 函数启动 monitor 与 dreamview 模块。

monitor 模块的启动过程暂且放在一边，下面从 dreamview 模块的 start 函数入手，分析一下 dreamview 的启动流程。scripts/dreamview.sh 文件的内容如下：

```
DIR="$( cd "$( dirname "${BASH_SOURCE[0]}" )" && pwd )"

cd "${DIR}/.."
```

```
source "${DIR}/apollo_base.sh"

# run function from apollo_base.sh
# run command_name module_name
run dreamview "$@"
```

里面压根没有 start 函数，但我们找到一个 apollo_base.sh 脚本文件，并且有一条调用语句：run dreamview "$@"（展开以后就是 run dreamview start）。

我们有理由判断，run 函数存在于 apollo_base.sh 脚本文件中，现在到里面一探究竟。不出意外，果然有一个 run 函数：

```
function run() {
  local module=$1
  shift
  run_customized_path $module $module "$@"
}
```

上述代码中，module 的值为 dreamview，$@的值为 start，因此后面继续调用 run_customized_path dreamview dreamview start。继续顺藤摸瓜，查看 run_customized_path 函数：

```
 function run_customized_path() {
  local module_path=$1
  local module=$2
  local cmd=$3
  shift 3
  case $cmd in
    start)
      start_customized_path $module_path $module "$@"
      ;;
    start_fe)
      start_fe_customized_path $module_path $module "$@"
      ;;
    start_gdb)
      start_gdb_customized_path $module_path $module "$@"
      ;;
    start_prof)
      start_prof_customized_path $module_path $module "$@"
      ;;
    stop)
      stop_customized_path $module_path $module
      ;;
    help)
      help
      ;;架构概述
```

```
    *)
      start_customized_path $module_path $module $cmd "$@"
    ;;
  esac
}实际调用的是 start_customized_path dreamview dreamview。
```

再来查看 start_customized_path 函数：

```
function start_customized_path() {
  MODULE_PATH=$1
  MODULE=$2
  shift 2

  is_stopped_customized_path "${MODULE_PATH}" "${MODULE}"
  if [ $? -eq 1 ]; then
    eval "nohup cyber_launch start
/apollo/modules/${MODULE_PATH}/launch/${MODULE}.launch &"
    sleep 0.5
    is_stopped_customized_path "${MODULE_PATH}" "${MODULE}"
    if [ $? -eq 0 ]; then
      echo "Launched module ${MODULE}."
      return 0
    else
      echo "Could not launch module ${MODULE}. Is it already built?"
      return 1
    fi
  else
    echo "Module ${MODULE} is already running - skipping."
    return 2
  fi
}
```

在 start_customized_path 函数内部，首先调用 is_stopped_customized_path 函数来判断（在内部实际上通过指令$(pgrep -c -f "modules/dreamview/launch/dreamview.launch")来判断）dreamview 模块是否已启动。

若该模块未启动，则使用指令：

```
nohup cyber_launch start/apollo/modules/dreamview/launch/dreamview.launch &
# 以非挂断方式启动后台进程模块 dreamview
```

cyber_launch 是 Cyber 平台提供的一个 Python 工具程序，完整路径为：

```
${APOLLO_HOME}/cyber/tools/cyber_launch/cyber_launch
```

可通过 sudo find / -name cyber_launch 查找，${APOLLO_HOME}表示 Apollo 项目的根目录。下面继续研究 cyber_launch 中的 main 函数：

```
def main():
    """
    Main function
    """
    if cyber_path is None:
        logger.error(
            'Error: environment variable CYBER_PATH not found, set environment
first.')
        sys.exit(1)
    os.chdir(cyber_path)
    parser = argparse.ArgumentParser(description='cyber launcher')
    subparsers = parser.add_subparsers(help='sub-command help')

    start_parser = subparsers.add_parser(
        'start', help='launch/benchmark.launch')
    start_parser.add_argument('file', nargs='?', action='store',
                              help='launch file, default is cyber.launch')

    stop_parser = subparsers.add_parser(
        'stop', help='stop all the module in launch file')
    stop_parser.add_argument('file', nargs='?', action='store',
                             help='launch file, default stop all the launcher')

    # restart_parser = subparsers.add_parser('restart', help='restart the
module')
    # restart_parser.add_argument('file', nargs='?', action='store',
help='launch file,
    #                             default is cyber.launch')

    params = parser.parse_args(sys.argv[1:])

    command = sys.argv[1]
    if command == 'start':
        start(params.file)
    elif command == 'stop':
        stop_launch(params.file)
    # elif command == 'restart':
    #    restart(params.file)
    else:
        logger.error('Invalid command %s' % command)
        sys.exit(1)
```

该函数无非进行一些命令行参数解析，然后调用

start(/apollo/modules/dreamview/launch/dreamview.launch)函数启动 dreamview 模块。

继续查看 start 函数，该函数的内容很长，不再详细解释，其主要功能是解析 XML 文件 /apollo/modules/dreamview/launch/dreamview.launch 中的各项元素：name、dag_conf、type、process_name、exception_handler。

其值分别为：dreamview、null、binary、/apollo/bazel-bin/modules/dreamview/dreamview --flagfile=/apollo/modules/common/data/global_flagfile.txt、respawn，首先调用 ProcessWrapper(process_name.split()[0], 0, [""], process_name, process_type,exception_handler)创建一个 ProcessWrapper 对象 pw，然后调用 pw.start()函数启动 dreamview 模块：

```
def start(launch_file=''):
    """
    Start all modules in xml config
    """
    pmon = ProcessMonitor()
    # Find launch file
    if launch_file[0] == '/':
        launch_file = launch_file
    elif launch_file == os.path.basename(launch_file):
        launch_file = os.path.join(cyber_path, 'launch', launch_file)
    else:
        if os.path.exists(os.path.join(g_pwd, launch_file)):
            launch_file = os.path.join(g_pwd, launch_file)
        else:
            logger.error('Cannot find launch file: %s ' % launch_file)
            sys.exit(1)
    logger.info('Launch file [%s]' % launch_file)
    logger.info('=' * 120)

    if not os.path.isfile(launch_file):
        logger.error('Launch xml file %s does not exist' % launch_file)
        sys.exit(1)

    try:
        tree = ET.parse(launch_file)
    except Exception:
        logger.error('Parse xml failed. illegal xml!')
        sys.exit(1)
    total_dag_num = 0
    dictionary = {}
    dag_dict = {}
    root1 = tree.getroot()
    for module in root1.findall('module'):
```

```
        dag_conf = module.find('dag_conf').text
        process_name = module.find('process_name').text
        process_type = module.find('type')
        if process_type is None:
            process_type = 'library'
        else:
            process_type = process_type.text
            if process_type is None:
                process_type = 'library'
            process_type = process_type.strip()
        if process_type != 'binary':
            if dag_conf is None or not dag_conf.strip():
                logger.error('Library dag conf is null')
                continue
            if process_name is None:
                process_name = 'mainboard_default_' + str(os.getpid())
            process_name = process_name.strip()
            if dictionary.has_key(str(process_name)):
                dictionary[str(process_name)] += 1
            else:
                dictionary[str(process_name)] = 1
            if not dag_dict.has_key(str(process_name)):
                dag_dict[str(process_name)] = [str(dag_conf)]
            else:
                dag_dict[str(process_name)].append(str(dag_conf))
            if dag_conf is not None:
                total_dag_num += 1

    process_list = []
    root = tree.getroot()
    for module in root.findall('module'):
        module_name = module.find('name').text
        dag_conf = module.find('dag_conf').text
        process_name = module.find('process_name').text
        sched_name = module.find('sched_name')
        process_type = module.find('type')
        exception_handler = module.find('exception_handler')
        if process_type is None:
            process_type = 'library'
        else:
            process_type = process_type.text
            if process_type is None:
                process_type = 'library'
```

```
            process_type = process_type.strip()

        if sched_name is None:
            sched_name = "CYBER_DEFAULT"
        else:
            sched_name = sched_name.text

        if process_name is None:
            process_name = 'mainboard_default_' + str(os.getpid())
        if dag_conf is None:
            dag_conf = ''
        if module_name is None:
            module_name = ''
        if exception_handler is None:
            exception_handler = ''
        else:
            exception_handler = exception_handler.text
        module_name = module_name.strip()
        dag_conf = dag_conf.strip()
        process_name = process_name.strip()
        sched_name = sched_name.strip()
        exception_handler = exception_handler.strip()

        logger.info('Load module [%s] %s: [%s] [%s] conf: [%s] exception_handler:
[%s]' %
                    (module_name, process_type, process_name, sched_name,
dag_conf,
                     exception_handler))

        if process_name not in process_list:
            if process_type == 'binary':
                if len(process_name) == 0:
                    logger.error(
                        'Start binary failed. Binary process_name is null.')
                    continue
                pw = ProcessWrapper(
                    process_name.split()[0], 0, [
                        ""], process_name, process_type,
                    exception_handler)
            # Default is library
            else:
                pw = ProcessWrapper(
                    g_binary_name, 0, dag_dict[
```

```
                    str(process_name)], process_name,
                process_type, sched_name, exception_handler)
            result = pw.start()
            if result != 0:
                logger.error(
                    'Start manager [%s] failed. Stop all!' % process_name)
                stop()
            pmon.register(pw)
            process_list.append(process_name)

    # No module in xml
    if not process_list:
        logger.error("No module was found in xml config.")
        return
    all_died = pmon.run()
    if not all_died:
        logger.info("Stop all processes...")
        stop()
    logger.info("Cyber exit.")
```

下面查看 ProcessWrapper 类里的 start 函数：

```
def start(self):
        """
        Start a manager in process name
        """
        if self.process_type == 'binary':
            args_list = self.name.split()
        else:
            args_list = [self.binary_path, '-d'] + self.dag_list
            if len(self.name) != 0:
                args_list.append('-p')
                args_list.append(self.name)
            if len(self.sched_name) != 0:
                args_list.append('-s')
                args_list.append(self.sched_name)

        self.args = args_list

        try:
            self.popen = subprocess.Popen(args_list, stdout=subprocess.PIPE,
                                          stderr=subprocess.STDOUT)
        except Exception as err:
            logger.error('Subprocess Popen exception: ' + str(err))
```

```
            return 2
        else:
            if self.popen.pid == 0 or self.popen.returncode is not None:
                logger.error('Start process [%s] failed.' % self.name)
                return 2

        th = threading.Thread(target=module_monitor, args=(self, ))
        th.setDaemon(True)
        th.start()
        self.started = True
        self.pid = self.popen.pid
        logger.info('Start process [%s] successfully. pid: %d' %
                    (self.name, self.popen.pid))
        logger.info('-' * 120)
        return 0
```

在该函数内部调用/apollo/bazel-bin/modules/dreamview/dreamview --flagfile=/apollo/modules/common/data/global_flagfile.txt 最终启动了 dreamview 进程。dreamview 进程的 main 函数位于/apollo/modules/dreamview/backend/main.cc 中，内容如下：

```
  int main(int argc, char *argv[]) {
    google::ParseCommandLineFlags(&argc, &argv, true);
    // add by caros for dv performance improve

apollo::cyber::GlobalData::Instance()->SetProcessGroup("dreamview_sched");
    apollo::cyber::Init(argv[0]);

    apollo::dreamview::Dreamview dreamview;
    const bool init_success = dreamview.Init().ok() && dreamview.Start().ok();
    if (!init_success) {
      AERROR << "Failed to initialize dreamview server";
      return -1;
    }
    apollo::cyber::WaitForShutdown();
    dreamview.Stop();
    apollo::cyber::Clear();
    return 0;
  }
```

该函数初始化 Cyber 环境，并调用 Dreamview::Init()和 Dreamview::Start()函数，启动 dreamview 后台监护进程。然后进入消息处理循环，直到等待 cyber::WaitForShutdown()返回，清理资源并退出 main 函数。

Apollo 3.5 使用 Cyber 启动 Localization、Perception、Prediction、Planning、Control 等功能模块。若只看各模块的 BUILD 文件，则肯定无法找到该模块的启动入口 main 函数（Apollo 3.5 之前的版

本均是如此处理）。

下面以 Prediction 模块为例具体阐述。

Prediction 模块 BUILD 文件中生成 binary 文件的配置项如下：

```
cc_binary(
    name = "libprediction_component.so",
    linkshared = True,
    linkstatic = False,
    deps = [":prediction_component_lib"],
)
```

该配置项中没有 source 文件，仅包含一个依赖项：prediction_component_lib。注意后者的定义如下：

```
cc_library(
    name = "prediction_component_lib",
    srcs = ["prediction_component.cc"],
    hdrs = [
        "prediction_component.h",
    ],
    copts = [
        "-DMODULE_NAME=\\\"prediction\\\"",
    ],
    deps = [
        "//cyber/common:file",
        "//modules/common/adapters:adapter_gflags",
        "//modules/prediction/common:message_process",
        "//modules/prediction/evaluator:evaluator_manager",
        "//modules/prediction/predictor:predictor_manager",
        "//modules/prediction/proto:offline_features_proto",
        "//modules/prediction/scenario:scenario_manager",
        "//modules/prediction/util:data_extraction",
    ],
)
```

在 srcs 文件 prediction_component.cc 以及 deps 文件中均找不到 main 函数。那么 main 函数被隐藏在哪里？如果没有 main 函数，binary 文件 libprediction_component.so 又是如何启动的？答案很简单，Prediction 模块的 binary 文件 libprediction_component.so 作为 Cyber 的一个组件启动，不需要 main 函数。

下面详细阐述在 DreamView 界面中启动 Prediction 模块的过程。DreamView 前端界面操作此处不表，后端的消息响应函数 HMI::RegisterMessageHandlers()位于/apollo/modules/dreamview/backend/hmi/hmi.cc 文件中：

```
void HMI::RegisterMessageHandlers() {
```

```
    // Broadcast HMIStatus to clients when status changed
    hmi_worker_->RegisterStatusUpdateHandler(
        [this](const bool status_changed, HMIStatus* status) {
          if (!status_changed) {
            // Status doesn't change, skip broadcasting.
            return;
          }
          websocket_->BroadcastData(
              JsonUtil::ProtoToTypedJson("HMIStatus", *status).dump());
          if (status->current_map().empty()) {
            monitor_log_buffer_.WARN("You haven't selected a map yet!");
          }
          if (status->current_vehicle().empty()) {
            monitor_log_buffer_.WARN("You haven't selected a vehicle yet!");
          }
        });

    // Send current status and vehicle param to newly joined client
    websocket_->RegisterConnectionReadyHandler(
        [this](WebSocketHandler::Connection* conn) {
          const auto status_json =
              JsonUtil::ProtoToTypedJson("HMIStatus",
hmi_worker_->GetStatus());
          websocket_->SendData(conn, status_json.dump());
          SendVehicleParam(conn);
        });

    websocket_->RegisterMessageHandler(
        "HMIAction",
        [this](const Json& json, WebSocketHandler::Connection* conn) {
          // Run HMIWorker::Trigger(action) if json is {action: "<action>"}
          // Run HMIWorker::Trigger(action, value) if "value" field is provided
          std::string action;
          if (!JsonUtil::GetStringFromJson(json, "action", &action)) {
            AERROR << "Truncated HMIAction request.";
            return;
          }
          HMIAction hmi_action;
          if (!HMIAction_Parse(action, &hmi_action)) {
            AERROR << "Invalid HMIAction string: " << action;
            return;
          }
          std::string value;
```

```
        if (JsonUtil::GetStringFromJson(json, "value", &value)) {
          hmi_worker_->Trigger(hmi_action, value);
        } else {
          hmi_worker_->Trigger(hmi_action);
        }

        // Extra works for current Dreamview
        if (hmi_action == HMIAction::CHANGE_MAP) {
          // Reload simulation map after changing map
          CHECK(map_service_->ReloadMap(true))
              << "Failed to load new simulation map: " << value;
        } else if (hmi_action == HMIAction::CHANGE_VEHICLE) {
          // Reload lidar params for point cloud service
          PointCloudUpdater::LoadLidarHeight(FLAGS_lidar_height_yaml);
          SendVehicleParam();
        } else if (hmi_action == HMIAction::CHANGE_MODE) {
          static constexpr char kCalibrationMode[] = "Vehicle Calibration";
          if (value == kCalibrationMode) {
            data_collection_monitor_->Start();
          } else {
            data_collection_monitor_->Stop();
          }
        }
      });

  // HMI client asks for adding new DriveEvent
  websocket_->RegisterMessageHandler(
      "SubmitDriveEvent",
      [this](const Json& json, WebSocketHandler::Connection* conn) {
        // json should contain event_time_ms and event_msg
        uint64_t event_time_ms;
        std::string event_msg;
        std::vector<std::string> event_types;
        bool is_reportable;
        if (JsonUtil::GetNumberFromJson(json, "event_time_ms",
                                        &event_time_ms) &&
            JsonUtil::GetStringFromJson(json, "event_msg", &event_msg) &&
            JsonUtil::GetStringVectorFromJson(json, "event_type",
                                              &event_types) &&
            JsonUtil::GetBooleanFromJson(json, "is_reportable",
                                         &is_reportable)) {
          hmi_worker_->SubmitDriveEvent(event_time_ms, event_msg, event_types,
                                        is_reportable);
```

```
        monitor_log_buffer_.INFO("Drive event added.");
      } else {
        AERROR << "Truncated SubmitDriveEvent request.";
        monitor_log_buffer_.WARN("Failed to submit a drive event.");
      }
    });
}
```

其中，HMIAction_Parse(action, &hmi_action)用于解析动作参数，hmi_worker_->Trigger(hmi_action,value)用于执行相关动作。对于 Prediction 模块的启动而言，hmi_action 的值为 HMIAction::START_MODULE，value 的值为 prediction。

实际上，DreamView 将操作模式分为多种 HMI Mode，这些模式位于目录/apollo/modules/dreamview/conf/hmi_modes 中，每一个配置文件均对应一种 HMI Mode。但无论处于哪种 HMI Mode，对于 Planning 模块的启动而言，hmi_action 的值均为 HMIAction::START_MODULE，value 的值均为 prediction。

当然，Standard Mode 与 Navigation Mode 对应的 dag_files 不一样，Standard Mode 的 dag_files 为/apollo/modules/prediction/dag/prediction.dag，Navigation Mode 的 dag_files 为/apollo/modules/prediction/dag/prediction_navi.dag。

HMIWorker::Trigger(const HMIAction action, const std::string& value)函数位于文件/apollo/modules/dreamview/backend/hmi/hmi_worker.cc 中，内容如下：

```
bool HMIWorker::Trigger(const HMIAction action, const std::string& value) {
  AINFO << "HMIAction " << HMIAction_Name(action) << "(" << value
        << ") was triggered!";
  switch (action) {
    case HMIAction::CHANGE_MODE:
      ChangeMode(value);
      break;
    case HMIAction::CHANGE_MAP:
      ChangeMap(value);
      break;
    case HMIAction::CHANGE_VEHICLE:
      ChangeVehicle(value);
      break;
    case HMIAction::START_MODULE:
      StartModule(value);
      break;
    case HMIAction::STOP_MODULE:
      StopModule(value);
      break;
    case HMIAction::RECORD_AUDIO:
      RecordAudio(value);
```

```
      break;
    default:
      AERROR << "HMIAction not implemented, yet!";
      return false;
  }
  return true;
}
```

上述函数中成员变量 current_mode_保存着当前 HMI Mode 对应配置文件包含的所有配置项。例如，modules/dreamview/conf/hmi_modes/mkz_standard_debug.pb.txt 里面就包含 MKZ 标准调试模式下所有的功能模块，该配置文件通过 HMIWorker::LoadMode(const std::string& mode_config_path) 函数读入成员变量 current_mode_中。如果基于字符串 module 查找到了对应的模块名以及对应的启动配置文件 dag_files，就调用 System 函数（内部实际调用 std::system 函数）基于命令 module_conf->start_command()启动一个进程。这个 start_command 从何而来？需进一步分析 HMIWorker::LoadMode(const std::string& mode_config_path)函数：

```
HMIMode HMIWorker::LoadMode(const std::string& mode_config_path) {
  HMIMode mode;
  CHECK(cyber::common::GetProtoFromFile(mode_config_path, &mode))
      << "Unable to parse HMIMode from file " << mode_config_path;
  // Translate cyber_modules to regular modules
  for (const auto& iter : mode.cyber_modules()) {
    const std::string& module_name = iter.first;
    const CyberModule& cyber_module = iter.second;
    // Each cyber module should have at least one dag file
    CHECK(!cyber_module.dag_files().empty())
        << "None dag file is provided for " << module_name << " module in "
        << mode_config_path;

    Module& module = LookupOrInsert(mode.mutable_modules(), module_name, {});
    module.set_required_for_safety(cyber_module.required_for_safety());

    // Construct start_command
    //     nohup mainboard -p <process_group> -d <dag> ... &
    module.set_start_command("nohup mainboard");
    const auto& process_group = cyber_module.process_group();
    if (!process_group.empty()) {
      StrAppend(module.mutable_start_command(), " -p ", process_group);
    }
    for (const std::string& dag : cyber_module.dag_files()) {
      StrAppend(module.mutable_start_command(), " -d ", dag);
    }
    StrAppend(module.mutable_start_command(), " &");
```

```
    // Construct stop_command: pkill -f '<dag[0]>'
    const std::string& first_dag = cyber_module.dag_files(0);
    module.set_stop_command(StrCat("pkill -f \"", first_dag, "\""));
    // Construct process_monitor_config

module.mutable_process_monitor_config()->add_command_keywords("mainboard");

module.mutable_process_monitor_config()->add_command_keywords(first_dag);
   }
   mode.clear_cyber_modules();
   AINFO << "Loaded HMI mode: " << mode.DebugString();
   return mode;
  }
```

通过该函数可以看到，构建出的 start_command 格式为 nohup mainboard -p-d... &。

其中，process_group 与 dag 均来自于当前 HMI Mode 对应的配置文件。以 modules/dreamview/conf/hmi_modes/mkz_close_loop.pb.txt 为例，它包含两个 cyber_modules 配置项，对于 Computer 模块而言，它包含 11 个 dag_files 文件（对应 11 个子功能模块），这些子功能模块全部属于名为 compute_sched 的 process_group。dag 自不必言，每个子功能模块对应一个 dag_files，Prediction 子功能模块对应的 dag_files 为/apollo/modules/prediction/dag/prediction.dag。

```
  module_config {
    module_library :
"/apollo/bazel-bin/modules/prediction/libprediction_component.so"
    components {
      class_name : "PredictionComponent"
      config {
        name: "prediction"
        flag_file_path: "/apollo/modules/prediction/conf/prediction.conf"
        readers: [
          {
            channel: "/apollo/perception/obstacles"
            qos_profile: {
              depth : 1
            }
          }
        ]
      }
    }
  }
```

至此，我们终于找到了 Prediction 功能模块的启动命令：

```
1nohup mainboard -p compute_sched -d /apollo/modules/prediction/dag/predict
ion.dag &
```

nohup 表示非挂断方式启动，mainboard 无疑是启动的主程序，入口 main 函数必定包含于其中。

process_group 的意义不是那么大，无非对功能模块分组而已，dag_files 才是我们启动相关功能模块的真正配置文件。查看 Cyber 模块的构建文件/apollo/cyber/BUILD，可发现如下内容：

```
cc_binary(
    name = "mainboard",
    srcs = [
        "mainboard/mainboard.cc",
        "mainboard/module_argument.cc",
        "mainboard/module_argument.h",
        "mainboard/module_controller.cc",
        "mainboard/module_controller.h",
    ],
    copts = [
        "-pthread",
    ],
    linkstatic = False,
    deps = [
        ":cyber_core",
        "//cyber/proto:dag_conf_cc_proto",
    ],
)
```

至此，可执行文件 mainboard 的踪迹水落石出。

果不其然，入口函数 main 位于文件 cyber/mainboard/mainboard.cc 中：

```
int main(int argc, char** argv) {
  google::SetUsageMessage("we use this program to load dag and run user apps.");

  // parse the argument
  ModuleArgument module_args;
  module_args.ParseArgument(argc, argv);

  // initialize cyber
  apollo::cyber::Init(argv[0]);

  // start module
  ModuleController controller(module_args);
  if (!controller.Init()) {
    controller.Clear();
    AERROR << "module start error.";
    return -1;
```

```
  }

  apollo::cyber::WaitForShutdown();
  controller.Clear();
  AINFO << "exit mainboard.";

  return 0;
}
```

main函数十分简单，首先是解析参数，初始化Cyber环境，接下来创建一个ModuleController类对象controller，后调用controller.Init()启动相关功能模块。最后，一直等待cyber::WaitForShutdown()返回，清理资源并退出main函数。ModuleController::Init()函数十分简单，内部调用了ModuleController::LoadAll()函数：

```
bool ModuleController::LoadAll() {
  const std::string work_root = common::WorkRoot();
  const std::string current_path = common::GetCurrentPath();
  const std::string dag_root_path = common::GetAbsolutePath(work_root, "dag");

  for (auto& dag_conf : args_.GetDAGConfList()) {
    std::string module_path = "";
    if (dag_conf == common::GetFileName(dag_conf)) {
      // case dag conf argument var is a filename
      module_path = common::GetAbsolutePath(dag_root_path, dag_conf);
    } else if (dag_conf[0] == '/') {
      // case dag conf argument var is an absolute path
      module_path = dag_conf;
    } else {
      // case dag conf argument var is a relative path
      module_path = common::GetAbsolutePath(current_path, dag_conf);
      if (!common::PathExists(module_path)) {
        module_path = common::GetAbsolutePath(work_root, dag_conf);
      }
    }
    AINFO << "Start initialize dag: " << module_path;
    if (!LoadModule(module_path)) {
      AERROR << "Failed to load module: " << module_path;
      return false;
    }
  }
  return true;
}
```

上述函数处理一个dag_conf配置文件循环，读取配置文件中的所有dag_conf，并逐一调用bool

ModuleController::LoadModule(const std::string& path)函数加载功能模块：

```
bool ModuleController::LoadModule(const std::string& path) {
  DagConfig dag_config;
  if (!common::GetProtoFromFile(path, &dag_config)) {
    AERROR << "Get proto failed, file: " << path;
    return false;
  }
  return LoadModule(dag_config);
}
```

上述函数从磁盘配置文件中读取配置信息，并调用 bool ModuleController::LoadModule(const DagConfig& dag_config)函数加载功能模块：

```
bool ModuleController::LoadModule(const DagConfig& dag_config) {
  const std::string work_root = common::WorkRoot();

  for (auto module_config : dag_config.module_config()) {
    std::string load_path;
    if (module_config.module_library().front() == '/') {
      load_path = module_config.module_library();
    } else {
      load_path =
          common::GetAbsolutePath(work_root, module_config.module_library());
    }

    if (!common::PathExists(load_path)) {
      AERROR << "Path does not exist: " << load_path;
      return false;
    }

    class_loader_manager_.LoadLibrary(load_path);

    for (auto& component : module_config.components()) {
      const std::string& class_name = component.class_name();
      std::shared_ptr<ComponentBase> base =
          class_loader_manager_.CreateClassObj<ComponentBase>(class_name);
      if (base == nullptr || !base->Initialize(component.config())) {
        return false;
      }
      component_list_.emplace_back(std::move(base));
    }

    for (auto& component : module_config.timer_components()) {
```

```
      const std::string& class_name = component.class_name();
      std::shared_ptr<ComponentBase> base =
          class_loader_manager_.CreateClassObj<ComponentBase>(class_name);
      if (base == nullptr || !base->Initialize(component.config())) {
        return false;
      }
      component_list_.emplace_back(std::move(base));
    }
  }
  return true;
}
```

上述函数看似很长，核心思想无非是调用 class_loader_manager_.LoadLibrary(load_path);加载功能模块，创建并初始化功能模块类对象，并将该功能模块加入 Cyber 的组件列表中统一调度管理。

10.3.2　Apollo 模块注册及动态创建

整个 Prediction 模块的启动过程已阐述完毕，但仍有一个问题需要解决：Prediction 模块是如何作为 Cyber 的一个组件注册并动态创建的？首先看组件注册过程。注意到 modules/prediction/prediction_component.h 的组件类 PredictionComponent 继承自 cyber::Component<perception::PerceptionObstacles>，同时，使用宏 CYBER_REGISTER_COMPONENT(PredictionComponent)将规划组件 PredictionComponent 注册到 Cyber 的组件类管理器。查看源代码可知：

```
#define CYBER_REGISTER_COMPONENT(name) \
CLASS_LOADER_REGISTER_CLASS(name, apollo::cyber::ComponentBase)
```

而后者的定义为：

```
#define CLASS_LOADER_REGISTER_CLASS(Derived, Base) \
CLASS_LOADER_REGISTER_CLASS_INTERNAL_1(Derived, Base, __COUNTER__)
```

继续展开得到：

```
#define CLASS_LOADER_REGISTER_CLASS_INTERNAL_1(Derived, Base, UniqueID) \
CLASS_LOADER_REGISTER_CLASS_INTERNAL(Derived, Base, UniqueID)
```

仍然需要进一步展开：

```
#define CLASS_LOADER_REGISTER_CLASS_INTERNAL(Derived, Base, UniqueID)
  namespace {
  struct ProxyType##UniqueID {
    ProxyType##UniqueID() {
      apollo::cyber::class_loader::utility::RegisterClass<Derived, Base>(
          #Derived, #Base);
    }
  };
```

```
  static ProxyType##UniqueID g_register_class_##UniqueID;
  }
```

将 PlanningComponent 代入上述宏，最终得到：

```
namespace {

 struct ProxyType__COUNTER__ {
ProxyType__COUNTER__() {
  apollo::cyber::class_loader::utility::RegisterClass(
      "PredictionComponent", "apollo::cyber::ComponentBase");

  }
  };
  static ProxyType__COUNTER__ g_register_class___COUNTER__;

 }
```

注意两点：

第一，上述定义位于 namespace apollo::planning 内。

第二，___COUNTER__是 C 语言的一个计数器宏，这里仅代表一个占位符，实际展开时可能就是 78 之类的数字，即 ProxyType__COUNTER__实际上应为 ProxyType78 之类的命名。

上述代码简洁明了，首先定义了一个结构体 ProxyType__COUNTER__，该结构体仅包含一个构造函数，在内部调用 apollo::cyber::class_loader::utility::RegisterClass 注册 apollo::cyber::ComponentBase 类的派生类 PredictionComponent，并定义了一个静态全局结构体 ProxyType__COUNTER__变量：g_register_class___COUNTER__。

继续观察 apollo::cyber::class_loader::utility::RegisterClass 函数：

```
template <typename Derived, typename Base>
void RegisterClass(const std::string& class_name,
                   const std::string& base_class_name) {
  AINFO << "registerclass:" << class_name << "," << base_class_name << ","
        << GetCurLoadingLibraryName();

  utility::AbstractClassFactory<Base>* new_class_factory_obj =
      new utility::ClassFactory<Derived, Base>(class_name, base_class_name);
  new_class_factory_obj->AddOwnedClassLoader(GetCurActiveClassLoader());

new_class_factory_obj->SetRelativeLibraryPath(GetCurLoadingLibraryName());

  GetClassFactoryMapMapMutex().lock();
  ClassClassFactoryMap& factory_map =
      GetClassFactoryMapByBaseClass(typeid(Base).name());
```

```
  factory_map[class_name] = new_class_factory_obj;
  GetClassFactoryMapMapMutex().unlock();
}
```

该函数创建一个模板类 utility::ClassFactory 对象 new_class_factory_obj，为其添加类加载器，设置加载库的路径，最后将工厂类对象加入 ClassClassFactoryMap 对象 factory_map 统一管理。通过该函数可以清楚地看到，Cyber 使用工厂方法模式完成产品类（例如 PredictionComponent）对象的创建。

具体的动态创建过程我们需要简单学习一下。功能模块类 PredictionComponent 对象在 bool ModuleController::LoadModule(const DagConfig& dag_config)函数内部创建：

```
bool ModuleController::LoadModule(const DagConfig& dag_config) {
  const std::string work_root = common::WorkRoot();

  for (auto module_config : dag_config.module_config()) {
    std::string load_path;
    if (module_config.module_library().front() == '/') {
      load_path = module_config.module_library();
    } else {
      load_path =
          common::GetAbsolutePath(work_root, module_config.module_library());
    }

    if (!common::PathExists(load_path)) {
      AERROR << "Path does not exist: " << load_path;
      return false;
    }

    class_loader_manager_.LoadLibrary(load_path);

    for (auto& component : module_config.components()) {
      const std::string& class_name = component.class_name();
      std::shared_ptr<ComponentBase> base =
          class_loader_manager_.CreateClassObj<ComponentBase>(class_name);
      if (base == nullptr || !base->Initialize(component.config())) {
        return false;
      }
      component_list_.emplace_back(std::move(base));
    }

    for (auto& component : module_config.timer_components()) {
      const std::string& class_name = component.class_name();
      std::shared_ptr<ComponentBase> base =
```

```
          class_loader_manager_.CreateClassObj<ComponentBase>(class_name);
      if (base == nullptr || !base->Initialize(component.config())) {
        return false;
      }
      component_list_.emplace_back(std::move(base));
    }
  }
  return true;
}
```

已经知道，PredictionComponent 对象是通过 class_loader_manager_.CreateClassObj(class_name) 创建出来的，而 class_loader_manager_是一个 class_loader::ClassLoaderManager 类对象。现在的问题是，class_loader::ClassLoaderManager 与前面提到的工厂类 utility::AbstractClassFactory 是如何联系起来的？先看 ClassLoaderManager::CreateClassObj 函数（位于文件 cyber/class_loader/class_loader_manager.h 中）：

```
  template <typename Base>
std::shared_ptr ClassLoaderManager::CreateClassObj(
    const std::string& class_name) {
  std::vector class_loaders = GetAllValidClassLoaders();
  for (auto class_loader : class_loaders) {
    if (class_loader->IsClassValid(class_name)) {
      return (class_loader->CreateClassObj(class_name));
    }
  }
  AERROR << "Invalid class name: " << class_name;
  return std::shared_ptr();
}
```

上述函数中，从所有 class_loaders 中找出一个正确的 class_loader，并调用 class_loader->CreateClassObj(class_name)（位于文件 cyber/class_loader/class_loader.h 中）。

创建功能模块组件类对象：

```
  template <typename Base>
  std::shared_ptr<Base> ClassLoader::CreateClassObj(
      const std::string& class_name) {
    if (!IsLibraryLoaded()) {
      LoadLibrary();
    }

    Base* class_object = utility::CreateClassObj<Base>(class_name, this);
    if (nullptr == class_object) {
      AWARN << "CreateClassObj failed, ensure class has been registered. "
            << "classname: " << class_name << ",lib: " << GetLibraryPath();
```

```
    return std::shared_ptr<Base>();
  }

  std::lock_guard<std::mutex> lck(classobj_ref_count_mutex_);
  classobj_ref_count_ = classobj_ref_count_ + 1;
  std::shared_ptr<Base> classObjSharePtr(
      class_object, std::bind(&ClassLoader::OnClassObjDeleter<Base>, this,
                              std::placeholders::_1));
  return classObjSharePtr;
}
```

上述函数继续调用 utility::CreateClassObj(class_name, this)（位于文件 cyber/class_loader/utility/class_loader_utility.h 中）。

创建功能模块组件类对象：

```
template <typename Base>
Base* CreateClassObj(const std::string& class_name, ClassLoader* loader) {
  GetClassFactoryMapMapMutex().lock();
  ClassClassFactoryMap& factoryMap =
      GetClassFactoryMapByBaseClass(typeid(Base).name());
  AbstractClassFactory<Base>* factory = nullptr;
  if (factoryMap.find(class_name) != factoryMap.end()) {
    factory = dynamic_cast<utility::AbstractClassFactory<Base>*>(
        factoryMap[class_name]);
  }
  GetClassFactoryMapMapMutex().unlock();

  Base* classobj = nullptr;
  if (factory && factory->IsOwnedBy(loader)) {
    classobj = factory->CreateObj();
  }

  return classobj;
}
```

上述函数使用 factory = dynamic_cast(factoryMap[class_name]);获取对应的工厂对象指针。至此，终于将 class_loader::ClassLoaderManager 与上文中的工厂类 utility::AbstractClassFactory 联系起来了。工厂对象指针找到后，使用 classobj = factory->CreateObj();就顺理成章地将功能模块类对象创建出来了。

参考文献

[1] Rich C, Ponsler B, Holroyd A, et al. [IEEE 2010 5th ACM/IEEE International Conference on Human-Robot Interaction (HRI) - Osaka, Japan (2010.03.2-2010.03.5)] 2010 5th ACM/IEEE International Conference on Human-Robot Interaction (HRI) - Recognizing engagement in human-robot interaction[J]. 2010:375-382.

[2] Lu D V, Lu D V. Contextualized Robot Navigation[J]. Dissertations & Theses - Gradworks, 2014(Oct).

[3] Hornung A, Dornbush A, Likhachev M, et al. Anytime Footstep Planning with Suboptimality Bounds[C]// IEEE-RAS International Conference on Humanoid Robots (Humanoids). IEEE, 2012.

[4] Foote T. tf: The Transform Library[C]// 2013 IEEE Conference on Technologies for Practical Robot Applications (TePRA). IEEE, 2013.

[5] Jones E G. Navigation in Three-Dimensional Cluttered Environments for Mobile Manipulation[C]// IEEE International Conference on Robotics & Automation. IEEE, 2012.

[6] Steder B, Rusu R B, Konolige K, et al. Point feature extraction on 3D range scans taking into account object boundaries[C]// IEEE International Conference on Robotics & Automation. IEEE, 2011.

第 11 章

无人驾驶地图技术

地图是我们驾驶过程中需要使用的重要工具，当驾驶过程升级为无人驾驶时，对地图的要求变得更加严格。这就是我们在无人驾驶中经常提到的高精地图（High Defination Map）。高精地图在无人驾驶系统运作的各个阶段都有重要的作用，在自动驾驶的定位、决策、规划和感知过程中都离不开高精地图的支持。高精地图和普通地图最大的区别在于高精地图可以达到厘米级别的精度，同时高精地图还包含大量的驾驶辅助信息。

本章主要讲解高精地图的概念和应用，特别是映射到 Apollo 无人驾驶系统中成为 Pnc 地图和相对地图及其相关应用。内容分为三部分，第一部分着重介绍高精地图的概念和应用，分析 OpenDrive 标准，并且对高精地图的解决方案的一些代表产品进行介绍；第二部分讲解高精地图映射到 Apollo 无人驾驶系统中成为 Pnc 地图及其相关子系统的应用接口；第三部分讲解高精地图映射到 Apollo 无人驾驶系统中成为相对地图及其相关子系统的应用接口。

11.1　高精地图

高精地图是无人驾驶软件输入的起点。正如它的名称中体现的一样，高精地图和普通地图最大的区别在于高精地图可以达到厘米级别的精度。同时高精地图还包含大量的驾驶辅助信息，最重要的就是道路网络的三维描述。比如交叉路口的布局、交通标志牌的位置等也包含环境的语义信息，比如交通信号灯的颜色含义、道路限速、从哪里开始转向和交通规则等。高精地图在无人驾驶系统运作的各个阶段都有重要的作用。

11.1.1　高精地图在自动驾驶子系统中的应用

在定位过程中，使用来自传感器收集到的 Landmarks 的信息与以后的高精地图里面预存的

Landmarks 进行匹配就能得到车辆自身的精确位置。高精地图提供标准的位置，现在有一种低成本的设计方案，采用单目相机拍摄虚线和实线，把采集的图像和高精地图进行比对，通过算法可以知道当前在道路的第几个车道。计算出车道后解决定位问题，这是横向定位。纵向定位可以借助交通信号灯、路灯、灯杆等实现定位。

在感知过程中，高精地图有两点应用：第一，高精地图为感知增加了系统的冗余度，每个传感器都有自己的测量距离、天气或光照条件的限制，在较远的区域或因为障碍物被遮挡的区域，高精地图都可以为传感器融合提供一个新的数据源；第二，高精地图可以通过 ROI（Region Of Interest，感兴趣区域）缩小传感器的探测范围，提高探测精度和速度，节省计算资源

在规划过程中，高精地图可以帮助找到合适的行驶区域，与运动物体的历史数据一起进行之后运动的预测。同时，如果车在路上发现前面有事故或者施工路面，这个时候就需要变道，高精地图提供了有利的支持。

在决策过程中，高精地图中的信息会成为决策的最有力保障。例如，车到了十字路口，高精地图会采集安全岛的信息，复杂十字路口有安全岛，车在决策过程中需要参考安全岛等要素，否则车辆冲上安全岛会导致发生交通事故。

综上所述，在自动驾驶的几个重要系统流程的定位、决策、规划和感知模型都离不开高精地图的支持，有了高精地图可以节省很多传感器的成本，成本的下降能够加速量产在行业中的推进，这是高精地图在目前最大的意义所在。同时，业界对于脱落高精地图的自动驾驶方案的研究也在持续进行，由于目前自动驾驶的主流是模仿人的驾驶行为，因此地图组件对于无人驾驶过程来说并不是完全必需的，但是目前阶段，高精地图还无法替代自动驾驶核心组件。高精地图的解决方案主要由地图提供商和自动驾驶独角兽企业提供。下面我们列举一些典型的代表。

11.1.2 高精地图的解决方案

1. HERE HD Live Map

HERE HD Live Map（高清实时地图）是 HERE 公司的核心产品，有很好的基础优势。作为一家传统图商，HERE 公司的用户基数可以保证地图以更快的速度和形式更新。由于 HERE 在基础地图研发和产品化过程中深耕多年，因此它的基础地图设计是比较完善的。HERE HD Live Map 的基础就是它的地图架构设计，在此基础上通过采集的数据基于激光雷达、相机建立 16 线程的 Base Map。为了满足高精地图制作的精度需要，HERE 公司采用了两种地图采集策略：自主采集和众包采集，当然这两种策略也是业界普遍使用的。自主采集需要拥有隶属于公司的专业采集车辆。针对 HERE 的采集车辆，这类车辆配有 GPS、激光雷达、相机等，每天采集 28TB 的数据量，精度可达厘米。众包采集利用众包车辆传感器采集行车路径、车道标志、道路边缘、路标、路面标志等，结合卫星图像等多种数据源，保持高清地图的新鲜感。HERE 的众包服务通过与汽车生产厂家合作，在车辆出厂前安装对应的感应器模块和传输功能模块。由于采集车辆只是在行驶过程中回传行驶数据，因此在众包地图采集中对数据的融合处理是十分重要的，HERE 采用在云计算中映射学习的融合方案，车辆大小、传感器设置和行驶路径能够更加准确地识别路边的同一个物体。机器学习将这些变化的传感器数据聚合起来，以确定路边工件的精确位置。在高精地图更新后，使用高精地图的

车辆也需要尽快更新，更新地图的具体流程是：创建地图对象并将其添加到地图数据库，然后将其发布到 HD Live 地图，并将必要的平面图发送回车辆，以便车辆能够准确和实时地表示道路网络。

在采集数据处理的过程中，HERE 有一整套处理方案，一般称作 HERE HD Live Map Learning 方案。不同于利用神经网络的图像处理方法，HERE 利用点云分割技术对 Features 进行分析。在多次采集后，可将同一区域的点云补齐。但目前的图像处理方法已较为成熟，而点云技术（点云 SLAM、点云分割、点云特征提取等）仍需完善发展。

当然，对于开发人员来说，我们会很感兴趣 HERE 公司是如何表述其地图产品的。HERE 公司对地图做了 4 个分层结构。

第 1 层是 Road & Lane Model Layer（车道边界和区分界线）：

```
A highly precise representation of road network.
```

第 2 层是 Localization Model Layer（基于 Camera 或点云）：

```
Help a vehicle to find its exact Position of lane it is driving in.
```

第 3 层是 Activity Layer（动态信息层、道路实时信息）：

```
Understanding dynamic events in the road network.
```

第 4 层是 Analytics Layer（司机驾驶习惯分析）：

```
Tell how humans actually behave in a piece of road.
```

在 HERE 的解决方案中，可以通过检测与定位约束纵向行驶信息，通过车道线约束横向行驶信息。

2. MobileEye

MobileEye 是业内非常知名的公司，是英特尔旗下的企业。MobileEye 号称为全球 25 家知名车厂合作商提供更安全的技术解决方案，有 2500 万辆车在使用他们的技术，13 家车厂正在使用 MobileEye 的技术在攻关自动驾驶。相比于 HERE，MobileEye 更侧重于使用 Camera，在图像处理方面也做得更好，使用视觉信息来辅助驾驶，是一种基于众包的视觉制图。

```
MobileEye-Pillars of Autonomous Driving
```

MobileEye 把技术层次分为感知、映射和驾驶策略 3 个层次。

- 感知：MobileEye 的软件可以进行传感器融合，从摄像机传感器、雷达和激光雷达传感器中解读数据。在图像处理方面，MobileEye 经验丰富，使用自己独有的算法来检测对象，确保安全行驶和系统决策。L3 以下的自动驾驶不需要高精地图，但是 L3 以上可以使用基于 LiDAR 和 Camera 的协同方案或者多传感器融合的高精地图方案。
- 映射：自动驾驶汽车需要大量的系统冗余来处理无法预料的情况。在所有条件下，车辆相对于道路边界和交叉口的精确定位都需要高精地图。MobileEye 提供基于 REM 的框架（REM™），它使用众包的策略，让用户能低成本地构建和快速更新高清地图。
- 驾驶策略：在 MobileEye 的驾驶策略中，他们认为，一旦一辆自动驾驶汽车能够感知周围的场景并在地图上进行定位，要解决的最后一件事情就是学习和共享人类司机的驾驶策略。

MobileEye 声称，传感、测绘和强大的计算能力赋予了自动驾驶车辆超人的视觉和反应时间。MobileEye 对驾驶策略的强化学习将提供多变量情况的分析方案，并且尽可能地逼近人类的行为和判断方式。这证明 MobileEye 对于复制人类的驾驶行为还是很看重的，至少把其单独地作为一个数据层去阐述处理。

说到 MobileEye，要重点提及其制图方案。MobileEye 的众包流程方案跟 HERE 的很像，只不过他们的方案更多地基于视觉来做，都是收集数据→上传云端→处理→下发车端。MobileEye 的 REM 系统（道路经验管理系统）非常知名，提供实时匿名众包的汽车数据，用于高精度地图的制作和使用。

MobileEye 的 REM 解决方案由 3 层组成：采集器（任何装有摄影机的车辆）、云端和自动驾驶车辆。相比 HERE 来说，MobileEye 基于视觉的方案，使用时最大的缺陷就是道路线的判断不连续。这会造成没有车道线了，车辆不知道怎么走了。在复杂的道路中，一旦出现红绿灯等难以识别的物体，MobileEye 所推崇的单靠视觉信息的解决方案将难以支撑全自动驾驶技术。

MobileEye 把 REM 采集、发送云端、处理、发回车端的过程称为“路书”。搭载 MobileEye 的车端首先会对环境进行识别，然后进行语义分析和几何形状提取，将其压缩后打包上传，这个过程称为道路段数据（Road Segment Data，RSD）。经过 REM 系统采集处理的 RSDs，其数据包大小可以达到 10KB/公里，并达到“高精度低延时”的效果。MobileEye 还会将不同路段的数据打断上传。这就是 MobileEye 的众包方案：所有的数据都在云端，大家一起来贡献相关数据，并且获得更好（高精度低延时）的数据回馈。正是由于激光雷达的解决方案存在诸多的限制：高成本、低规模化和点云算法尚不完善，在现行的网络条件下，MobileEye 的 RSD 方案至少看起来让自动驾驶这件事儿变得更加可行了。不过在 MobileEye 对外公布的演示视频中，我们可以看到其场景都是非城市的简单场景。在更复杂的环境中，其解决方案还是存在局限性的。

3. Google Waymo

谷歌 Waymo 业内知名，谷歌在业内做自动驾驶非常早，但是其对外披露的信息极少。这导致业界和开发者基本对于谷歌的解决方案“只能靠猜”。

在谷歌透露出来的地图解决方案中，我们可以发现在高精地图的层面上，大家对于道路信息的描述基本都是一致的。比如 Lane、路口虚拟线和道路停止线的理念，谷歌的解决方案本质上是为自动驾驶提供一个可运行的静态环境。

谷歌的地图解决方案中，谷歌将地图提供的静态环境和基于感知的动态环境（人物、车辆、道路标志）等信息结合在一起。这样融合高精地图使搭载 Waymo 的无人车完成对环境的感知。谷歌通常将红绿灯感知为框体，并且将人行横道的识别放在非常重要的位置。谷歌将根据地图提供的静态信息确定红绿灯的位置，基于感知到的红绿灯状态为其打上标签（红灯禁止或者绿灯通行），再为车辆决策提供依据，并且蓝色的预测轨迹为车辆规划行驶路径。

在谷歌对于高精地图的阐述中，他们的研发团队认为，仅有矢量数据是不够的。业界所推崇的矢量类型地图对于谷歌来说过于传统，他们更期待自己能够研发出栅格式的高精地图。

这种地图记录了所有道路上的物体信息，并且将不存在于静态地图中的动态物体自动过滤，由此降低车端感知识别的难度，达到更好的检测效果。至于谷歌所透露出来的环境地图，其红绿灯和停止线的设置跟业界的标准基本一致。

谷歌 Waymo 声称其降低了车载雷达的综合成本，谷歌 Waymo 的实验车大家应该非常熟悉。其车辆顶部可能搭载了激光雷达+视觉系统，车辆四周搭载了激光雷达。其整体方案也是激光雷达+视觉融合。谷歌自研的激光雷达据称可以检测到两个足球场（240 米）外的物体数据，并且整体的生产成本比 Velodyne 的 64 线激光雷达的售价（8 万美元）低 90%左右，这个价格对于开发者来说是非常诱人。

11.1.3　OpenDrive 地图格式简介

百度高精地图是基于 OpenDrive 地图格式的扩展版本，本节首先介绍 OpenDrive 地图文件。OpenDrive 地图文件以 XML 为文件标准格式，该 XML 文件中包含很多地图信息，例如 Road、Junction 等。图 11.1 所示为绘制地图的一个简单思路，读取 OpenDRIVE 文件（地图数据），构造路网，通过渲染展示给用户。

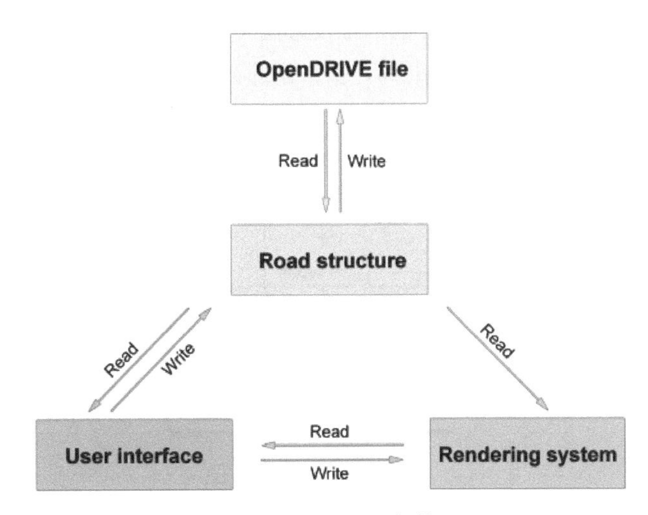

图 11.1　OpenDrive 架构

下面结合 OpenDRIVE 文件中的数据介绍如何构造路网。

1. 坐标系

首先需要考虑的是坐标系，我们平时看到的地图都是二维的，而地球是个椭球体，因此需要通过某种方式将椭球上的表示转换为二维平面。

在 GIS 中，一般使用两种常用的坐标系类型：一是全局坐标系或球坐标系，例如经纬度。这些坐标系通常称为地理坐标系（Geographic Coordinate System，GCS）。地理坐标系使用三维球面来定义地球上的位置，如图 11.2 所示。

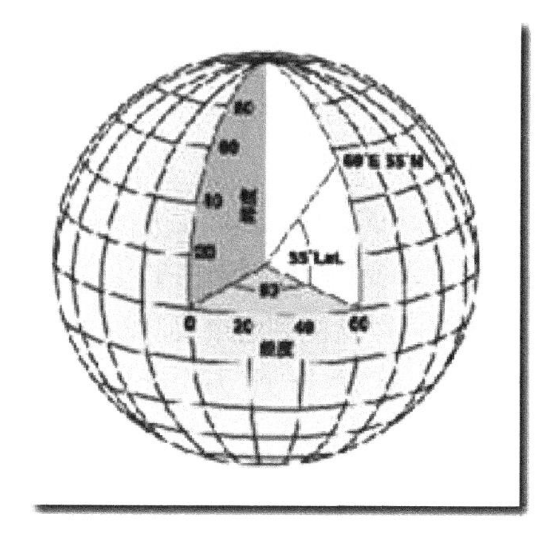

图 11.2　地理坐标系

二是基于横轴墨卡托、亚尔勃斯等积或罗宾森等地图投影的运算方案，地图投影（以及其他多种地图投影模型）提供了各种机制将地球球面的地图投影到二维笛卡尔坐标平面上。这种坐标系称为投影坐标系（Projection Coordinate System，PCS），且在二维平面中进行定义。

与地理坐标系不同，在二维空间范围内，投影坐标系的长度、角度和面积恒定。投影坐标系始终基于地理坐标系，而地理坐标系则是基于球体或旋转椭球体的。

图 11.3 所示为经过某种投影变换后的二维投影图（本文记作 x、y 坐标系），x 表示经度（正：东经，负：西经），y 表示纬度（正：北纬，负：南纬）。

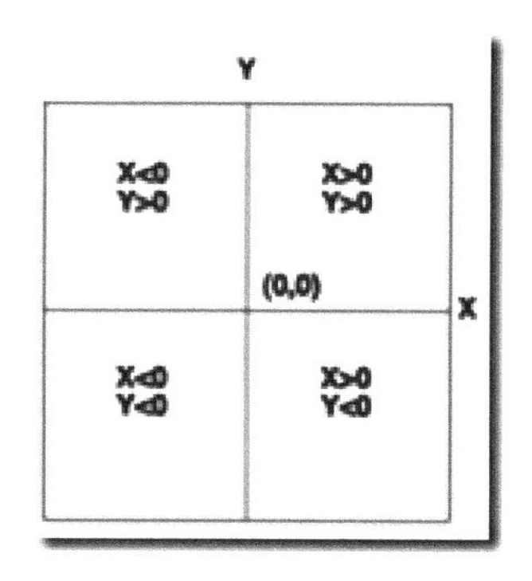

图 11.3　二维投影图

在 OpenDrive 数据中，具体体现如下：

```
<header revMajor="1" revMinor="4" name="OpenDRIVE TestFile" version="1"
date="Tue Aug 15 16:21:00 2017" north="5.4355531085039526e+06"
south="5.4143611839699000e+06" east="3.2681018217470363e+07"
west="3.2670519445542485e+07" vendor="AUTONAVI">
        <geoReference originLat="3.2675915701523371e+07"
originLong="5.4273604273430230e+06" originAlt="0.0000000000000000e+00"
originHdg="0.0000000000000000e+00">
            <![CDATA[PROJCS["WGS 84 / UTM zone 32N",GEOGCS["WGS
84",DATUM["WGS_1984",SPHEROID["WGS
84",6378137,298.257223563,AUTHORITY["EPSG","7030"]],AUTHORITY["EPSG","6326"]],
```

```
PRIMEM["Greenwich",0,AUTHORITY["EPSG","8901"]],UNIT["degree",0.01745329251994328,AUTHORITY["EPSG","9122"]],AUTHORITY["EPSG","4326"]],UNIT["metre",1,AUTHORITY["EPSG","9001"]],PROJECTION["Transverse_Mercator"],PARAMETER["latitude_of_origin",0],PARAMETER["central_meridian",117],PARAMETER["scale_factor",0.9996],PARAMETER["false_easting",500000],PARAMETER["false_northing",0],AUTHORITY["EPSG","32650"],AXIS["Easting",EAST],AXIS["Northing",NORTH]]]]>
        </geoReference>
    </header>
```

geoReference 元素定义了该文件使用的投影坐标系，其中地理坐标系为 WGS-84，而投影坐标系采用的是 Transverse Mercator（横轴墨卡托）投影。

在 OpenDrive 数据中大量使用的位置信息都是投影后的 x、y 坐标，而除了该投影坐标系外，还定义了一种轨迹坐标系，如图 11.4 所示。

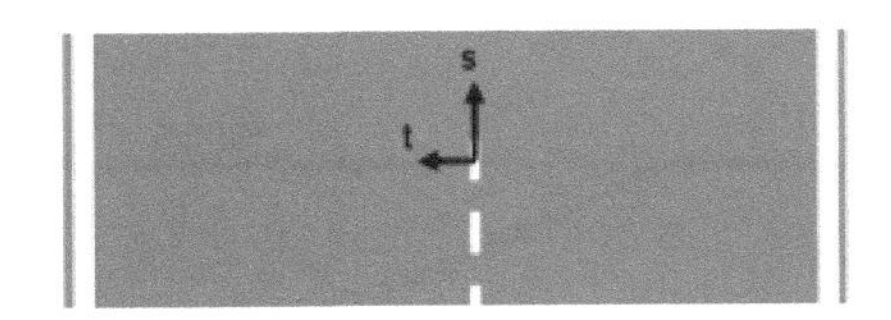

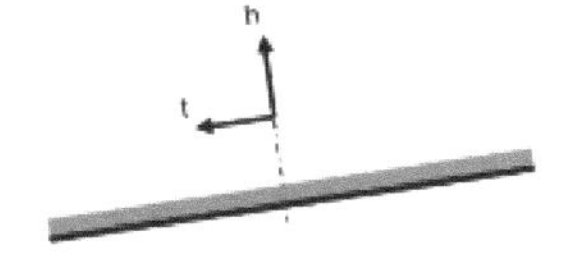

图 11.4　轨迹坐标系

s 坐标是沿着参考线的，关于参考线后面介绍。长度是在 x、y 坐标下计算的，应该是通过积分计算的（例如图 11.5 中的 123.45）。

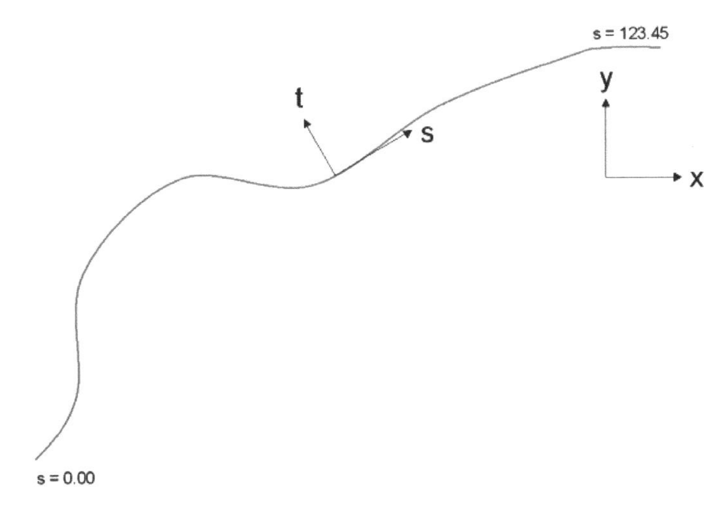

图 11.5　轨迹坐标系中的 s 和 t 坐标

t 坐标是相对于参考线的侧向位置，左为正，右为负。

2. 参考线

参考线是路网结构中一个很重要的概念，绘制地图的时候要先画参考线，参考线包含 x、y 位置坐标和路的形状属性，然后在参考线的基础上再去画其他元素。

图 11.6 是 OpenDRIVE 路网结构中的一条道路（Road），该道路由 3 部分组成：蓝色的参考线、车道线和车道线的其他属性（限速等）。

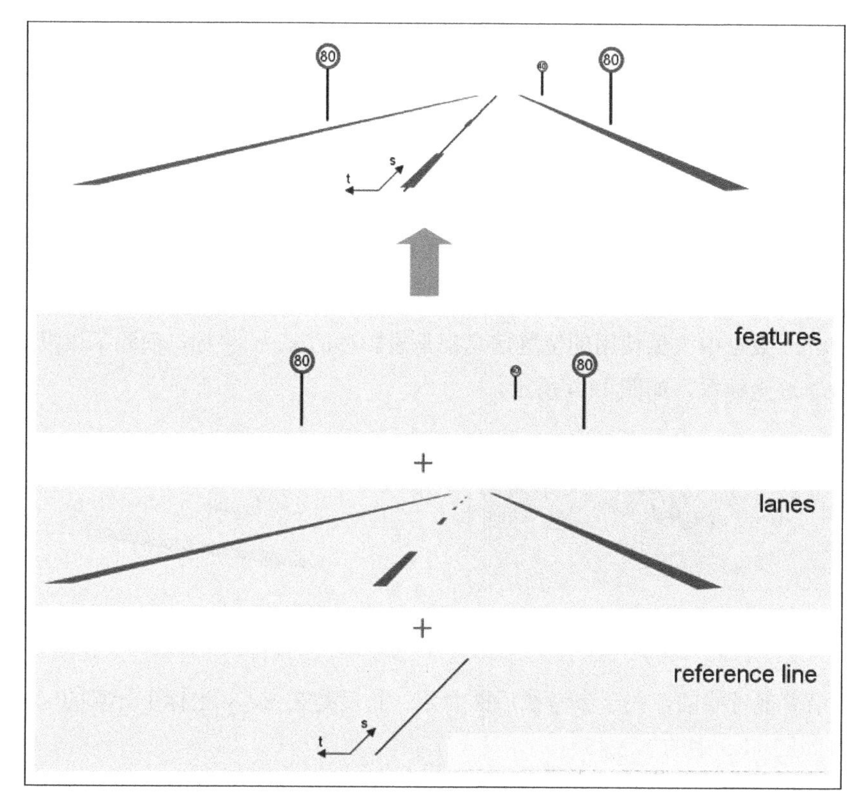

图 11.6 参考线

整个地图路网由很多道路构成，而每个道路中都会包含参考线，参考线就是一条线，它没有宽度。参考线的线条有好几种类型，如直线、螺旋线等，图 11.7 所示为几种常见的参考线，注意图中的两个坐标系：x、y 和 s、t。

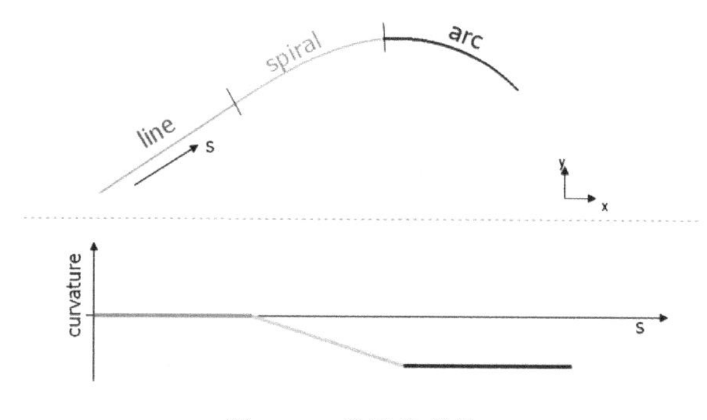

图 11.7 常见参考线

而在 OpenDrive 数据中，参考线体现在 planView 元素下的 geometry，顾名思义，是指俯视图下的道路的几何形状，该道路可以是直线、螺旋线等。

```
    <planView>
            <geometry s="0.0000000000000000e+00" x="-3.0024844302609563e+03"
y="2.7962779217697680e+03" hdg="2.8046409307224991e+00"
length="1.3647961224498889e+02">
                <arc curvature="-3.3912133325369736e-05" />
            </geometry>
            <geometry s="1.3647961224498889e+02" x="-3.1311844847723842e+03"
y="2.8416976011684164e+03" hdg="2.7974752903867355e+00"
length="9.9592435217850038e+01">
                <arc curvature="-1.2467775981482813e-03" />
            </geometry>
    </planView>
```

一条道路并不是只有一条参考线，因为假如一条道路的长度为 100 米，有可能这 100 米有些地方是直路，有些地方是拐弯的曲线，每一条都是一个 geometry 标签，通过 s（起始位置）和长度进行连接（后一个标签的 s 是前一个的 length）。

而属性中的 x、y、hdg 分别是投影坐标系 x、y 下的起始点位置以及起始点的角度（定义了曲线方程以及起始点坐标和长度，曲线肯定就能画出来了）。

3. 车道

一条道路中包含很多的车道（Lane），而车道本身有宽度（width）以及虚线、实线等属性参数（roadMark）。结合这些参数，我们就能在参考线的基础上将车道画出来，如图 11.8 所示。

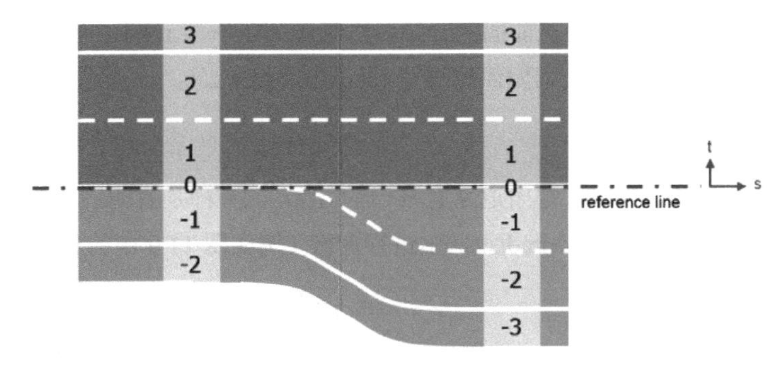

图 11.8　车道

下面是 OpenDrive 中对应的车道相关的元素。

```
    <lanes>
            <laneSection s="0.0000000000000000e+00">
                <center>
                    <lane id="0" type="driving" level="false">
                        <link />
                        <roadMark sOffset="0.0000000000000000e+00" type="solid"
weight="standard" color="standard" material="standard" LDM="none"
width="2.9999999999999999e-01" laneChange="both" />
                    </lane>
                </center>
```

```
                <right>
                    <lane id="-1" type="driving" level="false">
                        <link>
                            <predecessor id="-1" />
                            <successor id="-1" />
                        </link>
                        <width sOffset="0.0000000000000000e+00"
a="3.8890850467340541e+00" b="-1.4514389448175911e-03"
c="1.0899364495936138e-04" d="-1.3397356888919692e-06" />
                        <width sOffset="7.6500000000000000e+01"
a="3.8161122099921028e+00" b="1.6531839124687595e-03"
c="-3.0234314157904548e-06" d="-2.9791355887866248e-08" />
                        <roadMark sOffset="0.0000000000000000e+00" type="broken"
weight="standard" color="standard" material="standard" LDM="none"
width="1.4999999999999999e-01" laneChange="both" />
                    </lane>
                </right>
            </laneSection>
    </lanes>
```

在车道下还有一个 laneSection（车道横截面）的概念，一条道路包含数个 laneSection，每个 laneSection 中又包含车道。在一个 laneSection 中，车道顺着参考线分为 left、center、right。参考线是 center，没有宽度，只是一条线。left 的车道的 id 为正，right 的车道的 id 为负。图 11.8 中左边定义了 5 条车道：1、2、3、-1、-2，而右边多了一条-3。此外，在 lane 元素中，width 元素定义了车道的宽度，都是基于曲线进行拟合的。roadMark 元素定义了车道线的属性，OpenDrive 中规定的车道线的属性有实线、虚线等。通过前面的介绍，已经画完了一个 road，而前面提到的 OpenDrive 地图是由多条车道组成的，下面介绍如何将这些 road 连接起来。

4. 道路 road 连接

道路之间的连接定义了两种（每条道路有唯一的 ID），一种是有明确的连接关系，例如前后只有一条道路，那么通过 successor/predecessor 进行连接（例如图 11.9 中的 ROAD 1 和 ROAD 2）。

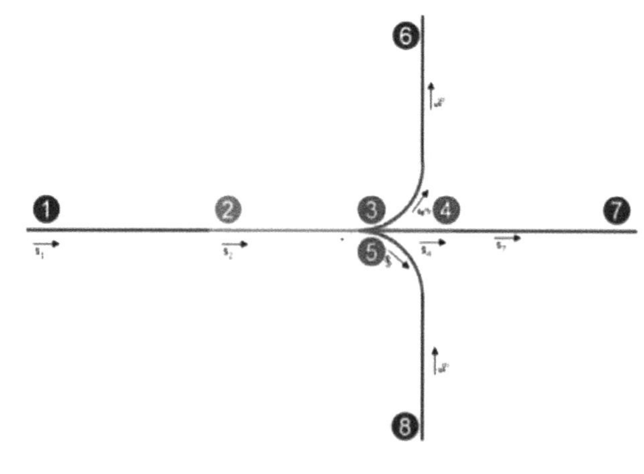

图 11.9　道路连接

如果前后的连接关系不是很明确，就需要一个道路连接关系（junction），如图 11.10 所示。图 11.9 中 ROAD 2 的后置节点 successor 就无法确定，在 OpenDrive 中，将图 11.10 中的 3、4、5 称为 junction，ROAD 3、4、5 称为 connectingRoad，而 2 称为 incomingRoad，6、7、8 称为 outgoingRoad。

Road	Predecessor	Successor
1	-	2
2	1	ambiguous
3	2	6
4	2	7
5	2	8
6	3	-
7	4	-
8	-	5

图 11.10 道路连接关系

具体体现在 OpenDrive 中，代码如下。junction 41 的 incomingRoad 为 ROAD 42，而 ROAD 55 属于 junction 中的 connectingRoad。

```
    <junction name="normal" id="41">
        <connection id="1" incomingRoad="42" connectingRoad="55"
contactPoint="start">
            <laneLink from="-3" to="-3" />
            <laneLink from="-2" to="-2" />
            <laneLink from="-1" to="-1" />
        </connection>
    </junction>
```

由于道路中又包含很多车道，因此需要将车道的连接关系表示清楚，上面的 laneLink 元素就是用来连接车道的，对 ROAD 42 和 55 的-3、-2、-1 进行连接（center 右边的车道）。

图 11.11 是一个完整的 OpenDrive 实例。

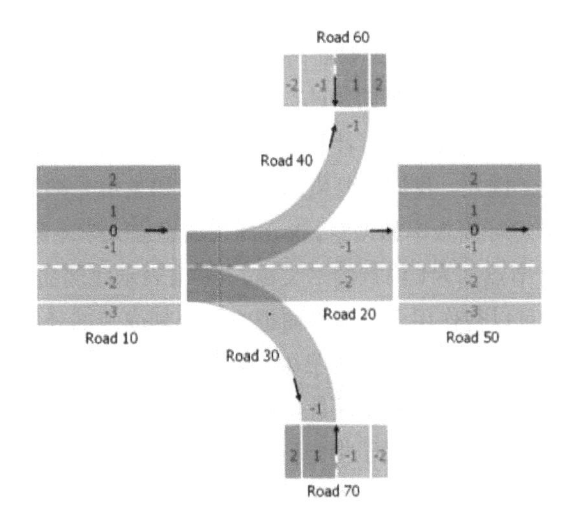

图 11.11 OpenDrive 实例

图 11.11 体现了 OpenDrive 中的路网连接关系，分为：

左边 Road 10，参考线黑线所在的位置，左边有 Lane 1 和 Lane 2，右边为 Lane-1、Lane-2 和 Lane-3。

上边 Road 60，参考线黑线所在的位置，左边有 Lane 1 和 Lane 2，右边为 Lane-1 和 Lane-2。

右边 Road 50，参考线黑线所在的位置，左边有 Lane 1 和 Lane 2，右边为 Lane-1、Lane-2 和 Lane -3。

下边 Road 70，参考线黑线所在的位置，左边有 Lane 1 和 Lane 2，右边为 Lane-1 和 Lane-2。

中间是一个 junction，包含 Road 40、Road 20 和 Road 30，都是 connectingRoad，对应的连接关系为 Road 10-1 Lane (incomingRoad)<—>Road 40(connectingRoad)-1 Lane，Road 10-2 Lane (incomingRoad)<—>Road 30(connectingRoad)-1 Lane，Road 10-1 Lane (incomingRoad)<—>Road 20(connectingRoad) -1 Lane，Road 10-2 Lane (incomingRoad)<—>Road 20(connectingRoad) -2 Lane，而 outgoingRoad 与之类似。

总之，对于一条道路来说，先确定参考线，有了参考线的几何形状和位置后，再确定参考线左右的车道，车道又有实线和虚线等属性。道路和道路之间通过普通连接和 junction 进行连接，同时还要将道路中的相关车道进行连接。

11.1.4 百度 Apollo 相关源代码分析

在 Apollo 系统中，高精地图相关代码主要在 modules\map\hdmap 文件夹下，hdmap.cc 文件为程序入口，实际上实现主要在 hdmap_impl.cc 文件中。在实现中最重要的有两部分：一部分是构建单一地图元素的集合，另一部分是利用算法快速查找符合条件的地图元素。如果我们使用面向对象的思路来分析这两个过程，其实这两个过程就是 Apollo 高精度地图的 set 方法集合和对应的 get 方法集合。在这个过程中，最重要的是能够快速完成某种子元素的查找过程，因此在设计这一模块的时候使用了 KDTree 的数据结构。

在 SIFT 图像特征匹配等应用中，需要在高维特征空间中快速找到距离目标图像特征最近邻的那个特征点，往往需要进行比较的特征向量的数量很大，如果进行朴素最近邻搜索，也就是依次计算目标点和每一个待匹配特征的距离，再计算最短距离这样的策略，那么特征匹配算法的时间复杂度将会高得令人难以接受。因此，我们需要借助一种存储和表示 k 维数据的数据结构，既能够方便地存储 k 维数据，又能够进行高效率的搜索。k-d 树由斯坦福大学本科生 Jon Louis Bentley 于 1975 年首次提出。k-d 树是每个节点都为 k 维点的二叉树，其中 k 表示存储的数据的维度，d 就是 dimension 的意思。所有非叶子节点可以视作用一个超平面把空间分割成两部分，在超平面左边的点代表节点的左子树，在超平面右边的点代表节点的右子树。超平面的方向可以用下述方法来选择：每个节点都与 k 维中垂直于超平面的那一维有关。因此，如果选择按照 x 轴划分，那么所有 x 值小于指定值的节点都会出现在左子树，所有 x 值大于指定值的节点都会出现在右子树。这样超平面可以用该 x 值来确定，其法矢为 x 轴的单位向量。一个三维空间内的 3-d 树如图 11.12 所示。

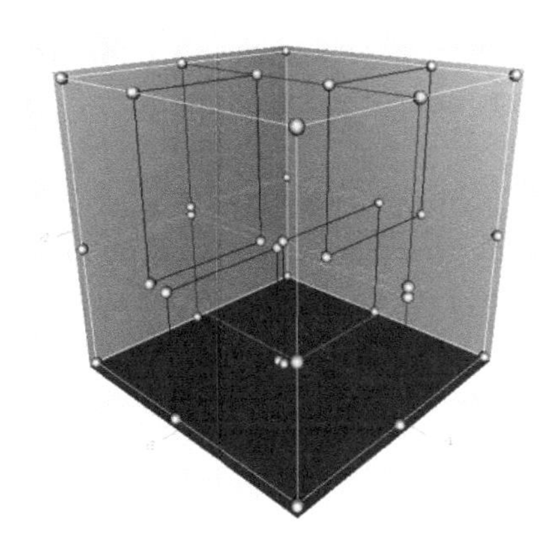

图 11.12　3-d 树示意图

构造 k-d 树的方法：首先构造根节点，根节点对应于整个 k 维空间，包含所有的实例点（至于如何选取划分点，有不同的策略。常用的一种方法是：对于所有的样本点，统计它们在每个维上的方差，挑选出方差中的最大值，对应的维就是要进行数据切分的维度。数据方差最大表明沿该维度数据点分散得比较开，这个方向上进行数据分割可以获得最好的分辨率；然后将所有样本点按切分维度的值进行排序，位于正中间的那个数据点选为分裂节点）。然后利用递归的方法分别构造 k-d 树根节点的左右子树。在超矩形区域上选择一个坐标轴（切分维度）和一个分裂节点，以通过此分裂节点且垂直于切分方向坐标轴的直线作为分隔线，将当前超矩形区域分隔成左右或者上下两个子超矩形区域，对应于分裂节点的左右子树的根节点。实例也就被分到两个不相交的区域中了。重复此过程，直到子区域内没有实例点时终止。终止时的节点为叶节点。

通常依次选择坐标轴对空间切分，选择训练实例点在选定坐标轴上的中位数为切分点，这样得到的 k-d 树是平衡的，但并不一定能保证检索的效率最优。

k-d 树最近邻搜索算法基本的思路很简单：首先通过二叉树搜索（比较待查询节点和分裂节点的分裂维的值，小于等于就进入左子树分支，等于就进入右子树分支，直到叶子节点），顺着“搜索路径”很快能找到最近邻的近似点，也就是与待查询点处于同一个子空间的叶子节点；然后回溯搜索路径，并判断搜索路径上的节点的其他子节点空间中是否可能有距离查询点更近的数据点，如果可能有，就需要跳到其他子节点空间中去搜索（将其他子节点加入搜索路径）。重复这个过程直到搜索路径为空。

Apollo 中的 KDTree 的实现是 AABoxKDTree2d，文件位于 modules\common\math 中。相关的构造核心代码如下：

```
    if (SplitToSubNodes(objects, params)) {
        std::vector<ObjectPtr> left_subnode_objects;
        std::vector<ObjectPtr> right_subnode_objects;
        PartitionObjects(objects, &left_subnode_objects,
&right_subnode_objects);

        // Split to sub-nodes.
        if (!left_subnode_objects.empty()) {
```

```
        left_subnode_.reset(new AABoxKDTree2dNode<ObjectType>(
            left_subnode_objects, params, depth + 1));
      }
      if (!right_subnode_objects.empty()) {
        right_subnode_.reset(new AABoxKDTree2dNode<ObjectType>(
            right_subnode_objects, params, depth + 1));
      }
    } else {
      InitObjects(objects);
  }
```

在搜索核心代码中，可以清楚地看到递归过程的应用。比如其中最重要的 GetObjects 函数，实现如下：

```
std::vector<ObjectPtr> GetObjects(const Vec2d &point,
                                  const double distance) const {
    if (root_ == nullptr) {
      return {};
    }
    return root_->GetObjects(point, distance);
  }
```

有了 KDTree 的相关知识准备，下面使用注释的方式详细分析一下 hdmap_impl.cc 的源代码。

```
// 地图文件读取功能
int HDMapImpl::LoadMapFromFile(const std::string& map_filename) {
  // 读取之前清除内存中的地图数据和相关的关系（包括道路信息、路口信息、交通标识信息等）
  Clear();
  // OpenDrive 格式的地图标准形式是 XML，在 Apollo 系统中的基础文件是 Proto 标准格式的，
在 HDmap 文件中支持载入这两种格式的 OpenDrive Apollo 地图
  if (apollo::common::util::EndWith(map_filename, ".xml")) {
    if (!adapter::OpendriveAdapter::LoadData(map_filename, &map_)) {
      return -1;
    }
  } else if (!cyber::common::GetProtoFromFile(map_filename, &map_)) {
    return -1;
  }

  return LoadMapFromProto(map_);
}
// 从 Proto 文件中载入地图数据
int HDMapImpl::LoadMapFromProto(const Map& map_proto) {
  if (&map_proto != &map_) {
//避免不必要的内存浪费，在载入数据之前对 map 对象进行清理
    Clear();
```

```
    map_ = map_proto;
  }
  //遍历读取车道线信息
  for (const auto& lane : map_.lane()) {
    lane_table_[lane.id().id()].reset(new LaneInfo(lane));
  }
  //遍历读取路口信息
  for (const auto& junction : map_.junction()) {
    junction_table_[junction.id().id()].reset(new JunctionInfo(junction));
  }
  //遍历读取交通标志信息
  for (const auto& signal : map_.signal()) {
    signal_table_[signal.id().id()].reset(new SignalInfo(signal));
  }
  //遍历读取人行横道信息
  for (const auto& crosswalk : map_.crosswalk()) {
    crosswalk_table_[crosswalk.id().id()].reset(new
CrosswalkInfo(crosswalk));
  }
  //遍历读取停车标志信息
  for (const auto& stop_sign : map_.stop_sign()) {
    stop_sign_table_[stop_sign.id().id()].reset(new
StopSignInfo(stop_sign));
  }
  //遍历读取停车标志信息
  for (const auto& yield_sign : map_.yield()) {
    yield_sign_table_[yield_sign.id().id()].reset(
        new YieldSignInfo(yield_sign));
  }
  //遍历读取清洁区域信息
  for (const auto& clear_area : map_.clear_area()) {
    clear_area_table_[clear_area.id().id()].reset(
        new ClearAreaInfo(clear_area));
  }
  //遍历读取减速带地图信息
  for (const auto& speed_bump : map_.speed_bump()) {
    speed_bump_table_[speed_bump.id().id()].reset(
        new SpeedBumpInfo(speed_bump));
  }
  //遍历读取停车位置地图信息
  for (const auto& parking_space : map_.parking_space()) {
    parking_space_table_[parking_space.id().id()].reset(
        new ParkingSpaceInfo(parking_space));
```

```
  }
//遍历读取 PNC 路口信息
  for (const auto& pnc_junction : map_.pnc_junction()) {
    pnc_junction_table_[pnc_junction.id().id()].reset(
        new PNCJunctionInfo(pnc_junction));
  }
//遍历读取 PNC 路口信息
  for (const auto& overlap : map_.overlap()) {
    overlap_table_[overlap.id().id()].reset(new OverlapInfo(overlap));
  }
//遍历道路地图信息
  for (const auto& road : map_.road()) {
    road_table_[road.id().id()].reset(new RoadInfo(road));
  }
//遍历道路、道路行车区域和车道线的地图关联关系信息
  for (const auto& road_ptr_pair : road_table_) {
    const auto& road_id = road_ptr_pair.second->id();
    for (const auto& road_section : road_ptr_pair.second->sections()) {
      const auto& section_id = road_section.id();
      for (const auto& lane_id : road_section.lane_id()) {
        auto iter = lane_table_.find(lane_id.id());
        if (iter != lane_table_.end()) {
          iter->second->set_road_id(road_id);
          iter->second->set_section_id(section_id);
        } else {
          AFATAL << "Unknown lane id: " << lane_id.id();
        }
      }
    }
  }
//遍历车道信息提取对应关系
  for (const auto& lane_ptr_pair : lane_table_) {
    lane_ptr_pair.second->PostProcess(*this);
  }
//遍历路口信息提取对应关系
  for (const auto& junction_ptr_pair : junction_table_) {
    junction_ptr_pair.second->PostProcess(*this);
  }
//遍历停车标志信息提取对应关系
  for (const auto& stop_sign_ptr_pair : stop_sign_table_) {
    stop_sign_ptr_pair.second->PostProcess(*this);
  }
//建立上面解析出的车道线、路口、交通信号等地图对象对应的 KDTree
```

```
  BuildLaneSegmentKDTree();
  BuildJunctionPolygonKDTree();
  BuildSignalSegmentKDTree();
  BuildCrosswalkPolygonKDTree();
  BuildStopSignSegmentKDTree();
  BuildYieldSignSegmentKDTree();
  BuildClearAreaPolygonKDTree();
  BuildSpeedBumpSegmentKDTree();
  BuildParkingSpacePolygonKDTree();
  BuildPNCJunctionPolygonKDTree();
  return 0;
}
//通过车道线对象的 id 来得到车道线对象
LaneInfoConstPtr HDMapImpl::GetLaneById(const Id& id) const {
  LaneTable::const_iterator it = lane_table_.find(id.id());
  return it != lane_table_.end() ? it->second : nullptr;
}
//通过路口对象的 id 来得到路口对象
JunctionInfoConstPtr HDMapImpl::GetJunctionById(const Id& id) const {
  JunctionTable::const_iterator it = junction_table_.find(id.id());
  return it != junction_table_.end() ? it->second : nullptr;
}
//通过交通信号对象的 id 来得到交通信号对象
SignalInfoConstPtr HDMapImpl::GetSignalById(const Id& id) const {
  SignalTable::const_iterator it = signal_table_.find(id.id());
  return it != signal_table_.end() ? it->second : nullptr;
}
//通过人行横道对象的 id 来得到人行横道对象
CrosswalkInfoConstPtr HDMapImpl::GetCrosswalkById(const Id& id) const {
  CrosswalkTable::const_iterator it = crosswalk_table_.find(id.id());
  return it != crosswalk_table_.end() ? it->second : nullptr;
}
//通过停车标志对象的 id 来得到对应的停车标志对象
StopSignInfoConstPtr HDMapImpl::GetStopSignById(const Id& id) const {
  StopSignTable::const_iterator it = stop_sign_table_.find(id.id());
  return it != stop_sign_table_.end() ? it->second : nullptr;
}
//通过让路标志对象的 id 来得到对应的让路标志对象
YieldSignInfoConstPtr HDMapImpl::GetYieldSignById(const Id& id) const {
  YieldSignTable::const_iterator it = yield_sign_table_.find(id.id());
  return it != yield_sign_table_.end() ? it->second : nullptr;
}
//通过清洁区对象的 id 来得到对应的清洁区对象
```

```
ClearAreaInfoConstPtr HDMapImpl::GetClearAreaById(const Id& id) const {
  ClearAreaTable::const_iterator it = clear_area_table_.find(id.id());
  return it != clear_area_table_.end() ? it->second : nullptr;
}
//通过减速带对象的 id 来得到对应的减速带对象
SpeedBumpInfoConstPtr HDMapImpl::GetSpeedBumpById(const Id& id) const {
  SpeedBumpTable::const_iterator it = speed_bump_table_.find(id.id());
  return it != speed_bump_table_.end() ? it->second : nullptr;
}
//通过覆盖区对象的 id 来得到对应的覆盖区对象
OverlapInfoConstPtr HDMapImpl::GetOverlapById(const Id& id) const {
  OverlapTable::const_iterator it = overlap_table_.find(id.id());
  return it != overlap_table_.end() ? it->second : nullptr;
}
//通过道路对象的 id 来得到对应的道路对象
RoadInfoConstPtr HDMapImpl::GetRoadById(const Id& id) const {
  RoadTable::const_iterator it = road_table_.find(id.id());
  return it != road_table_.end() ? it->second : nullptr;
}
//通过停车位对象的 id 来得到对应的停车位对象
ParkingSpaceInfoConstPtr HDMapImpl::GetParkingSpaceById(const Id& id) const {
  ParkingSpaceTable::const_iterator it = parking_space_table_.find(id.id());
  return it != parking_space_table_.end() ? it->second : nullptr;
}
//通过 PNC 路口对象的 id 来得到对应的 PNC 路口对象
PNCJunctionInfoConstPtr HDMapImpl::GetPNCJunctionById(const Id& id) const {
  PNCJunctionTable::const_iterator it = pnc_junction_table_.find(id.id());
  return it != pnc_junction_table_.end() ? it->second : nullptr;
}
//通过起点和距离来得到符合标准的车道线对象集合
int HDMapImpl::GetLanes(const PointENU& point, double distance,
                        std::vector<LaneInfoConstPtr>* lanes) const {
  return GetLanes({point.x(), point.y()}, distance, lanes);
}

int HDMapImpl::GetLanes(const Vec2d& point, double distance,
                        std::vector<LaneInfoConstPtr>* lanes) const {
  if (lanes == nullptr || lane_segment_kdtree_ == nullptr) {
    return -1;
  }
  lanes->clear();
  std::vector<std::string> ids;
  //得到车道线的算法依赖于在对象构建环节建立的车道线 KDtree
```

```
  const int status =
      SearchObjects(point, distance, *lane_segment_kdtree_, &ids);
  if (status < 0) {
    return status;
  }
  for (const auto& id : ids) {
    lanes->emplace_back(GetLaneById(CreateHDMapId(id)));
  }
  return 0;
}
//通过起点和距离来得到符合标准的道路对象集合
int HDMapImpl::GetRoads(const PointENU& point, double distance,
                        std::vector<RoadInfoConstPtr>* roads) const {
  return GetRoads({point.x(), point.y()}, distance, roads);
}

int HDMapImpl::GetRoads(const Vec2d& point, double distance,
                        std::vector<RoadInfoConstPtr>* roads) const {
  std::vector<LaneInfoConstPtr> lanes;
//因为道路中包含车道线，所以调用筛选车道线的方法来得到道路对应的车道线
  if (GetLanes(point, distance, &lanes) != 0) {
    return -1;
  }
//从符合条件的车道线对象中找到道路的 ID
  std::unordered_set<std::string> road_ids;
  for (auto& lane : lanes) {
    if (!lane->road_id().id().empty()) {
      road_ids.insert(lane->road_id().id());
    }
  }
//通过利用 ID 找到对应道路对象的方法找到对应的道路对象集合
  for (auto& road_id : road_ids) {
    RoadInfoConstPtr road = GetRoadById(CreateHDMapId(road_id));
    CHECK_NOTNULL(road);
    roads->push_back(road);
  }

  return 0;
}
//通过起点和距离来得到符合标准的路口对象集合
int HDMapImpl::GetJunctions(
    const PointENU& point, double distance,
    std::vector<JunctionInfoConstPtr>* junctions) const {
```

```
  return GetJunctions({point.x(), point.y()}, distance, junctions);
}

int HDMapImpl::GetJunctions(
    const Vec2d& point, double distance,
    std::vector<JunctionInfoConstPtr>* junctions) const {
  if (junctions == nullptr || junction_polygon_kdtree_ == nullptr) {
    return -1;
  }
  junctions->clear();
  std::vector<std::string> ids;
//得到路口对象的算法依赖于在对象构建环节建立的路口对象的 KDtree
  const int status =
      SearchObjects(point, distance, *junction_polygon_kdtree_, &ids);
  if (status < 0) {
    return status;
  }
  for (const auto& id : ids) {
    junctions->emplace_back(GetJunctionById(CreateHDMapId(id)));
  }
  return 0;
}
//通过起点和距离来得到符合标准的交通标志对象集合
int HDMapImpl::GetSignals(const PointENU& point, double distance,
                          std::vector<SignalInfoConstPtr>* signals) const {
  return GetSignals({point.x(), point.y()}, distance, signals);
}

int HDMapImpl::GetSignals(const Vec2d& point, double distance,
                          std::vector<SignalInfoConstPtr>* signals) const {
  if (signals == nullptr || signal_segment_kdtree_ == nullptr) {
    return -1;
  }
  signals->clear();
  std::vector<std::string> ids;
//得到交通标志对象的算法依赖于在对象构建环节建立的交通标志对象的 KDtree
  const int status =
      SearchObjects(point, distance, *signal_segment_kdtree_, &ids);
  if (status < 0) {
    return status;
  }
  for (const auto& id : ids) {
    signals->emplace_back(GetSignalById(CreateHDMapId(id)));
```

```
  }
  return 0;
}
//通过起点和距离来得到符合标准的人行横道对象集合
int HDMapImpl::GetCrosswalks(
    const PointENU& point, double distance,
    std::vector<CrosswalkInfoConstPtr>* crosswalks) const {
  return GetCrosswalks({point.x(), point.y()}, distance, crosswalks);
}

int HDMapImpl::GetCrosswalks(
    const Vec2d& point, double distance,
    std::vector<CrosswalkInfoConstPtr>* crosswalks) const {
  if (crosswalks == nullptr || crosswalk_polygon_kdtree_ == nullptr) {
    return -1;
  }
  crosswalks->clear();
  std::vector<std::string> ids;
//得到人行横道对象的算法依赖于在对象构建环节建立的人行横道对象的 KDtree
  const int status =
      SearchObjects(point, distance, *crosswalk_polygon_kdtree_, &ids);
  if (status < 0) {
    return status;
  }
  for (const auto& id : ids) {
    crosswalks->emplace_back(GetCrosswalkById(CreateHDMapId(id)));
  }
  return 0;
}
//通过起点和距离来得到符合标准的停车标志对象集合
int HDMapImpl::GetStopSigns(
    const PointENU& point, double distance,
    std::vector<StopSignInfoConstPtr>* stop_signs) const {
  return GetStopSigns({point.x(), point.y()}, distance, stop_signs);
}

int HDMapImpl::GetStopSigns(
    const Vec2d& point, double distance,
    std::vector<StopSignInfoConstPtr>* stop_signs) const {
  if (stop_signs == nullptr || stop_sign_segment_kdtree_ == nullptr) {
    return -1;
  }
  stop_signs->clear();
```

```
  std::vector<std::string> ids;
//得到停车标志对象的算法依赖于在对象构建环节建立的停车标志对象的 KDtree
  const int status =
      SearchObjects(point, distance, *stop_sign_segment_kdtree_, &ids);
  if (status < 0) {
    return status;
  }
  for (const auto& id : ids) {
    stop_signs->emplace_back(GetStopSignById(CreateHDMapId(id)));
  }
  return 0;
}
//通过起点和距离来得到符合标准的让路标志对象集合
int HDMapImpl::GetYieldSigns(
    const PointENU& point, double distance,
    std::vector<YieldSignInfoConstPtr>* yield_signs) const {
  return GetYieldSigns({point.x(), point.y()}, distance, yield_signs);
}

int HDMapImpl::GetYieldSigns(
    const Vec2d& point, double distance,
    std::vector<YieldSignInfoConstPtr>* yield_signs) const {
  if (yield_signs == nullptr || yield_sign_segment_kdtree_ == nullptr) {
    return -1;
  }
  yield_signs->clear();
  std::vector<std::string> ids;
//得到让路标志对象的算法依赖于在对象构建环节建立的让路标志对象的 KDtree
  const int status =
      SearchObjects(point, distance, *yield_sign_segment_kdtree_, &ids);
  if (status < 0) {
    return status;
  }
  for (const auto& id : ids) {
    yield_signs->emplace_back(GetYieldSignById(CreateHDMapId(id)));
  }

  return 0;
}
//通过起点和距离来得到符合标准的畅通区对象集合
int HDMapImpl::GetClearAreas(
    const PointENU& point, double distance,
    std::vector<ClearAreaInfoConstPtr>* clear_areas) const {
```

```
  return GetClearAreas({point.x(), point.y()}, distance, clear_areas);
}

int HDMapImpl::GetClearAreas(
    const Vec2d& point, double distance,
    std::vector<ClearAreaInfoConstPtr>* clear_areas) const {
  if (clear_areas == nullptr || clear_area_polygon_kdtree_ == nullptr) {
    return -1;
  }
  clear_areas->clear();
  std::vector<std::string> ids;
//得到畅通区域对象的算法依赖于在对象构建环节建立的畅通区域对象的 KDtree
  const int status =
      SearchObjects(point, distance, *clear_area_polygon_kdtree_, &ids);
  if (status < 0) {
    return status;
  }
  for (const auto& id : ids) {
    clear_areas->emplace_back(GetClearAreaById(CreateHDMapId(id)));
  }

  return 0;
}
//通过起点和距离来得到符合标准的减速带对象集合
int HDMapImpl::GetSpeedBumps(
    const PointENU& point, double distance,
    std::vector<SpeedBumpInfoConstPtr>* speed_bumps) const {
  return GetSpeedBumps({point.x(), point.y()}, distance, speed_bumps);
}

int HDMapImpl::GetSpeedBumps(
    const Vec2d& point, double distance,
    std::vector<SpeedBumpInfoConstPtr>* speed_bumps) const {
  if (speed_bumps == nullptr || speed_bump_segment_kdtree_ == nullptr) {
    return -1;
  }
  speed_bumps->clear();
  std::vector<std::string> ids;
//得到减速带对象的算法依赖于在对象构建环节建立的减速带对象的 KDtree
  const int status =
      SearchObjects(point, distance, *speed_bump_segment_kdtree_, &ids);
  if (status < 0) {
    return status;
```

```
  }
  for (const auto& id : ids) {
    speed_bumps->emplace_back(GetSpeedBumpById(CreateHDMapId(id)));
  }

  return 0;
}
//通过起点和距离来得到符合标准的停车区域对象集合
int HDMapImpl::GetParkingSpaces(
    const PointENU& point, double distance,
    std::vector<ParkingSpaceInfoConstPtr>* parking_spaces) const {
  return GetParkingSpaces({point.x(), point.y()}, distance, parking_spaces);
}

int HDMapImpl::GetParkingSpaces(
    const Vec2d& point, double distance,
    std::vector<ParkingSpaceInfoConstPtr>* parking_spaces) const {
  if (parking_spaces == nullptr || parking_space_polygon_kdtree_ == nullptr)
{
    return -1;
  }
  parking_spaces->clear();
  std::vector<std::string> ids;
//得到停车区域对象的算法依赖于在对象构建环节建立的停车区域对象的 KDtree
  const int status =
      SearchObjects(point, distance, *parking_space_polygon_kdtree_, &ids);
  if (status < 0) {
    return status;
  }
  for (const auto& id : ids) {
    parking_spaces->emplace_back(GetParkingSpaceById(CreateHDMapId(id)));
  }

  return 0;
}
//通过起点和距离来得到符合标准的 PNC 路口集合
int HDMapImpl::GetPNCJunctions(
    const apollo::common::PointENU& point, double distance,
    std::vector<PNCJunctionInfoConstPtr>* pnc_junctions) const {
  return GetPNCJunctions({point.x(), point.y()}, distance, pnc_junctions);
}

int HDMapImpl::GetPNCJunctions(
```

```
    const apollo::common::math::Vec2d& point, double distance,
    std::vector<PNCJunctionInfoConstPtr>* pnc_junctions) const {
  if (pnc_junctions == nullptr || pnc_junction_polygon_kdtree_ == nullptr) {
    return -1;
  }
  pnc_junctions->clear();

  std::vector<std::string> ids;
//得到 PNC 路口对象的算法依赖于在对象构建环节建立的 PNC 路口对象的 KDtree
  const int status =
      SearchObjects(point, distance, *pnc_junction_polygon_kdtree_, &ids);
  if (status < 0) {
    return status;
  }

  for (const auto& id : ids) {
    pnc_junctions->emplace_back(GetPNCJunctionById(CreateHDMapId(id)));
  }

  return 0;
}
//通过给定点得到距离给定点最近的车道线
int HDMapImpl::GetNearestLane(const PointENU& point,
                              LaneInfoConstPtr* nearest_lane, double* nearest_s,
                              double* nearest_l) const {
  return GetNearestLane({point.x(), point.y()}, nearest_lane, nearest_s,
                        nearest_l);
}

int HDMapImpl::GetNearestLane(const Vec2d& point,
                              LaneInfoConstPtr* nearest_lane, double* nearest_s,
                              double* nearest_l) const {
  CHECK_NOTNULL(nearest_lane);
  CHECK_NOTNULL(nearest_s);
  CHECK_NOTNULL(nearest_l);
//通过 KDTree 的遍历来得到离最近点最近的车道线
  const auto* segment_object = lane_segment_kdtree_->GetNearestObject(point);
  if (segment_object == nullptr) {
    return -1;
  }
  const Id& lane_id = segment_object->object()->id();
  *nearest_lane = GetLaneById(lane_id);
  CHECK(*nearest_lane);
```

```
  const int id = segment_object->id();
  const auto& segment = (*nearest_lane)->segments()[id];
  Vec2d nearest_pt;
  segment.DistanceTo(point, &nearest_pt);
  *nearest_s = (*nearest_lane)->accumulate_s()[id] +
             nearest_pt.DistanceTo(segment.start());
  *nearest_l = segment.unit_direction().CrossProd(point - segment.start());

  return 0;
}
//通过给定点的位置信息和朝向信息得到与给定点距离最近而且方向一致的车道线
int HDMapImpl::GetNearestLaneWithHeading(
    const PointENU& point, const double distance, const double central_heading,
    const double max_heading_difference, LaneInfoConstPtr* nearest_lane,
    double* nearest_s, double* nearest_l) const {
  return GetNearestLaneWithHeading({point.x(), point.y()}, distance,
                                  central_heading, max_heading_difference,
                                  nearest_lane, nearest_s, nearest_l);
}

int HDMapImpl::GetNearestLaneWithHeading(
    const Vec2d& point, const double distance, const double central_heading,
    const double OpenDrive 地图格式简介
max_heading_difference, LaneInfoConstPtr* nearest_lane,
    double* nearest_s, double* nearest_l) const {
  CHECK_NOTNULL(nearest_lane);
  CHECK_NOTNULL(nearest_s);
  CHECK_NOTNULL(nearest_l);

  std::vector<LaneInfoConstPtr> lanes;
//需要首先调用获得符合朝向的车道线函数，获得所有符合朝向和位置的车道线集合
  if (GetLanesWithHeading(point, distance, central_heading,
                         max_heading_difference, &lanes) != 0) {
    return -1;
  }

  double s = 0;
  size_t s_index = 0;
  Vec2d map_point;
  double min_distance = distance;
//接下来需要遍历得到的车道线集合，选择出距离最近的一条车道线
  for (const auto& lane : lanes) {
    double s_offset = 0.0;
```

```
    int s_offset_index = 0;
    double distance =
        lane->DistanceTo(point, &map_point, &s_offset, &s_offset_index);
    if (distance < min_distance) {
      min_distance = distance;
      *nearest_lane = lane;
      s = s_offset;
      s_index = s_offset_index;
    }
  }

  if (*nearest_lane == nullptr) {
    return -1;
  }

  *nearest_s = s;
  int segment_index = static_cast<int>(
      std::min(s_index, (*nearest_lane)->segments().size() - 1));
  const auto& segment_2d = (*nearest_lane)->segments()[segment_index];
  *nearest_l =
      segment_2d.unit_direction().CrossProd(point - segment_2d.start());

  return 0;
}
//通过给定点的位置信息距离和朝向信息选出符合条件的车道线集合
int HDMapImpl::GetLanesWithHeading(const PointENU& point, const double
distance,
                                   const double central_heading,
                                   const double max_heading_difference,
                                   std::vector<LaneInfoConstPtr>* lanes) const {
  return GetLanesWithHeading({point.x(), point.y()}, distance,
central_heading,
                             max_heading_difference, lanes);
}

int HDMapImpl::GetLanesWithHeading(const Vec2d& point, const double distance,
                                   const double central_heading,
                                   const double max_heading_difference,
                                   std::vector<LaneInfoConstPtr>* lanes) const {
  CHECK_NOTNULL(lanes);
  std::vector<LaneInfoConstPtr> all_lanes;
//首先调用取得车道线的函数
  const int status = GetLanes(point, distance, &all_lanes);
```

```
  if (status < 0 || all_lanes.size() <= 0) {
    return -1;
  }

  lanes->clear();
//然后需要遍历检测每个车道线的朝向，选出朝向在指定朝向误差范围的车道线集合
  for (auto& lane : all_lanes) {
    Vec2d proj_pt(0.0, 0.0);
    double s_offset = 0.0;
    int s_offset_index = 0;
    double dis = lane->DistanceTo(point, &proj_pt, &s_offset, &s_offset_index);
    if (dis <= distance) {
      double heading_diff =
          fabs(lane->headings()[s_offset_index] - central_heading);
      if (fabs(apollo::common::math::NormalizeAngle(heading_diff)) <=
          max_heading_difference) {
        lanes->push_back(lane);
      }
    }
  }

  return 0;
}
//通过给定点的位置信息距离选出符合条件的道路边界对象集合
int HDMapImpl::GetRoadBoundaries(
    const PointENU& point, double radius,
    std::vector<RoadROIBoundaryPtr>* road_boundaries,
    std::vector<JunctionBoundaryPtr>* junctions) const {
  CHECK_NOTNULL(road_boundaries);
  CHECK_NOTNULL(junctions);

  road_boundaries->clear();
  junctions->clear();

  std::vector<LaneInfoConstPtr> lanes;
//首先调用取得车道线集合的函数
  if (GetLanes(point, radius, &lanes) != 0 || lanes.size() <= 0) {
    return -1;
  }

  std::unordered_set<std::string> junction_id_set;
  std::unordered_set<std::string> road_section_id_set;
//遍历车道线集合
```

```
  for (const auto& lane : lanes) {
//通过车道线信息中的道路 ID 得到对应的道路信息
    const auto road_id = lane->road_id();
    const auto section_id = lane->section_id();
    std::string unique_id = road_id.id() + section_id.id();
    if (road_section_id_set.count(unique_id) > 0) {
      continue;
    }
road_section_id_set.insert(unique_id);
    const auto road_ptr = GetRoadById(road_id);
    CHECK_NOTNULL(road_ptr);
    if (road_ptr->has_junction_id()) {//如果道路信息中存在路口信息，就提取路口信息
      const Id junction_id = road_ptr->junction_id();
      if (junction_id_set.count(junction_id.id()) > 0) {
        continue;
      }
      junction_id_set.insert(junction_id.id());
      JunctionBoundaryPtr junction_boundary_ptr(new JunctionBoundary());
      junction_boundary_ptr->junction_info = GetJunctionById(junction_id);
      CHECK_NOTNULL(junction_boundary_ptr->junction_info);
      junctions->push_back(junction_boundary_ptr);
    } else {//如果没有路口信息，就需要使用 ROI 过滤器从道路旁边的路口信息提取路口信息
      RoadROIBoundaryPtr road_boundary_ptr(new RoadROIBoundary());
      road_boundary_ptr->mutable_id()->CopyFrom(road_ptr->id());
      for (const auto& section : road_ptr->sections()) {
        if (section.id().id() == section_id.id()) {
          road_boundary_ptr->add_road_boundaries()->CopyFrom(
              section.boundary());
        }
      }
      road_boundaries->push_back(road_boundary_ptr);
    }
  }

  return 0;
}

int HDMapImpl::GetRoadBoundaries(
    const PointENU& point, double radius,
    std::vector<RoadRoiPtr>* road_boundaries,
    std::vector<JunctionInfoConstPtr>* junctions) const {
  if (road_boundaries == nullptr || junctions == nullptr) {
    AERROR << "the pointer in parameter is null";
```

```
    return -1;
  }
  road_boundaries->clear();
  junctions->clear();
  std::set<std::string> junction_id_set;
  std::vector<RoadInfoConstPtr> roads;
  //首先调用取得道路集合的函数
if (GetRoads(point, radius, &roads) != 0) {
    AERROR << "can not get roads in the range.";
    return -1;
  }
  //遍历查询出的道路集合
for (const auto& road_ptr : roads) {
    if (road_ptr->has_junction_id()) {//如果道路信息中存在路口信息，就提取路口信息
      JunctionInfoConstPtr junction_ptr =
          GetJunctionById(road_ptr->junction_id());
      if (junction_id_set.find(junction_ptr->id().id()) ==
          junction_id_set.end()) {
        junctions->push_back(junction_ptr);
        junction_id_set.insert(junction_ptr->id().id());
      }
    } else {//如果没有路口信息，就需要使用 ROI 过滤器从道路旁边的路口信息提取路口信息
      RoadRoiPtr road_boundary_ptr(new RoadRoi());
      const std::vector<apollo::hdmap::RoadBoundary>& temp_road_boundaries =
          road_ptr->GetBoundaries();
      road_boundary_ptr->id = road_ptr->id();
      for (const auto& temp_road_boundary : temp_road_boundaries) {
        apollo::hdmap::BoundaryPolygon boundary_polygon =
            temp_road_boundary.outer_polygon();
        for (const auto& edge : boundary_polygon.edge()) {
          if (edge.type() == apollo::hdmap::BoundaryEdge::LEFT_BOUNDARY) {
            for (const auto& s : edge.curve().segment()) {
              for (const auto& p : s.line_segment().point()) {
                road_boundary_ptr->left_boundary.line_points.push_back(p);
              }
            }
          }
          if (edge.type() == apollo::hdmap::BoundaryEdge::RIGHT_BOUNDARY) {
            for (const auto& s : edge.curve().segment()) {
              for (const auto& p : s.line_segment().point()) {
                road_boundary_ptr->right_boundary.line_points.push_back(p);
              }
            }
```

```
          }
        }
        if (temp_road_boundary.hole_size() != 0) {
          for (const auto& hole : temp_road_boundary.hole()) {
            PolygonBoundary hole_boundary;
            for (const auto& edge : hole.edge()) {
              if (edge.type() == apollo::hdmap::BoundaryEdge::NORMAL) {
                for (const auto& s : edge.curve().segment()) {
                  for (const auto& p : s.line_segment().point()) {
                    hole_boundary.polygon_points.push_back(p);
                  }
                }
              }
            }
            road_boundary_ptr->holes_boundary.push_back(hole_boundary);
          }
        }
      }
      road_boundaries->push_back(road_boundary_ptr);
    }
  }
  return 0;
}
//查询给定点附近的 ROI（路口和道路边界）信息
int HDMapImpl::GetRoi(const apollo::common::PointENU& point, double radius,
                      std::vector<RoadRoiPtr>* roads_roi,
                      std::vector<PolygonRoiPtr>* polygons_roi) {
  if (roads_roi == nullptr || polygons_roi == nullptr) {
    AERROR << "the pointer in parameter is null";
    return -1;
  }
  roads_roi->clear();
  polygons_roi->clear();
  std::set<std::string> polygon_id_set;
  std::vector<RoadInfoConstPtr> roads;
  std::vector<LaneInfoConstPtr> lanes;
 //查询给定点的道路对象集合
  if (GetRoads(point, radius, &roads) != 0) {
    AERROR << "can not get roads in the range.";
    return -1;
  }
 //查询给定点的车道线对象集合
  if (GetLanes(point, radius, &lanes) != 0) {
```

```
    AERROR << "can not get lanes in the range.";
    return -1;
  }
  for (const auto& road_ptr : roads) {
    // 获取路口多边形信息
    if (road_ptr->has_junction_id()) {
      JunctionInfoConstPtr junction_ptr =
          GetJunctionById(road_ptr->junction_id());
      if (polygon_id_set.find(junction_ptr->id().id()) ==
          polygon_id_set.end()) {
        PolygonRoiPtr polygon_roi_ptr(new PolygonRoi());
        polygon_roi_ptr->polygon = junction_ptr->polygon();
        polygon_roi_ptr->attribute.type = PolygonType::JUNCTION_POLYGON;
        polygon_roi_ptr->attribute.id = junction_ptr->id();
        polygons_roi->push_back(polygon_roi_ptr);
        polygon_id_set.insert(junction_ptr->id().id());
      }
    } else {
      // 获取道路边界信息
      RoadRoiPtr road_boundary_ptr(new RoadRoi());
      std::vector<apollo::hdmap::RoadBoundary> temp_roads_roi;
      temp_roads_roi = road_ptr->GetBoundaries();
      if (!temp_roads_roi.empty()) {
        road_boundary_ptr->id = road_ptr->id();
        for (const auto& temp_road_boundary : temp_roads_roi) {
          apollo::hdmap::BoundaryPolygon boundary_polygon =
              temp_road_boundary.outer_polygon();
          for (const auto& edge : boundary_polygon.edge()) {
            if (edge.type() == apollo::hdmap::BoundaryEdge::LEFT_BOUNDARY) {
              for (const auto& s : edge.curve().segment()) {
                for (const auto& p : s.line_segment().point()) {
                  road_boundary_ptr->left_boundary.line_points.push_back(p);
                }
              }
            }
            if (edge.type() == apollo::hdmap::BoundaryEdge::RIGHT_BOUNDARY) {
              for (const auto& s : edge.curve().segment()) {
                for (const auto& p : s.line_segment().point()) {
                  road_boundary_ptr->right_boundary.line_points.push_back(p);
                }
              }
            }
          }
```

```
        if (temp_road_boundary.hole_size() != 0) {
          for (const auto& hole : temp_road_boundary.hole()) {
            PolygonBoundary hole_boundary;
            for (const auto& edge : hole.edge()) {
              if (edge.type() == apollo::hdmap::BoundaryEdge::NORMAL) {
                for (const auto& s : edge.curve().segment()) {
                  for (const auto& p : s.line_segment().point()) {
                    hole_boundary.polygon_points.push_back(p);
                  }
                }
              }
            }
            road_boundary_ptr->holes_boundary.push_back(hole_boundary);
          }
        }
      }
      roads_roi->push_back(road_boundary_ptr);
    }
  }
}

for (const auto& lane_ptr : lanes) {
  // 获取停车区域信息
  for (const auto& overlap_id : lane_ptr->lane().overlap_id()) {
    OverlapInfoConstPtr overlap_ptr = GetOverlapById(overlap_id);
    for (int i = 0; i < overlap_ptr->overlap().object_size(); ++i) {
      if (overlap_ptr->overlap().object(i).id().id() == lane_ptr->id().id())
{
        continue;
      } else {
        ParkingSpaceInfoConstPtr parkingspace_ptr =
            GetParkingSpaceById(overlap_ptr->overlap().object(i).id());
        if (parkingspace_ptr != nullptr) {
          if (polygon_id_set.find(parkingspace_ptr->id().id()) ==
              polygon_id_set.end()) {
            PolygonRoiPtr polygon_roi_ptr(new PolygonRoi());
            polygon_roi_ptr->polygon = parkingspace_ptr->polygon();
            polygon_roi_ptr->attribute.type =
                PolygonType::PARKINGSPACE_POLYGON;
            polygon_roi_ptr->attribute.id = parkingspace_ptr->id();
            polygons_roi->push_back(polygon_roi_ptr);
            polygon_id_set.insert(parkingspace_ptr->id().id());
          }
```

```
          }
        }
      }
    }
  }
  return 0;
}
//查询给定点的前方最近的一个交通标志（在车道线上）
int HDMapImpl::GetForwardNearestSignalsOnLane(
    const apollo::common::PointENU& point, const double distance,
    std::vector<SignalInfoConstPtr>* signals) const {
  CHECK_NOTNULL(signals);

  signals->clear();
  LaneInfoConstPtr lane_ptr = nullptr;
  double nearest_s = 0.0;
  double nearest_l = 0.0;

  std::vector<LaneInfoConstPtr> temp_surrounding_lanes;
  std::vector<LaneInfoConstPtr> surrounding_lanes;
  int s_index = 0;
  apollo::common::math::Vec2d car_point;
  car_point.set_x(point.x());
  car_point.set_y(point.y());
  apollo::common::math::Vec2d map_point;
//利用给定点查询车道线集合
  if (GetLanes(point, kLanesSearchRange, &temp_surrounding_lanes) == -1) {
    AINFO << "Can not find lanes around car.";
    return -1;
  }
//通过遍历检测车辆在哪条车道线上，把符合条件的车道线存储在 surrounding_lanes 中
  for (const auto& surround_lane : temp_surrounding_lanes) {
    if (surround_lane->IsOnLane(car_point)) {
      surrounding_lanes.push_back(surround_lane);
    }
  }
//对 surrounding_lanes 进行非空判断，如果为空，就证明车并不在车道线上行驶
  if (surrounding_lanes.empty()) {
    AINFO << "Car is not on lane.";
    return -1;
  }
//遍历车辆所在的车道线的集合，筛选出有交通标识的车道线
  for (const auto& lane : surrounding_lanes) {
```

```
    if (!lane->signals().empty()) {
      lane_ptr = lane;
      nearest_l =
          lane_ptr->DistanceTo(car_point, &map_point, &nearest_s, &s_index);
      break;
    }
  }
 //通过 GetNearestLane 函数获取最近的一条车道线
 if (lane_ptr == nullptr) {
    GetNearestLane(point, &lane_ptr, &nearest_s, &nearest_l);
    if (lane_ptr == nullptr) {
      return -1;
    }
  }
 //取得车道线的交通标志位置集合和车道线覆盖区，通过比较两者的距离得到最近的交通标志
  double unused_distance = distance + kBackwardDistance;
  double back_distance = kBackwardDistance;
  double s = nearest_s;
  while (s < back_distance) {
    for (const auto& predecessor_lane_id : lane_ptr->lane().predecessor_id())
{
      lane_ptr = GetLaneById(predecessor_lane_id);
      if (lane_ptr->lane().turn() == apollo::hdmap::Lane::NO_TURN) {
        break;
      }
    }
    back_distance = back_distance - s;
    s = lane_ptr->total_length();
  }
  double s_start = s - back_distance;
  while (lane_ptr != nullptr) {
    double signal_min_dist = std::numeric_limits<double>::infinity();
    std::vector<SignalInfoConstPtr> min_dist_signal_ptr;
    for (const auto& overlap_id : lane_ptr->lane().overlap_id()) {
      OverlapInfoConstPtr overlap_ptr = GetOverlapById(overlap_id);
      double lane_overlap_offset_s = 0.0;
      SignalInfoConstPtr signal_ptr = nullptr;
      for (int i = 0; i < overlap_ptr->overlap().object_size(); ++i) {
        if (overlap_ptr->overlap().object(i).id().id() == lane_ptr->id().id())
{
          lane_overlap_offset_s =
              overlap_ptr->overlap().object(i).lane_overlap_info().start_s() -
              s_start;
```

```
          continue;
        }
        signal_ptr = GetSignalById(overlap_ptr->overlap().object(i).id());
        if (signal_ptr == nullptr || lane_overlap_offset_s < 0.0) {
          break;
        }
        if (lane_overlap_offset_s < signal_min_dist) {
          signal_min_dist = lane_overlap_offset_s;
          min_dist_signal_ptr.clear();
          min_dist_signal_ptr.push_back(signal_ptr);
        } else if (lane_overlap_offset_s < (signal_min_dist + 0.1) &&
                   lane_overlap_offset_s > (signal_min_dist - 0.1)) {
          min_dist_signal_ptr.push_back(signal_ptr);
        }
      }
    }
    if (!min_dist_signal_ptr.empty() && unused_distance >= signal_min_dist) {
      *signals = min_dist_signal_ptr;
      break;
    }
    unused_distance = unused_distance - (lane_ptr->total_length() - s_start);
    if (unused_distance <= 0) {
      break;
    }
    LaneInfoConstPtr tmp_lane_ptr = nullptr;
    for (const auto& successor_lane_id : lane_ptr->lane().successor_id()) {
      tmp_lane_ptr = GetLaneById(successor_lane_id);
      if (tmp_lane_ptr->lane().turn() == apollo::hdmap::Lane::NO_TURN) {
        break;
      }
    }
    lane_ptr = tmp_lane_ptr;
    s_start = 0;
  }
  return 0;
}

int HDMapImpl::GetStopSignAssociatedStopSigns(
    const Id& id, std::vector<StopSignInfoConstPtr>* stop_signs) const {
  CHECK_NOTNULL(stop_signs);

  const auto& stop_sign = GetStopSignById(id);
  if (stop_sign == nullptr) {
```

```
    return -1;
  }

  std::vector<Id> associate_stop_sign_ids;
  const auto junction_ids = stop_sign->OverlapJunctionIds();
  for (const auto& junction_id : junction_ids) {
    const auto& junction = GetJunctionById(junction_id);
    if (junction == nullptr) {
      continue;
    }
    const auto stop_sign_ids = junction->OverlapStopSignIds();
    std::copy(stop_sign_ids.begin(), stop_sign_ids.end(),
              std::back_inserter(associate_stop_sign_ids));
  }

  std::vector<Id> associate_lane_ids;
  for (const auto& stop_sign_id : associate_stop_sign_ids) {
    if (stop_sign_id.id() == id.id()) {
      // exclude current stop sign
      continue;
    }
    const auto& stop_sign = GetStopSignById(stop_sign_id);
    if (stop_sign == nullptr) {
      continue;
    }
    stop_signs->push_back(stop_sign);
  }

  return 0;
}
//得到给定车道集合中的给定停车标志（参数为停车标志 ID）所在的车道线集合
int HDMapImpl::GetStopSignAssociatedLanes(
    const Id& id, std::vector<LaneInfoConstPtr>* lanes) const {
  CHECK_NOTNULL(lanes);
//通过停车标志 ID 得到对应的停车标志
  const auto& stop_sign = GetStopSignById(id);
  if (stop_sign == nullptr) {
    return -1;
  }
//通过查询停车标志得到与标志相关的路口信息
  std::vector<Id> associate_stop_sign_ids;
  const auto junction_ids = stop_sign->OverlapJunctionIds();
  for (const auto& junction_id : junction_ids) {
```

```
      const auto& junction = GetJunctionById(junction_id);
      if (junction == nullptr) {
        continue;
    }
    //得到路口对应的停止标志 ID 集合
      const auto stop_sign_ids = junction->OverlapStopSignIds();
      std::copy(stop_sign_ids.begin(), stop_sign_ids.end(),
                std::back_inserter(associate_stop_sign_ids));
    }
    //通过遍历所有停止标志得到与停止标志相关的车道线 ID
    std::vector<Id> associate_lane_ids;
    for (const auto& stop_sign_id : associate_stop_sign_ids) {
      if (stop_sign_id.id() == id.id()) {
        // exclude current stop sign
        continue;
      }
      const auto& stop_sign = GetStopSignById(stop_sign_id);
      if (stop_sign == nullptr) {
        continue;
      }
      const auto lane_ids = stop_sign->OverlapLaneIds();
      std::copy(lane_ids.begin(), lane_ids.end(),
                std::back_inserter(associate_lane_ids));
    }
    //通过遍历车道线 ID 列表得到车道线集合
    for (const auto lane_id : associate_lane_ids) {
      const auto& lane = GetLaneById(lane_id);
      if (lane == nullptr) {
        continue;
      }
      lanes->push_back(lane);
    }

    return 0;
  }
```

//得到给定点附近的地图信息，分为两个步骤：首先，查询出附近的各种地图元素信息，之后，存入 Map 对象中

```
  int HDMapImpl::GetLocalMap(const apollo::common::PointENU& point,
                             const std::pair<double, double>& range,
                             Map* local_map) const {
    CHECK_NOTNULL(local_map);
  //查询出附近的各种地图元素信息，包括车道线、路口、交通标志等
    double distance = std::max(range.first, range.second);
```

```
CHECK_GT(distance, 0.0);

std::vector<LaneInfoConstPtr> lanes;
GetLanes(point, distance, &lanes);

std::vector<JunctionInfoConstPtr> junctions;
GetJunctions(point, distance, &junctions);

std::vector<CrosswalkInfoConstPtr> crosswalks;
GetCrosswalks(point, distance, &crosswalks);

std::vector<SignalInfoConstPtr> signals;
GetSignals(point, distance, &signals);

std::vector<StopSignInfoConstPtr> stop_signs;
GetStopSigns(point, distance, &stop_signs);

std::vector<YieldSignInfoConstPtr> yield_signs;
GetYieldSigns(point, distance, &yield_signs);

std::vector<ClearAreaInfoConstPtr> clear_areas;
GetClearAreas(point, distance, &clear_areas);

std::vector<SpeedBumpInfoConstPtr> speed_bumps;
GetSpeedBumps(point, distance, &speed_bumps);

std::vector<RoadInfoConstPtr> roads;
GetRoads(point, distance, &roads);

std::vector<ParkingSpaceInfoConstPtr> parking_spaces;
GetParkingSpaces(point, distance, &parking_spaces);

std::unordered_set<std::string> map_element_ids;
std::vector<Id> overlap_ids;
//遍历查询出的各个元素集合，存入 Map 对象中
for (auto& lane_ptr : lanes) {
  map_element_ids.insert(lane_ptr->id().id());
  std::copy(lane_ptr->lane().overlap_id().begin(),
          lane_ptr->lane().overlap_id().end(),
          std::back_inserter(overlap_ids));
  *local_map->add_lane() = lane_ptr->lane();
}
```

```
for (auto& crosswalk_ptr : crosswalks) {
  map_element_ids.insert(crosswalk_ptr->id().id());
  std::copy(crosswalk_ptr->crosswalk().overlap_id().begin(),
          crosswalk_ptr->crosswalk().overlap_id().end(),
          std::back_inserter(overlap_ids));
  *local_map->add_crosswalk() = crosswalk_ptr->crosswalk();
}

for (auto& junction_ptr : junctions) {
  map_element_ids.insert(junction_ptr->id().id());
  std::copy(junction_ptr->junction().overlap_id().begin(),
          junction_ptr->junction().overlap_id().end(),
          std::back_inserter(overlap_ids));
  *local_map->add_junction() = junction_ptr->junction();
}

for (auto& signal_ptr : signals) {
  map_element_ids.insert(signal_ptr->id().id());
  std::copy(signal_ptr->signal().overlap_id().begin(),
          signal_ptr->signal().overlap_id().end(),
          std::back_inserter(overlap_ids));
  *local_map->add_signal() = signal_ptr->signal();
}

for (auto& stop_sign_ptr : stop_signs) {
  map_element_ids.insert(stop_sign_ptr->id().id());
  std::copy(stop_sign_ptr->stop_sign().overlap_id().begin(),
          stop_sign_ptr->stop_sign().overlap_id().end(),
          std::back_inserter(overlap_ids));
  *local_map->add_stop_sign() = stop_sign_ptr->stop_sign();
}

for (auto& yield_sign_ptr : yield_signs) {
  std::copy(yield_sign_ptr->yield_sign().overlap_id().begin(),
          yield_sign_ptr->yield_sign().overlap_id().end(),
          std::back_inserter(overlap_ids));
  map_element_ids.insert(yield_sign_ptr->id().id());
  *local_map->add_yield() = yield_sign_ptr->yield_sign();
}

for (auto& clear_area_ptr : clear_areas) {
  map_element_ids.insert(clear_area_ptr->id().id());
  std::copy(clear_area_ptr->clear_area().overlap_id().begin(),
```

```
          clear_area_ptr->clear_area().overlap_id().end(),
          std::back_inserter(overlap_ids));
    *local_map->add_clear_area() = clear_area_ptr->clear_area();
  }

  for (auto& speed_bump_ptr : speed_bumps) {
    map_element_ids.insert(speed_bump_ptr->id().id());
    std::copy(speed_bump_ptr->speed_bump().overlap_id().begin(),
          speed_bump_ptr->speed_bump().overlap_id().end(),
          std::back_inserter(overlap_ids));
    *local_map->add_speed_bump() = speed_bump_ptr->speed_bump();
  }

  for (auto& road_ptr : roads) {
    map_element_ids.insert(road_ptr->id().id());
    *local_map->add_road() = road_ptr->road();
  }

  for (auto& parking_space_ptr : parking_spaces) {
    map_element_ids.insert(parking_space_ptr->id().id());
    std::copy(parking_space_ptr->parking_space().overlap_id().begin(),
          parking_space_ptr->parking_space().overlap_id().end(),
          std::back_inserter(overlap_ids));
    *local_map->add_parking_space() = parking_space_ptr->parking_space();
  }

  for (auto& overlap_id : overlap_ids) {
    auto overlap_ptr = GetOverlapById(overlap_id);
    CHECK_NOTNULL(overlap_ptr);

    bool need_delete = false;
    for (auto& overlap_object : overlap_ptr->overlap().object()) {
      if (map_element_ids.count(overlap_object.id().id()) <= 0) {
        need_delete = true;
      }
    }

    if (!need_delete) {
      *local_map->add_overlap() = overlap_ptr->overlap();
    }
  }

  return 0;
}
```

11.2 PncMap

PncMap 是 Apollo 中用于规划的地图封装，相关代码主要位于 apollo\modules\map\pnc_map 文件夹下，其中 pnc_map.cc 是其中的核心文件。我们可以简单地理解 PncMap 实际上就是 routing 和 planning 模块的地图接口中间件，因为 routing 和 planning 模块需要由一些处理过的带有一定关联关系的高精地图封装，这个工作就是由 PncMap 完成的。

下面我们用注释的方式来分析一下百度 Apollo PncMap 源码中的核心逻辑。

相关源码分析如下：

```
//定义了车辆后向距离
DEFINE_double(
    look_backward_distance, 50,
    "look backward this distance when creating 参考线 from routing");
//定义了车辆前向较近距离
DEFINE_double(look_forward_short_distance, 180,
              "short look forward this distance when creating 参考线 "
              "from routing when ADC is slow");
//定义了车辆前向较远距离
DEFINE_double(
    look_forward_long_distance, 250,
    "look forward this distance when creating 参考线 from routing");

namespace apollo {
namespace hdmap {

using apollo::common::PointENU;
using apollo::common::VehicleState;
using apollo::routing::RoutingResponse;
using common::util::MakePointENU;

namespace {

// 轨迹近似计算中允许的最大横向误差值
const double kTrajectoryApproximationMaxError = 2.0;

}  // namespace
// Pnc 地图中含有的高精地图对象
PncMap::PncMap(const HDMap *hdmap) : hdmap_(hdmap) {}

const hdmap::HDMap *PncMap::hdmap() const { return hdmap_; }
```

```
// 根据给定的 LaneWaypoint 对象得到对应的车道线对象，LaneWaypoint 对象定义位于 apollo\
modules\routing\proto\routing.proto 文件中，定义如下
  message LaneWaypoint {
    optional string id = 1;
    optional double s = 2;
    optional apollo.common.PointENU pose = 3;
  }
  LaneWaypoint PncMap::ToLaneWaypoint(
  const routing::LaneWaypoint &waypoint) const {
  // 调用高精地图通过 ID 获得车道线的方法得到对应车道线对象
    auto lane = hdmap_->GetLaneById(hdmap::MakeMapId(waypoint.id()));
    CHECK(lane) << "Invalid lane id: " << waypoint.id();
    return LaneWaypoint(lane, waypoint.s());
  }
  // 按照当前车速评估 8 秒（FLAGS_look_forward_time_sec）后车辆一共行驶的距离，如果小于
180 米，就使用 180 米，如果大于 180 米，就使用 250 米，因此系统默认的速度分界值是 22.5 m/s (81km/h)
  FLAGS_look_forward_time_sec 定义在
apollo\modules\common\configs\config_gflags.cc 文件中
  DEFINE_double(look_forward_time_sec, 8.0,
              "look forward time times adc speed to calculate this distance "
              "when creating 参考线 from routing");
  double PncMap::LookForwardDistance(double velocity) {
    auto forward_distance = velocity * FLAGS_look_forward_time_sec;
    return forward_distance > FLAGS_look_forward_short_distance
          ? FLAGS_look_forward_long_distance
          : FLAGS_look_forward_short_distance;
  }
  // 根据 LaneSegment 对象得到对应的车道线对象
  LaneSegment PncMap::ToLaneSegment(const routing::LaneSegment &segment) const
{
    auto lane = hdmap_->GetLaneById(hdmap::MakeMapId(segment.id()));
    CHECK(lane) << "Invalid lane id: " << segment.id();
    return LaneSegment(lane, segment.start_s(), segment.end_s());
  }
  // 更新下一次路由道路点的索引
  void PncMap::UpdateNextRoutingWaypointIndex(int cur_index) {
    if (cur_index < 0) {
      next_routing_waypoint_index_ = 0;
      return;
    }
    if (cur_index >= static_cast<int>(route_indices_.size())) {
```

```
    next_routing_waypoint_index_ = routing_waypoint_index_.size() - 1;
    return;
  }
  // 在车辆倒车的情况下向后搜索
  while (next_routing_waypoint_index_ != 0 &&
         next_routing_waypoint_index_ < routing_waypoint_index_.size() &&
         routing_waypoint_index_[next_routing_waypoint_index_].index >
             cur_index) {
    --next_routing_waypoint_index_;
  }
  while (next_routing_waypoint_index_ != 0 &&
         next_routing_waypoint_index_ < routing_waypoint_index_.size() &&
         routing_waypoint_index_[next_routing_waypoint_index_].index ==
             cur_index &&
         adc_waypoint_.s <
             routing_waypoint_index_[next_routing_waypoint_index_].waypoint.s)
{
    --next_routing_waypoint_index_;
  }
  // 当车辆前进的时候向前搜索
  while (next_routing_waypoint_index_ < routing_waypoint_index_.size() &&
         routing_waypoint_index_[next_routing_waypoint_index_].index <
             cur_index) {
    ++next_routing_waypoint_index_;
  }
  while (next_routing_waypoint_index_ < routing_waypoint_index_.size() &&
         cur_index ==
             routing_waypoint_index_[next_routing_waypoint_index_].index &&
         adc_waypoint_.s >=
             routing_waypoint_index_[next_routing_waypoint_index_].waypoint.s)
{
    ++next_routing_waypoint_index_;
  }
  if (next_routing_waypoint_index_ >= routing_waypoint_index_.size()) {
    next_routing_waypoint_index_ = routing_waypoint_index_.size() - 1;
  }
}
// 读取下一次路由道路点的集合
std::vector<routing::LaneWaypoint> PncMap::FutureRouteWaypoints() const {
  const auto &waypoints = routing_.routing_request().waypoint();
  return std::vector<routing::LaneWaypoint>(
      waypoints.begin() + next_routing_waypoint_index_, waypoints.end());
}
```

```
// 更新路由的范围
void PncMap::UpdateRoutingRange(int adc_index) {
  // 跟踪路由的范围
  if (range_start_ > adc_index || range_end_ < adc_index) {
    range_lane_ids_.clear();
    range_start_ = std::max(0, adc_index - 1);
    range_end_ = range_start_;
  }
  while (range_start_ + 1 < adc_index) {

range_lane_ids_.erase(route_indices_[range_start_].segment.lane->id().id());
    ++range_start_;
  }
  while (range_end_ < static_cast<int>(route_indices_.size())) {
    const auto &lane_id = route_indices_[range_end_].segment.lane->id().id();
    if (range_lane_ids_.count(lane_id) != 0) {
      break;
    }
    range_lane_ids_.insert(lane_id);
    ++range_end_;
  }
}
// 更新车辆的状态
bool PncMap::UpdateVehicleState(const VehicleState &vehicle_state) {
  if (!ValidateRouting(routing_)) {
    AERROR << "The routing is invalid when updatting vehicle state.";
    return false;
  }
  if (!adc_state_.has_x() ||
      (common::util::DistanceXY(adc_state_, vehicle_state) >
       FLAGS_replan_lateral_distance_threshold +
           FLAGS_replan_longitudinal_distance_threshold)) {
    // 首先需要重置道路位置点
    next_routing_waypoint_index_ = 0;
    adc_route_index_ = -1;
    stop_for_destination_ = false;
  }

  adc_state_ = vehicle_state;
  if (!GetNearestPointFromRouting(vehicle_state, &adc_waypoint_)) {
    AERROR << "Failed to get waypoint from routing with point: "
           << "(" << vehicle_state.x() << ", " << vehicle_state.y() << ", "
           << vehicle_state.z() << ").";
```

```
    return false;
  }
  int route_index = GetWaypointIndex(adc_waypoint_);
  if (route_index < 0 ||
      route_index >= static_cast<int>(route_indices_.size())) {
    AERROR << "Cannot find waypoint: " << adc_waypoint_.DebugString();
    return false;
  }

  // 记录当前车辆位置映射的所有得路由请求信息
  UpdateNextRoutingWaypointIndex(route_index);
  adc_route_index_ = route_index;
  UpdateRoutingRange(adc_route_index_);

  if (routing_waypoint_index_.empty()) {
    AERROR << "No routing waypoint index.";
    return false;
  }

  int last_index = GetWaypointIndex(routing_waypoint_index_.back().waypoint);
  if (next_routing_waypoint_index_ == routing_waypoint_index_.size() - 1 ||
      (!stop_for_destination_ &&
       last_index == routing_waypoint_index_.back().index)) {
    stop_for_destination_ = true;
  }
  return true;
}
// 判断路由的结果是否与之前的路由结果重复
bool PncMap::IsNewRouting(const routing::RoutingResponse &routing) const {
  return IsNewRouting(routing_, routing);
}

bool PncMap::IsNewRouting(const routing::RoutingResponse &prev,
                          const routing::RoutingResponse &routing) {
  if (!ValidateRouting(routing)) {
    ADEBUG << "The provided routing is invalid.";
    return false;
  }
// 使用 Proto 对象比较方法来完成是否一致的分析
  return !common::util::IsProtoEqual(prev, routing);
}
// 更新路由结果的应答
bool PncMap::UpdateRoutingResponse(const routing::RoutingResponse &routing) {
```

```
range_lane_ids_.clear();
route_indices_.clear();
all_lane_ids_.clear();
for (int road_index = 0; road_index < routing.road_size(); ++road_index) {
  const auto &road_segment = routing.road(road_index);
  for (int passage_index = 0; passage_index < road_segment.passage_size();
       ++passage_index) {
    const auto &passage = road_segment.passage(passage_index);
    for (int lane_index = 0; lane_index < passage.segment_size();
         ++lane_index) {
      all_lane_ids_.insert(passage.segment(lane_index).id());
      route_indices_.emplace_back();
      route_indices_.back().segment =
          ToLaneSegment(passage.segment(lane_index));
      if (route_indices_.back().segment.lane == nullptr) {
        AERROR << "Failed to get lane segment from passage.";
        return false;
      }
      route_indices_.back().index = {road_index, passage_index, lane_index};
    }
  }
}

range_start_ = 0;
range_end_ = 0;
adc_route_index_ = -1;
next_routing_waypoint_index_ = 0;
UpdateRoutingRange(adc_route_index_);

routing_waypoint_index_.clear();
const auto &request_waypoints = routing.routing_request().waypoint();
if (request_waypoints.empty()) {
  AERROR << "Invalid routing: no request waypoints.";
  return false;
}
int i = 0;
for (size_t j = 0; j < route_indices_.size(); ++j) {
  while (i < request_waypoints.size() &&
         RouteSegments::WithinLaneSegment(route_indices_[j].segment,
                                          request_waypoints.Get(i))) {
    routing_waypoint_index_.emplace_back(
        LaneWaypoint(route_indices_[j].segment.lane,
                     request_waypoints.Get(i).s()),
```

```
            j);
        ++i;
      }
    }
    routing_ = routing;
    adc_waypoint_ = LaneWaypoint();
    stop_for_destination_ = false;
    return true;
  }

  const routing::RoutingResponse &PncMap::routing_response() const {
    return routing_;
  }
  // 验证路由相应结果是否正确
  bool PncMap::ValidateRouting(const RoutingResponse &routing) {
    const int num_road = routing.road_size();
    if (num_road == 0) {
      AERROR << "Route is empty.";
      return false;
    }
    if (!routing.has_routing_request() ||
        routing.routing_request().waypoint_size() < 2) {
      AERROR << "Routing does not have request.";
      return false;
    }
    for (const auto &waypoint : routing.routing_request().waypoint()) {
      if (!waypoint.has_id() || !waypoint.has_s()) {
        AERROR << "Routing waypoint has no lane_id or s.";
        return false;
      }
    }
    return true;
  }
  // 向前行驶的道路点的索引
  int PncMap::SearchForwardWaypointIndex(int start,
                                         const LaneWaypoint &waypoint) const {
    int i = std::max(start, 0);
    while (
        i < static_cast<int>(route_indices_.size()) &&
        !RouteSegments::WithinLaneSegment(route_indices_[i].segment, waypoint))
{
      ++i;
    }
```

```
    return i;
  }
  // 向后行驶的道路点的索引
  int PncMap::SearchBackwardWaypointIndex(int start,
                                          const LaneWaypoint &waypoint) const {
    int i = std::min(static_cast<int>(route_indices_.size() - 1), start);
    while (i >= 0
&& !RouteSegments::WithinLaneSegment(route_indices_[i].segment,
                                     waypoint)) {
      --i;
    }
    return i;
  }

  // 返回下一个道路点的索引
  int PncMap::NextWaypointIndex(int index) const {
    if (index >= static_cast<int>(route_indices_.size() - 1)) {
      return static_cast<int>(route_indices_.size()) - 1;
    } else if (index < 0) {
      return 0;
    } else {
      return index + 1;
    }
  }
  // 通过当前道路点得到下面的道路点集合
  int PncMap::GetWaypointIndex(const LaneWaypoint &waypoint) const {
    int forward_index = SearchForwardWaypointIndex(adc_route_index_, waypoint);
    if (forward_index >= static_cast<int>(route_indices_.size())) {
      return SearchBackwardWaypointIndex(adc_route_index_, waypoint);
    }
    if (forward_index == adc_route_index_ ||
      forward_index == adc_route_index_ + 1) {
      return forward_index;
    }
    auto backward_index = SearchBackwardWaypointIndex(adc_route_index_,
waypoint);
    if (backward_index < 0) {
      return forward_index;
    }

    return (backward_index + 1 == adc_route_index_) ? backward_index
                                                     : forward_index;
  }
```

```
// 向后行驶的道路点的索引
bool PncMap::PassageToSegments(routing::Passage passage,
                               RouteSegments *segments) const {
  CHECK_NOTNULL(segments);
  segments->clear();
  for (const auto &lane : passage.segment()) {
    auto lane_ptr = hdmap_->GetLaneById(hdmap::MakeMapId(lane.id()));
    if (!lane_ptr) {
      AERROR << "Failed to find lane: " << lane.id();
      return false;
    }
    segments->emplace_back(lane_ptr, std::max(0.0, lane.start_s()),
                         std::min(lane_ptr->total_length(), lane.end_s()));
  }
  return !segments->empty();
}
// 得到最近的通道，主要用于车道变换的路由应用中
std::vector<int> PncMap::GetNeighborPassages(const routing::RoadSegment
&road,
                                             int start_passage) const {
  CHECK_GE(start_passage, 0);
  CHECK_LE(start_passage, road.passage_size());
  std::vector<int> result;
  const auto &source_passage = road.passage(start_passage);
  result.emplace_back(start_passage);
  if (source_passage.change_lane_type() == routing::FORWARD) {
    return result;
  }
  if (source_passage.can_exit()) {  //不需要变换车道
    return result;
  }
  RouteSegments source_segments;
  if (!PassageToSegments(source_passage, &source_segments)) {
    AERROR << "Failed to convert passage to segments";
    return result;
  }
  if (next_routing_waypoint_index_ < routing_waypoint_index_.size() &&
      source_segments.IsWaypointOnSegment(
          routing_waypoint_index_[next_routing_waypoint_index_].waypoint)) {
    ADEBUG << "Need to pass next waypoint[" << next_routing_waypoint_index_
           << "] before change lane";
    return result;
  }
```

```
  std::unordered_set<std::string> neighbor_lanes;
  if (source_passage.change_lane_type() == routing::LEFT) {
    for (const auto &segment : source_segments) {
      for (const auto &left_id :
           segment.lane->lane().left_neighbor_forward_lane_id()) {
        neighbor_lanes.insert(left_id.id());
      }
    }
  } else if (source_passage.change_lane_type() == routing::RIGHT) {
    for (const auto &segment : source_segments) {
      for (const auto &right_id :
           segment.lane->lane().right_neighbor_forward_lane_id()) {
        neighbor_lanes.insert(right_id.id());
      }
    }
  }

  for (int i = 0; i < road.passage_size(); ++i) {
    if (i == start_passage) {
      continue;
    }
    const auto &target_passage = road.passage(i);
    for (const auto &segment : target_passage.segment()) {
      if (neighbor_lanes.count(segment.id())) {
        result.emplace_back(i);
        break;
      }
    }
  }
  return result;
}
// 得到车辆需要的分段路由信息，使用默认的后向观测距离和根据车速选定的前向观测距离
bool PncMap::GetRouteSegments(const VehicleState &vehicle_state,
                              std::list<RouteSegments> *const route_segments) {
  double look_forward_distance =
      LookForwardDistance(vehicle_state.linear_velocity());
  double look_backward_distance = FLAGS_look_backward_distance;
  return GetRouteSegments(vehicle_state, look_backward_distance,
                          look_forward_distance, route_segments);
}

// 得到车辆需要的分段路由信息，输入参数中有前向观测距离和后向观测距离
bool PncMap::GetRouteSegments(const VehicleState &vehicle_state,
```

```
                                const double backward_length,
                                const double forward_length,
                                std::list<RouteSegments> *const route_segments) {
    if (!UpdateVehicleState(vehicle_state)) {
      AERROR << "Failed to update vehicle state in pnc_map.";
      return false;
    }
    // 我们的变道操作有一个默认的前提，就是车辆在目前车道线中的行驶方式是稳定的，换句话说，车辆的位置与车道线的中心线是非常接近的
    if (!adc_waypoint_.lane || adc_route_index_ < 0 ||
        adc_route_index_ >= static_cast<int>(route_indices_.size())) {
      AERROR << "Invalid vehicle state in pnc_map, update vehicle state first.";
      return false;
    }
    const auto &route_index = route_indices_[adc_route_index_].index;
    const int road_index = route_index[0];
    const int passage_index = route_index[1];
    const auto &road = routing_.road(road_index);
    // 首先通过 GetNeighborPassages 方法得到比较接近的可变道通道集合
    auto drive_passages = GetNeighborPassages(road, passage_index);
  // 遍历取得的可变道通道集合
    for (const int index : drive_passages) {
      const auto &passage = road.passage(index);
      RouteSegments segments;
      if (!PassageToSegments(passage, &segments)) {
        ADEBUG << "Failed to convert passage to lane segments.";
        continue;
      }
  // 得到某个通道中最近的道路点
      PointENU nearest_point =
          MakePointENU(adc_state_.x(), adc_state_.y(), adc_state_.z());
  // 车辆运动学模型优化道路点
      if (index == passage_index) {
        nearest_point = adc_waypoint_.lane->GetSmoothPoint(adc_waypoint_.s);
      }
      common::SLPoint sl;
      LaneWaypoint segment_waypoint;
  // 确认得到的道路点是合理的
      if (!segments.GetProjection(nearest_point, &sl, &segment_waypoint)) {
        ADEBUG << "Failed to get projection from point: "
               << nearest_point.ShortDebugString();
        continue;
      }
```

```
    if (index != passage_index) {
      if (!segments.CanDriveFrom(adc_waypoint_)) {
        ADEBUG << "You cannot drive from current waypoint to passage: "
               << index;
        continue;
      }
    }
  //根据结果更新分段路由的结果
    route_segments->emplace_back();
    const auto last_waypoint = segments.LastWaypoint();
    if (!ExtendSegments(segments, sl.s() - backward_length,
                        sl.s() + forward_length, &route_segments->back())) {
      AERROR << "Failed to extend segments with s=" << sl.s()
             << ", backward: " << backward_length
             << ", forward: " << forward_length;
      return false;
    }
    if (route_segments->back().IsWaypointOnSegment(last_waypoint)) {
      route_segments->back().SetRouteEndWaypoint(last_waypoint);
    }
    route_segments->back().SetCanExit(passage.can_exit());
    route_segments->back().SetNextAction(passage.change_lane_type());
    std::string route_segment_id =
        std::to_string(road_index) + "_" + std::to_string(index);
    route_segments->back().SetId(route_segment_id);
    route_segments->back().SetStopForDestination(stop_for_destination_);
    if (index == passage_index) {
      route_segments->back().SetIsOnSegment(true);
      route_segments->back().SetPreviousAction(routing::FORWARD);
    } else if (sl.l() > 0) {
      route_segments->back().SetPreviousAction(routing::RIGHT);
    } else {
      route_segments->back().SetPreviousAction(routing::LEFT);
    }
  }
  return !route_segments->empty();
}
// 通过路由算法得到最近的道路点
bool PncMap::GetNearestPointFromRouting(const VehicleState &state,
                                        LaneWaypoint *waypoint) const {
  const double kMaxDistance = 10.0;  // meters.
  waypoint->lane = nullptr;
  std::vector<LaneInfoConstPtr> lanes;
```

```
// 通过车辆状态得到车辆所在的道路点
 auto point = common::util::MakePointENU(state.x(), state.y(), state.z());
// 通过道路点得到符合条件的车道线
 const int status = hdmap_->GetLanesWithHeading(
     point, kMaxDistance, state.heading(), M_PI / 2.0, &lanes);
 if (status < 0) {
   AERROR << "Failed to get lane from point: " << point.ShortDebugString();
   return false;
 }
 if (lanes.empty()) {
   AERROR << "No valid lane found within " << kMaxDistance
          << " meters with heading " << state.heading();
   return false;
 }
 std::vector<LaneInfoConstPtr> valid_lanes;
 std::copy_if(lanes.begin(), lanes.end(), std::back_inserter(valid_lanes),
              [&](LaneInfoConstPtr ptr) {
                return range_lane_ids_.count(ptr->lane().id().id()) > 0;
              });
 if (valid_lanes.empty()) {
   std::copy_if(lanes.begin(), lanes.end(), std::back_inserter(valid_lanes),
                [&](LaneInfoConstPtr ptr) {
                  return all_lane_ids_.count(ptr->lane().id().id()) > 0;
                });
 }

 // 得到关联点附近的道路点
 double min_distance = std::numeric_limits<double>::infinity();
 for (const auto &lane : valid_lanes) {
   if (range_lane_ids_.count(lane->id().id()) == 0) {
     continue;
   }
   {
     double s = 0.0;
     double l = 0.0;
     if (!lane->GetProjection({point.x(), point.y()}, &s, &l)) {
       return false;
     }
     // 使用较大的 ε 来控制投影误差
     constexpr double kEpsilon = 0.5;
     if (s > (lane->total_length() + kEpsilon) || (s + kEpsilon) < 0.0) {
       continue;
     }
```

```
    }
    double distance = 0.0;
    common::PointENU map_point =
        lane->GetNearestPoint({point.x(), point.y()}, &distance);
    if (distance < min_distance) {
      min_distance = distance;
      double s = 0.0;
      double l = 0.0;
      if (!lane->GetProjection({map_point.x(), map_point.y()}, &s, &l)) {
        AERROR << "Failed to get projection for map_point: "
               << map_point.DebugString();
        return false;
      }
      waypoint->lane = lane;
      waypoint->s = s;
    }
  }
  if (waypoint->lane == nullptr) {
    AERROR << "Failed to find nearest point: " << point.ShortDebugString();
  }
  return waypoint->lane != nullptr;
}

LaneInfoConstPtr PncMap::GetRouteSuccessor(LaneInfoConstPtr lane) const {
  if (lane->lane().successor_id_size() == 0) {
    return nullptr;
  }
  hdmap::Id preferred_id = lane->lane().successor_id(0);
  for (const auto &lane_id : lane->lane().successor_id()) {
    if (range_lane_ids_.count(lane_id.id()) != 0) {
      preferred_id = lane_id;
      break;
    }
  }
  return hdmap_->GetLaneById(preferred_id);
}
// 通过车道线信息来计算路由的趋势
LaneInfoConstPtr PncMap::GetRoutePredecessor(LaneInfoConstPtr lane) const {
  if (lane->lane().predecessor_id_size() == 0) {
    return nullptr;
  }
  hdmap::Id preferred_id = lane->lane().predecessor_id(0);
  for (const auto &lane_id : lane->lane().predecessor_id()) {
```

```
      if (range_lane_ids_.count(lane_id.id()) != 0) {
        preferred_id = lane_id;
        break;
      }
    }
    return hdmap_->GetLaneById(preferred_id);
  }
  // 根据已有的道路路由分段结果扩展道路路由分段结果集合
  bool PncMap::ExtendSegments(const RouteSegments &segments,
                              const common::PointENU &point, double look_backward,
                              double look_forward,
                              RouteSegments *extended_segments) {
    common::SLPoint sl;
    LaneWaypoint waypoint;
    if (!segments.GetProjection(point, &sl, &waypoint)) {
      AERROR << "point: " << point.ShortDebugString() << " is not on segment";
      return false;
    }
    return ExtendSegments(segments, sl.s() - look_backward, sl.s() +
look_forward,
                          extended_segments);
  }

  bool PncMap::ExtendSegments(const RouteSegments &segments, double start_s,
                              double end_s,
                              RouteSegments *const truncated_segments) const {
    if (segments.empty()) {
      AERROR << "The input segments is empty";
      return false;
    }
    CHECK_NOTNULL(truncated_segments);
    truncated_segments->SetProperties(segments);

    if (start_s >= end_s) {
      AERROR << "start_s(" << start_s << " >= end_s(" << end_s << ")";
      return false;
    }
    std::unordered_set<std::string> unique_lanes;
    constexpr double kRouteEpsilon = 1e-3;
    // 从轨迹起点开始扩展轨迹
    if (start_s < 0) {
      const auto &first_segment = *segments.begin();
      auto lane = first_segment.lane;
```

```
      double s = first_segment.start_s;
      double extend_s = -start_s;
      std::vector<LaneSegment> extended_lane_segments;
      while (extend_s > kRouteEpsilon) {
        if (s <= kRouteEpsilon) {
          lane = GetRoutePredecessor(lane);
          if (lane == nullptr ||
              unique_lanes.find(lane->id().id()) != unique_lanes.end()) {
            break;
          }
          s = lane->total_length();
        } else {
          const double length = std::min(s, extend_s);
          extended_lane_segments.emplace_back(lane, s - length, s);
          extend_s -= length;
          s -= length;
          unique_lanes.insert(lane->id().id());
        }
      }
      truncated_segments->insert(truncated_segments->begin(),
                                 extended_lane_segments.rbegin(),
                                 extended_lane_segments.rend());
    }
    bool found_loop = false;
    double router_s = 0;
    for (const auto &lane_segment : segments) {
      const double adjusted_start_s = std::max(
          start_s - router_s + lane_segment.start_s, lane_segment.start_s);
      const double adjusted_end_s =
          std::min(end_s - router_s + lane_segment.start_s, lane_segment.end_s);
      if (adjusted_start_s < adjusted_end_s) {
        if (!truncated_segments->empty() &&
            truncated_segments->back().lane->id().id() ==
                lane_segment.lane->id().id()) {
          truncated_segments->back().end_s = adjusted_end_s;
        } else if (unique_lanes.find(lane_segment.lane->id().id()) ==
                   unique_lanes.end()) {
          truncated_segments->emplace_back(lane_segment.lane, adjusted_start_s,
                                           adjusted_end_s);
          unique_lanes.insert(lane_segment.lane->id().id());
        } else {
          found_loop = true;
          break;
```

```
      }
    }
    router_s += (lane_segment.end_s - lane_segment.start_s);
    if (router_s > end_s) {
      break;
    }
  }
  if (found_loop) {
    return true;
  }
  // 从轨迹终点开始扩展轨迹
  if (router_s < end_s && !truncated_segments->empty()) {
    auto &back = truncated_segments->back();
    if (back.lane->total_length() > back.end_s) {
      double origin_end_s = back.end_s;
      back.end_s =
          std::min(back.end_s + end_s - router_s, back.lane->total_length());
      router_s += back.end_s - origin_end_s;
    }
  }
  auto last_lane = segments.back().lane;
  while (router_s < end_s - kRouteEpsilon) {
    last_lane = GetRouteSuccessor(last_lane);
    if (last_lane == nullptr ||
        unique_lanes.find(last_lane->id().id()) != unique_lanes.end()) {
      break;
    }
    const double length = std::min(end_s - router_s, last_lane->total_length());
    truncated_segments->emplace_back(last_lane, 0, length);
    unique_lanes.insert(last_lane->id().id());
    router_s += length;
  }
  return true;
}
//将车道线对象转化成对应的道路点集合
void PncMap::AppendLaneToPoints(LaneInfoConstPtr lane, const double start_s,
                                const double end_s,
                                std::vector<MapPathPoint> *const points) {
  if (points == nullptr || start_s >= end_s) {
    return;
  }
  double accumulate_s = 0.0;
  for (size_t i = 0; i < lane->points().size(); ++i) {
```

```
    if (accumulate_s >= start_s && accumulate_s <= end_s) {
      points->emplace_back(lane->points()[i], lane->headings()[i],
                           LaneWaypoint(lane, accumulate_s));
    }
    if (i < lane->segments().size()) {
      const auto &segment = lane->segments()[i];
      const double next_accumulate_s = accumulate_s + segment.length();
      if (start_s > accumulate_s && start_s < next_accumulate_s) {
        points->emplace_back(segment.start() + segment.unit_direction() *
                                                   (start_s - accumulate_s),
                             lane->headings()[i], LaneWaypoint(lane, start_s));
      }
      if (end_s > accumulate_s && end_s < next_accumulate_s) {
        points->emplace_back(
            segment.start() + segment.unit_direction() * (end_s - accumulate_s),
            lane->headings()[i], LaneWaypoint(lane, end_s));
      }
      accumulate_s = next_accumulate_s;
    }
    if (accumulate_s > end_s) {
      break;
    }
  }
}

}  // namespace hdmap
}  // namespace apollo
```

11.3　Relative Map 模块

相对地图（Relative Map）模块是连接 HDMAP/感知模块和规划模块的中间层。该模块生成车身坐标系中的实时相对地图和规划参考线。相对地图模块的输入分为离线和在线两部分。离线部分是导航线（人工驾驶路径）和导航线附近的 HDMAP 信息。在线部分是感知模块提供的与交通标志相关的信息，如车道标志、人行横道、交通灯等，相对地图的生成既可以利用在线部分，又可以利用离线部分。

百度 Apollo 相关源码分析如下。系统输入为 DreamView 模块中的导航信息，感知模块中的道路标识和定位模块中的定位信息。系统输出为相对映射，映射格式遵循模块/map/proto/map.proto 中的定义。

```
syntax = "proto2";
```

```
package apollo.hdmap;

import "modules/map/proto/map_clear_area.proto";
import "modules/map/proto/map_crosswalk.proto";
import "modules/map/proto/map_junction.proto";
import "modules/map/proto/map_lane.proto";
import "modules/map/proto/map_overlap.proto";
import "modules/map/proto/map_signal.proto";
import "modules/map/proto/map_speed_bump.proto";
import "modules/map/proto/map_stop_sign.proto";
import "modules/map/proto/map_yield_sign.proto";
import "modules/map/proto/map_road.proto";
import "modules/map/proto/map_parking_space.proto";
import "modules/map/proto/map_pnc_junction.proto";

// 这条信息定义了我们如何将椭球面投影到平面上
message Projection {
  // PROJ.4 setting:
  // "+proj=tmerc +lat_0={origin.lat} +lon_0={origin.lon} +k={scale_factor}
  // +ellps=WGS84 +no_defs"
  optional string proj = 1;
}
//定义地图消息的消息头信息
message Header {
  optional bytes version = 1;
  optional bytes date = 2;
  optional Projection projection = 3;
  optional bytes district = 4;
  optional bytes generation = 5;
  optional bytes rev_major = 6;
  optional bytes rev_minor = 7;
  optional double left = 8;
  optional double top = 9;
  optional double right = 10;
  optional double bottom = 11;
  optional bytes vendor = 12;
}
//定义地图消息的消息信息，也就是消息头加上各种地图组件（车道线、路口、停车标志等）的相关信息，具体定义参考/map/proto 目录下对应名词的 proto 文件
message Map {
  optional Header header = 1;


  repeated Crosswalk crosswalk = 2;
```

```
  repeated Junction junction = 3;
  repeated Lane lane = 4;
  repeated StopSign stop_sign = 5;
  repeated Signal signal = 6;
  repeated YieldSign yield = 7;
  repeated Overlap overlap = 8;
  repeated ClearArea clear_area = 9;
  repeated SpeedBump speed_bump = 10;
  repeated Road road = 11;
  repeated ParkingSpace parking_space = 12;
  repeated PNCJunction pnc_junction = 13;
}
```

相对地图的入口组件是 RelativeMapComponent，对应的文件是 modules/map/relative_map/relative_map_component.h 和 relative_map_component.cc。在 RelativeMapComponent（相对地图组件）中会调用 RelativeMap 对象的 OnNavigationInfo 方法，OnNavigationInfo 实际调用了 NavigationLane 对象的 UpdateNavigationInfo 方法。至此，我们可以明确相对地图的具体实现在 NavigationLane 对象的代码中，代码位于 modules/map/relative_map/navigation_lane.cc 中。下面我们使用注释的方式具体分析 NavigationLane 对象。

```
namespace apollo {
namespace relative_map {

using apollo::common::VehicleStateProvider;
using apollo::common::math::Vec2d;
using apollo::common::util::DistanceXY;
using apollo::hdmap::Lane;
using apollo::common::util::operator+;
using apollo::common::util::IsFloatEqual;

namespace {
/**
 * @brief Create a single lane map.
 * @param navi_path_tuple A navigation path tuple.
 * @param map_config Map generation configuration information.
 * @param perception_obstacles The Perceived obstacle information and the lane
 * markings are used here.
 * @param hdmap The output single lane map in high-definition map format in the
 * relative map.
 * @param navigation_path The ouput navigation path map in the relative map.
* @return True if the map is created; false otherwise.
通过地图要素创建一个只有单一车道线的地图对象
 */
```

```
  bool CreateSingleLaneMap(
      const NaviPathTuple &navi_path_tuple, const MapGenerationParam
&map_config,
      const perception::PerceptionObstacles &perception_obstacles,
      hdmap::Map *const hdmap,
      google::protobuf::Map<std::string, NavigationPath> *const navigation_path)
{
    CHECK_NOTNULL(hdmap);
    CHECK_NOTNULL(navigation_path);

    const auto &navi_path = std::get<3>(navi_path_tuple);
    const auto &path = navi_path->path();
    if (path.path_point_size() < 2) {
      AERROR << "The path length of line index is invalid";
      return false;
    }
    auto *lane = hdmap->add_lane();
    lane->mutable_id()->set_id(std::to_string(navi_path->path_priority()) +
"_" +
                              path.name());
    (*navigation_path)[lane->id().id()] = *navi_path;
    // 设定车道线的类型为城市公路
    lane->set_type(Lane::CITY_DRIVING);
    lane->set_turn(Lane::NO_TURN);

    // 根据地图中的限速信息设定车道线的限速
    lane->set_speed_limit(map_config.default_speed_limit());

    // 设定车道线的中心线位置
    auto *curve_segment = lane->mutable_central_curve()->add_segment();
    curve_segment->set_heading(path.path_point(0).theta());
    auto *line_segment = curve_segment->mutable_line_segment();

    // 设定车道线的左边界
    hdmap::LineSegment *left_segment = nullptr;
    if (FLAGS_relative_map_generate_left_boundray) {
      auto *left_boundary = lane->mutable_left_boundary();
      auto *left_boundary_type = left_boundary->add_boundary_type();
      left_boundary->set_virtual_(false);
      left_boundary_type->set_s(0.0);
      left_boundary_type->add_types(
          perception_obstacles.lane_marker().left_lane_marker().lane_type());
      left_segment =
```

```
left_boundary->mutable_curve()->add_segment()->mutable_line_segment();
    }

    // 设定车道线的右边界
    auto *right_boundary = lane->mutable_right_boundary();
    auto *right_boundary_type = right_boundary->add_boundary_type();
    right_boundary->set_virtual_(false);
    right_boundary_type->set_s(0.0);
    right_boundary_type->add_types(
        perception_obstacles.lane_marker().right_lane_marker().lane_type());
    auto *right_segment =

right_boundary->mutable_curve()->add_segment()->mutable_line_segment();

    const double lane_left_width = std::get<1>(navi_path_tuple);
    const double lane_right_width = std::get<2>(navi_path_tuple);

    for (const auto &path_point : path.path_point()) {
      auto *point = line_segment->add_point();
      point->set_x(path_point.x());
      point->set_y(path_point.y());
      point->set_z(path_point.z());
      if (FLAGS_relative_map_generate_left_boundray) {
        auto *left_sample = lane->add_left_sample();
        left_sample->set_s(path_point.s());
        left_sample->set_width(lane_left_width);
        left_segment->add_point()->CopyFrom(
            *point + lane_left_width *
                     Vec2d::CreateUnitVec2d(path_point.theta() + M_PI_2));
      }

      auto *right_sample = lane->add_right_sample();
      right_sample->set_s(path_point.s());
      right_sample->set_width(lane_right_width);
      right_segment->add_point()->CopyFrom(
          *point +
          lane_right_width * Vec2d::CreateUnitVec2d(path_point.theta() -
M_PI_2));
    }
    return true;
  }
  }  // namespace
```

```
// NavigationLane 构造方法
NavigationLane::NavigationLane(const NavigationLaneConfig &config)
    : config_(config) {}
// NavigationLane 对象设定配置参数
void NavigationLane::SetConfig(const NavigationLaneConfig &config) {
  config_ = config;
}
// 更新导航车道线中的导航信息
void NavigationLane::UpdateNavigationInfo(
    const NavigationInfo &navigation_path) {
  navigation_info_ = navigation_path;
  last_project_index_map_.clear();
  navigation_path_list_.clear();
  current_navi_path_tuple_ = std::make_tuple(-1, -1.0, -1.0, nullptr);
  if (FLAGS_enable_cyclic_rerouting) {
    UpdateStitchIndexInfo();
  }
}
// 生成导航路径
bool NavigationLane::GeneratePath() {
  navigation_path_list_.clear();
  current_navi_path_tuple_ = std::make_tuple(-1, -1.0, -1.0, nullptr);

  // 确定在世界坐标系中的位置
  original_pose_ = VehicleStateProvider::Instance()->original_pose();

  int navigation_line_num = navigation_info_.navigation_path_size();
  const auto &lane_marker = perception_obstacles_.lane_marker();

  auto generate_path_on_perception_func = [this, &lane_marker]() {
    auto current_navi_path = std::make_shared<NavigationPath>();
    auto *path = current_navi_path->mutable_path();
    ConvertLaneMarkerToPath(lane_marker, path);
    current_navi_path->set_path_priority(0);
    double left_width = perceived_left_width_ > 0.0 ? perceived_left_width_
                                                     : default_left_width_;
    double right_width = perceived_right_width_ > 0.0 ? perceived_right_width_
                                                       : default_right_width_;
    current_navi_path_tuple_ =
        std::make_tuple(0, left_width, right_width, current_navi_path);
  };

  ADEBUG << "Beginning of NavigationLane::GeneratePath().";
```

```
ADEBUG << "navigation_line_num: " << navigation_line_num;

// priority: merge > navigation line > perception lane marker
if (config_.lane_source() == NavigationLaneConfig::OFFLINE_GENERATED &&
    navigation_line_num > 0) {
  // 根据导航线生成多条导航路径，导航路径的总数不会超过 10 条，因此程序运行效率是有保障的
  for (int i = 0; i < navigation_line_num; ++i) {
    auto current_navi_path = std::make_shared<NavigationPath>();
    auto *path = current_navi_path->mutable_path();
    if (ConvertNavigationLineToPath(i, path)) {
      current_navi_path->set_path_priority(
          navigation_info_.navigation_path(i).path_priority());
      navigation_path_list_.emplace_back(
          i, default_left_width_, default_right_width_, current_navi_path);
    }
  }

  //如果没有基于导航线生成导航路径，就会生成根据感知到的车道标记确定车辆位置的位置
  if (navigation_path_list_.empty()) {
    generate_path_on_perception_func();
    return true;
  }

  // 根据车辆的方向按照从左到右的顺序排列导航路径，在车辆坐标系中，y 坐标在车辆的左侧定义
为正，右侧定义为负。因此，下面的排序操作的依据是 y 坐标从左向右的顺序
  navigation_path_list_.sort(
      [](const NaviPathTuple &left, const NaviPathTuple &right) {
        double left_y = std::get<3>(left)->path().path_point(0).y();
        double right_y = std::get<3>(right)->path().path_point(0).y();
        return left_y > right_y;
      });

  // Get which navigation path the vehicle is currently on.
  double min_d = std::numeric_limits<double>::max();
  for (const auto &navi_path_tuple : navigation_path_list_) {
    int current_line_index = std::get<0>(navi_path_tuple);
    ADEBUG << "Current navigation path index is: " << current_line_index;
    double current_d = std::numeric_limits<double>::max();
    auto item_iter = last_project_index_map_.find(current_line_index);
    if (item_iter != last_project_index_map_.end()) {
      current_d = item_iter->second.second;
    }
    if (current_d < min_d) {
```

```
      min_d = current_d;
      current_navi_path_tuple_ = navi_path_tuple;
    }
  }

  //合并车辆所在的当前导航路径中的车道标记
  auto *path = std::get<3>(current_navi_path_tuple_)->mutable_path();
  MergeNavigationLineAndLaneMarker(std::get<0>(current_navi_path_tuple_),
                                    path);


  //设置车辆当前所在导航路径的宽度
  double left_width = perceived_left_width_ > 0.0 ? perceived_left_width_
                                                  : default_left_width_;
  double right_width = perceived_right_width_ > 0.0 ? perceived_right_width_
                                                    : default_right_width_;
  if (!IsFloatEqual(left_width, default_left_width_) &&
      !IsFloatEqual(right_width, default_right_width_)) {
    left_width = left_width > default_left_width_ ? left_width - min_d
                                                  : left_width + min_d;
    right_width = right_width > default_right_width_ ? right_width - min_d
                                                     : right_width + min_d;
  }

  ADEBUG << "The left width of current lane is: " << left_width
         << " and the right width of current lane is: " << right_width;

  std::get<1>(current_navi_path_tuple_) = left_width;
  std::get<2>(current_navi_path_tuple_) = right_width;
  auto curr_navi_path_iter = std::find_if(
      std::begin(navigation_path_list_), std::end(navigation_path_list_),
      [this](const NaviPathTuple &item) {
        return std::get<0>(item) == std::get<0>(current_navi_path_tuple_);
      });
  if (curr_navi_path_iter != std::end(navigation_path_list_)) {
    std::get<1>(*curr_navi_path_iter) = left_width;
    std::get<2>(*curr_navi_path_iter) = right_width;
  }
```

//设置每个导航路径及其相邻路径之间的宽度。使用多点平均值的原因是防止过多来自奇点的干扰。如果当前导航路径是车辆当前所在的路径，当前车道宽度就使用感知的宽度

```
  int average_point_size = 5;
  for (auto iter = navigation_path_list_.begin();
```

```
      iter != navigation_path_list_.end(); ++iter) {
    const auto &curr_path = std::get<3>(*iter)->path();

    // 左侧的邻居
    auto prev_iter = std::prev(iter);
    if (prev_iter != navigation_path_list_.end()) {
      const auto &prev_path = std::get<3>(*prev_iter)->path();
      average_point_size = std::min(
          average_point_size,
          std::min(curr_path.path_point_size(),
prev_path.path_point_size()));
      double lateral_distance_sum = 0.0;
      for (int i = 0; i < average_point_size; ++i) {
        lateral_distance_sum +=
            fabs(curr_path.path_point(i).y() - prev_path.path_point(i).y());
      }
      double width = lateral_distance_sum /
                     static_cast<double>(average_point_size) / 2.0;
      width = common::math::Clamp(width, config_.min_lane_half_width(),
                                  config_.max_lane_half_width());

      auto &curr_left_width = std::get<1>(*iter);
      auto &prev_right_width = std::get<2>(*prev_iter);
      if (std::get<0>(*iter) == std::get<0>(current_navi_path_tuple_)) {
        prev_right_width = 2.0 * width - curr_left_width;
      } else {
        curr_left_width = width;
        prev_right_width = width;
      }
    }
    // 右侧的邻居
    auto next_iter = std::next(iter);
    if (next_iter != navigation_path_list_.end()) {
      const auto &next_path = std::get<3>(*next_iter)->path();
      average_point_size = std::min(
          average_point_size,
          std::min(curr_path.path_point_size(),
next_path.path_point_size()));
      double lateral_distance_sum = 0.0;
      for (int i = 0; i < average_point_size; ++i) {
        lateral_distance_sum +=
            fabs(curr_path.path_point(i).y() - next_path.path_point(i).y());
      }
```

```
        double width = lateral_distance_sum /
                       static_cast<double>(average_point_size) / 2.0;
        width = common::math::Clamp(width, config_.min_lane_half_width(),
                                    config_.max_lane_half_width());

        auto &curr_right_width = std::get<2>(*iter);
        auto &next_left_width = std::get<1>(*next_iter);
        if (std::get<0>(*iter) == std::get<0>(current_navi_path_tuple_)) {
          next_left_width = 2.0 * width - curr_right_width;
        } else {
          next_left_width = width;
          curr_right_width = width;
        }
      }
    }

    return true;
  }

  // 根据感知生成车辆所在的导航路径车道标记
  generate_path_on_perception_func();
  return true;
}
//求三次多项式的简单封装
double NavigationLane::EvaluateCubicPolynomial(const double c0, const double
c1,
                                               const double c2, const double c3,
                                               const double x) const {
  return ((c3 * x + c2) * x + c1) * x + c0;
}
//合并导航线和道路标志
void NavigationLane::MergeNavigationLineAndLaneMarker(
    const int line_index, common::Path *const path) {
  CHECK_NOTNULL(path);

  // 如果“路径”点的大小小于 2，就表示需要首先生成导航路径
  if (path->path_point_size() < 2) {
    path->Clear();
    ConvertNavigationLineToPath(line_index, path);
  }

  //如果“路径”点的大小仍然小于 2，那么只需根据感知到的车道标记生成一条导航路径
```

```
  if (path->path_point_size() < 2) {
    path->Clear();
    ConvertLaneMarkerToPath(perception_obstacles_.lane_marker(), path);
    return;
  }

  common::Path lane_marker_path;
  ConvertLaneMarkerToPath(perception_obstacles_.lane_marker(),
                          &lane_marker_path);

  // 如果车道标志路径点的大小小于 2，就不需要合并
  if (lane_marker_path.path_point_size() < 2) {
    return;
  }

  int lane_marker_index = 0;
  double navigation_line_weight = 1.0 - config_.lane_marker_weight();
  for (int i = 0; i < path->path_point_size(); ++i) {
    auto *point = path->mutable_path_point(i);
    double s = point->s();
    auto lane_maker_point = GetPathPointByS(lane_marker_path,
lane_marker_index,s, &lane_marker_index);
    // 对于超出感知路径的导航路径的开始和结束部分，仅使用车辆坐标系中的 Y 坐标进行合并

    const int marker_size = lane_marker_path.path_point_size();
    if (lane_marker_index < 0 || lane_marker_index > (marker_size - 1)) {
      point->set_y(navigation_line_weight * point->y() +
                   (1 - navigation_line_weight) * lane_maker_point.y());
      lane_marker_index = 0;
      continue;
    }
    *point = common::util::GetWeightedAverageOfTwoPathPoints(
        *point, lane_maker_point, navigation_line_weight,
        1 - navigation_line_weight);
  }
 }
 // 获取一段路径中的路径点
 common::PathPoint NavigationLane::GetPathPointByS(const common::Path &path,
                                                   const int start_index,
                                                   const double s,
                                                   int *const matched_index) {
  CHECK_NOTNULL(matched_index);
```

```
  const int size = path.path_point_size();

  if (start_index < 0 || s < path.path_point(start_index).s()) {
    *matched_index = -1;
    return path.path_point(0);
  }

  if (s > path.path_point(size - 1).s() || start_index > (size - 1)) {
    *matched_index = size;
    return path.path_point(size - 1);
  }

  constexpr double kEpsilon = 1e-9;
  if (std::fabs(path.path_point(start_index).s() - s) < kEpsilon) {
    *matched_index = start_index;
    return path.path_point(start_index);
  }
  int i = start_index;
  while (i + 1 < size && path.path_point(i + 1).s() < s) {
    ++i;
  }
  *matched_index = i;

  const double r = (s - path.path_point(i).s()) /
                   (path.path_point(i + 1).s() - path.path_point(i).s());
  auto p = common::util::GetWeightedAverageOfTwoPathPoints(
      path.path_point(i), path.path_point(i + 1), 1 - r, r);
  return p;
}
// 将导航线转换成路径对象
bool NavigationLane::ConvertNavigationLineToPath(const int line_index,
                                                common::Path *const path) {
  CHECK_NOTNULL(path);
  if (!navigation_info_.navigation_path(line_index).has_path() ||
      navigation_info_.navigation_path(line_index).path().path_point_size()
==
          0) {
    // 导航线中的路径为空
    return false;
  }
  path->set_name("Path from navigation line index " +
                 std::to_string(line_index));
  const auto &navigation_path =
```

```
      navigation_info_.navigation_path(line_index).path();
  auto proj_index_pair = UpdateProjectionIndex(navigation_path, line_index);
  //根据当前车辆位置在“线路索引”车道上找不到合适的投影索引
  int current_project_index = proj_index_pair.first;
  if (current_project_index < 0 ||
      current_project_index >= navigation_path.path_point_size()) {
    AERROR << "Invalid projection index " << current_project_index
           << " in line " << line_index;
    last_project_index_map_.erase(line_index);
    return false;
  } else {
    last_project_index_map_[line_index] = proj_index_pair;
  }

  // 计算当前车辆状态与导航线之间的偏移量

  const double dx = -original_pose_.position().x();
  const double dy = -original_pose_.position().y();
  auto enu_to_flu_func = [this, dx, dy](const double enu_x, const double enu_y,
                                        const double enu_theta, double *flu_x,
                                        double *flu_y, double *flu_theta) {
    if (flu_x != nullptr && flu_y != nullptr) {
      Eigen::Vector2d flu_coordinate = common::math::RotateVector2d(
          {enu_x + dx, enu_y + dy}, -original_pose_.heading());

      *flu_x = flu_coordinate.x();
      *flu_y = flu_coordinate.y();
    }

    if (flu_theta != nullptr) {
      *flu_theta = common::math::NormalizeAngle(
          common::math::NormalizeAngle(enu_theta) - original_pose_.heading());
    }
  };

  auto gen_navi_path_loop_func =
      [this, &navigation_path, &enu_to_flu_func](
          const int start, const int end, const double ref_s_base,
          const double max_length, common::Path *path) {
        CHECK_NOTNULL(path);
        const double ref_s = navigation_path.path_point(start).s();
        for (int i = start; i < end; ++i) {
          auto *point = path->add_path_point();
```

```
          point->CopyFrom(navigation_path.path_point(i));

          double flu_x = 0.0;
          double flu_y = 0.0;
          double flu_theta = 0.0;
          enu_to_flu_func(point->x(), point->y(), point->theta(), &flu_x,
                          &flu_y, &flu_theta);

          point->set_x(flu_x);
          point->set_y(flu_y);
          point->set_theta(flu_theta);
          const double accumulated_s =
              navigation_path.path_point(i).s() - ref_s + ref_s_base;
          point->set_s(accumulated_s);

          if (accumulated_s > max_length) {
            break;
          }
        }
      };

  double dist = navigation_path.path_point().rbegin()->s() -
                navigation_path.path_point(current_project_index).s();
  // 将当前位置整合到循环/循环路线的起点
  if (FLAGS_enable_cyclic_rerouting &&
      dist < config_.max_len_from_navigation_line()) {
    auto item_iter = stitch_index_map_.find(line_index);
    if (item_iter != stitch_index_map_.end()) {
      int stitch_start_index =
          std::max(item_iter->second.first, item_iter->second.second);
      stitch_start_index = std::max(current_project_index,
stitch_start_index);
      stitch_start_index =
          std::min(navigation_path.path_point_size() - 1, stitch_start_index);

      int stitch_end_index =
          std::min(item_iter->second.first, item_iter->second.second);
      stitch_end_index = std::max(0, stitch_end_index);
      stitch_end_index = std::min(current_project_index, stitch_end_index);

      ADEBUG << "The stitch_start_index is: " << stitch_start_index << "; "
             << "the stitch_end_index is: " << stitch_end_index << "; "
             << "the current_project_index is: " << current_project_index
```

```
            << " for the navigation line: " << line_index;

    double length = navigation_path.path_point(stitch_start_index).s() -
                    navigation_path.path_point(current_project_index).s();
    gen_navi_path_loop_func(std::max(0, current_project_index - 3),
                            stitch_start_index + 1, 0.0, length, path);
    if (length > config_.max_len_from_navigation_line()) {
      return true;
    }
    gen_navi_path_loop_func(stitch_end_index,
                            navigation_path.path_point_size(), length,
                            config_.max_len_from_navigation_line(), path);
    return true;
  }
}

if (dist < 20) {
  return false;
}
gen_navi_path_loop_func(std::max(0, current_project_index - 3),
                        navigation_path.path_point_size(), 0.0,
                        config_.max_len_from_navigation_line(), path);
return true;
}

// 根据 path 计算投影状态
ProjIndexPair NavigationLane::UpdateProjectionIndex(const common::Path &path,
                                                    const int line_index) {
  if (path.path_point_size() < 2) {
    return std::make_pair(-1, std::numeric_limits<double>::max());
  }

  double min_d = std::numeric_limits<double>::max();
  const int path_size = path.path_point_size();
  int current_project_index = 0;
  auto item_iter = last_project_index_map_.find(line_index);
  if (item_iter != last_project_index_map_.end()) {
    current_project_index = std::max(0, item_iter->second.first);
  }

  // 用于检查车辆初始位置和当前导航线起点之间距离的 lambda 表达式

  auto check_distance_func = [this, &path, &path_size](
```

```
                                 const int project_index,
                                 double *project_distance) {
      // 将当前导航行的起始点从 ENU 坐标转化为车辆坐标。对于多车道情况，y 轴方向的距离可以适
当扩大，但 x 轴方向的距离应该很小

      // flu_x = (enu_x - x_shift) * cos(angle) + (enu_y - y_shift) *
      //  sin(angle)
      // flu_y = (enu_y - y_shift) * cos(angle) - (enu_x - x_shift) *
      //  sin(angle)
      if (project_index < 0 || project_index > path_size - 1) {
        return false;
      }
      double enu_x = path.path_point(project_index).x();
      double enu_y = path.path_point(project_index).y();
      double x_shift = original_pose_.position().x();
      double y_shift = original_pose_.position().y();
      double cos_angle = std::cos(original_pose_.heading());
      double sin_angle = std::sin(original_pose_.heading());
      double flu_x =
          (enu_x - x_shift) * cos_angle + (enu_y - y_shift) * sin_angle;
      double flu_y =
          (enu_y - y_shift) * cos_angle - (enu_x - x_shift) * sin_angle;

      if (project_distance != nullptr) {
        *project_distance = std::fabs(flu_y);
      }

      if (std::fabs(flu_x) < config_.max_distance_to_navigation_line() / 2.0 &&
          std::fabs(flu_y) < config_.max_distance_to_navigation_line() * 2.0) {
        return true;
      }
      return false;
    };

    int index = 0;
    for (int i = current_project_index; i + 1 < path_size; ++i) {
      const double d = DistanceXY(original_pose_.position(),
path.path_point(i));
      if (d < min_d) {
        min_d = d;
        index = i;
      }
```

```
    const double kMaxDistance = 50.0;
    if (current_project_index != 0 && d > kMaxDistance) {
      break;
    }
  }

  if (check_distance_func(index, &min_d)) {
    if (FLAGS_enable_cyclic_rerouting) {
      // 我们在此创建一个条件，将“当前项目索引”设置为 0，如果车辆到达循环/循环路线的终点。对于循环/循环导航线，其起点和终点之间的距离非常小，每次重新发送导航线都很烦琐且不必要

      auto item_iter = stitch_index_map_.find(line_index);
      if (item_iter != stitch_index_map_.end()) {
        int start_index =
            std::max(item_iter->second.first, item_iter->second.second);
        int end_index =
            std::min(item_iter->second.first, item_iter->second.second);
        int index_diff = index - start_index;
        if (index_diff >= 0) {
          index = std::min(end_index + index_diff, start_index);
          min_d = DistanceXY(original_pose_.position(),
path.path_point(index));
        }
      }
    }
    return std::make_pair(index, min_d);
  }

  return std::make_pair(-1, std::numeric_limits<double>::max());
}
//kappa 系数是一种衡量分类精度的指标。它是通过把所有地表真实分类中的像元总数（N）乘以混淆矩阵对角线（Xkk）的和，再减去某一类地表真实像元总数与该类中被分类像元总数之积对所有类别求和的结果，再除以总像元数的平方减去某一类地表真实像元总数与该类中被分类像元总数之积对所有类别求和的结果所得到的
double NavigationLane::GetKappa(const double c1, const double c2,
                                const double c3, const double x) {
  const double dy = 3 * c3 * x * x + 2 * c2 * x + c1;
  const double d2y = 6 * c3 * x + 2 * c2;
  return d2y / std::pow((1 + dy * dy), 1.5);
}
//将道路标志转换成路径信息
void NavigationLane::ConvertLaneMarkerToPath(
    const perception::LaneMarkers &lane_marker, common::Path *const path) {
```

```
    CHECK_NOTNULL(path);

    path->set_name("Path from lane markers.");
    const auto &left_lane = lane_marker.left_lane_marker();
    const auto &right_lane = lane_marker.right_lane_marker();

    double path_c0 = (left_lane.c0_position() + right_lane.c0_position()) / 2.0;

    double left_quality = left_lane.quality() + 0.001;
    double right_quality = right_lane.quality() + 0.001;

    double quality_divider = left_quality + right_quality;

    double path_c1 = (left_lane.c1_heading_angle() * left_quality +
                      right_lane.c1_heading_angle() * right_quality) /
                     quality_divider;

    double path_c2 = (left_lane.c2_curvature() * left_quality +
                      right_lane.c2_curvature() * right_quality) /
                     quality_divider;

    double path_c3 = (left_lane.c3_curvature_derivative() * left_quality +
                      right_lane.c3_curvature_derivative() * right_quality) /
                     quality_divider;

    const double current_speed =
        VehicleStateProvider::Instance()->vehicle_state().linear_velocity();
    double path_range =
        current_speed * config_.ratio_navigation_lane_len_to_speed();
    if (path_range <= config_.min_len_for_navigation_lane()) {
      path_range = config_.min_len_for_navigation_lane();
    } else {
      path_range = config_.max_len_for_navigation_lane();
    }

    const double unit_z = 1.0;
    const double start_s = -2.0;
    double accumulated_s = start_s;
    for (double z = start_s; z <= path_range; z += unit_z) {
      double x1 = z;
      double y1 = 0;
      if (left_lane.view_range()> config_.min_view_range_to_use_lane_marker() ||
```

```
      right_lane.view_range() > config_.min_view_range_to_use_lane_marker())
{
        y1 = EvaluateCubicPolynomial(path_c0, path_c1, path_c2, path_c3, z);
      }
      auto *point = path->add_path_point();
      point->set_x(x1);
      point->set_y(y1);

      if (path->path_point_size() > 1) {
        auto &pre_point = path->path_point(path->path_point_size() - 2);
        accumulated_s += std::hypot(x1 - pre_point.x(), y1 - pre_point.y());
      }
      point->set_s(accumulated_s);
      point->set_theta(
          std::atan2(3 * path_c3 * x1 * x1 + 2 * path_c2 * x1 + path_c1, 1));
      point->set_kappa(GetKappa(path_c1, path_c2, path_c3, x1));

      const double k1 = GetKappa(path_c1, path_c2, path_c3, x1 - 0.0001);
      const double k2 = GetKappa(path_c1, path_c2, path_c3, x1 + 0.0001);
      point->set_dkappa((k2 - k1) / 0.0002);
    }

    perceived_left_width_ = std::fabs(left_lane.c0_position());
    perceived_right_width_ = std::fabs(right_lane.c0_position());
    // 如果感知到的车道宽度不正确，就直接使用默认的车道宽度

    double perceived_lane_width = perceived_left_width_ +
perceived_right_width_;
    if (perceived_lane_width < 2.0 * config_.min_lane_half_width() ||
        perceived_lane_width > 2.0 * config_.max_lane_half_width()) {
      perceived_left_width_ = default_left_width_;
      perceived_right_width_ = default_right_width_;
    }
  }

  bool NavigationLane::CreateMap(const MapGenerationParam &map_config,
                                 MapMsg *const map_msg) const {
    auto *navigation_path = map_msg->mutable_navigation_path();
    auto *hdmap = map_msg->mutable_hdmap();
    auto *lane_marker = map_msg->mutable_lane_marker();

    lane_marker->CopyFrom(perception_obstacles_.lane_marker());
```

```
//如果没有基于导航线生成导航路径，我们就尝试使用“当前导航路径元组”创建地图，其生成基于
感知到的车道标志

    if (navigation_path_list_.empty()) {
      if (std::get<3>(current_navi_path_tuple_) != nullptr) {
        FLAGS_relative_map_generate_left_boundray = true;
        return CreateSingleLaneMap(current_navi_path_tuple_, map_config,
                                   perception_obstacles_, hdmap, navigation_path);
      } else {
        return false;
      }
    }

    int fail_num = 0;
    FLAGS_relative_map_generate_left_boundray = true;
    for (auto iter = navigation_path_list_.cbegin();
         iter != navigation_path_list_.cend(); ++iter) {
      std::size_t index = std::distance(navigation_path_list_.cbegin(), iter);
      if (!CreateSingleLaneMap(*iter, map_config, perception_obstacles_, hdmap,
                               navigation_path)) {
        AWARN << "Failed to generate lane: " << index;
        ++fail_num;
        FLAGS_relative_map_generate_left_boundray = true;
        continue;
      }
      FLAGS_relative_map_generate_left_boundray = false;

      //中间车道的左边界使用左车道的右边界

      int lane_index = static_cast<int>(index) - fail_num;
      if (lane_index > 0) {
        auto *left_boundary =
            hdmap->mutable_lane(lane_index)->mutable_left_boundary();
        left_boundary->CopyFrom(hdmap->lane(lane_index - 1).right_boundary());
        auto *left_sample =
            hdmap->mutable_lane(lane_index)->mutable_left_sample();
        left_sample->CopyFrom(hdmap->lane(lane_index - 1).right_sample());
      }
    }

    int lane_num = hdmap->lane_size();
    ADEBUG << "The Lane number is: " << lane_num;
```

```
    // 设置道路边界
    auto *road = hdmap->add_road();
    road->mutable_id()->set_id("road_" + hdmap->lane(0).id().id());
    auto *section = road->add_section();
    for (int i = 0; i < lane_num; ++i) {
      auto *lane_id = section->add_lane_id();
      lane_id->CopyFrom(hdmap->lane(0).id());
    }
    auto *outer_polygon = section->mutable_boundary()->mutable_outer_polygon();
    auto *left_edge = outer_polygon->add_edge();
    left_edge->set_type(apollo::hdmap::BoundaryEdge::LEFT_BOUNDARY);

left_edge->mutable_curve()->CopyFrom(hdmap->lane(0).left_boundary().curve());

    auto *right_edge = outer_polygon->add_edge();
    right_edge->set_type(apollo::hdmap::BoundaryEdge::RIGHT_BOUNDARY);
    right_edge->mutable_curve()->CopyFrom(
        hdmap->lane(lane_num - 1).right_boundary().curve());

    // 为每条车道线设定邻近的车道线
    if (lane_num < 2) {
      return true;
    }
    for (int i = 0; i < lane_num; ++i) {
      auto *lane = hdmap->mutable_lane(i);
      if (i > 0) {
        lane->add_left_neighbor_forward_lane_id()->CopyFrom(
            hdmap->lane(i - 1).id());
        ADEBUG << "Left neighbor is: " << hdmap->lane(i - 1).id().id();
      }
      if (i < lane_num - 1) {
        lane->add_right_neighbor_forward_lane_id()->CopyFrom(
            hdmap->lane(i + 1).id());
        ADEBUG << "Right neighbor is: " << hdmap->lane(i + 1).id().id();
      }
    }
    return true;
  }
  //更新车道融合后的索引信息
  void NavigationLane::UpdateStitchIndexInfo() {
    stitch_index_map_.clear();

    int navigation_line_num = navigation_info_.navigation_path_size();
```

```
if (navigation_line_num <= 0) {
  return;
}

constexpr int kMinPathPointSize = 10;
for (int i = 0; i < navigation_line_num; ++i) {
  const auto &navigation_path = navigation_info_.navigation_path(i).path();
  if (!navigation_info_.navigation_path(i).has_path() ||
      navigation_path.path_point_size() < kMinPathPointSize) {
    continue;
  }

  double min_distance = std::numeric_limits<double>::max();
  StitchIndexPair min_index_pair = std::make_pair(-1, -1);

  int path_size = navigation_path.path_point_size();
  const double start_s = navigation_path.path_point(0).s();
  const double end_s = navigation_path.path_point(path_size - 1).s();
  for (int m = 0; m < path_size; ++m) {
    double forward_s = navigation_path.path_point(m).s() - start_s;
    if (forward_s > config_.max_len_from_navigation_line()) {
      break;
    }

    for (int n = path_size - 1; n >= 0; --n) {
      double reverse_s = end_s - navigation_path.path_point(n).s();
      if (reverse_s > config_.max_len_from_navigation_line()) {
        break;
      }
      if (m == n) {
        break;
      }

      double current_distance = DistanceXY(navigation_path.path_point(m),
                                           navigation_path.path_point(n));
      if (current_distance < min_distance) {
        min_distance = current_distance;
        min_index_pair = std::make_pair(m, n);
      }
    }
  }

  if (min_distance < config_.min_lane_half_width()) {
```

```
      AINFO << "The stitching pair is: (" << min_index_pair.first << ", "
            << min_index_pair.second << ") for the navigation line: " << i;
      stitch_index_map_[i] = min_index_pair;
    }
  }
}

}  // namespace relative_map
}  // namespace apollo
```

参考文献

[1] OpenDRIVEFormatSpecRev www.opendrive.org/docs/OpenDRIVEFormatSpecRev1.4H.

[2] Xiaofang W U, Zhiyong X U, Zhongliang C, et al. Design and Implementation of High Precision Map Symbol Library Based on GDI[J]. Editorial Board of Geomatics & Information Science of Wuhan University, 2004, 29(10):928-932.

[3] Wei W. Designing and Implementation of Map Symbol Database Based on COM[J]. Editoral Board of Geomatics and Information Science of Wuhan University, 2002.

[4] Zhao R, Chen J, Wang D, et al. The design and implementation of geo-spatial database updating system based on digital map generalization[J]. Proceedings of SPIE - The International Society for Optical Engineering, 2007, 6795:67955S-67955S-6.

[5] Cui-Hua X, Yun L, Jie S, et al. A Constructing Algorithm of Concept Lattice with Attribute Generalization Based on Cloud Models[C]// Computer and Information Technology, 2005. CIT 2005. The Fifth International Conference on. IEEE Computer Society, 2005.

[6] Wang M L, Xu A J. Design and Implementation of Point Symbol Database Based on ArcGIS Engine[J]. Applied Mechanics and Materials, 2012, 263-266:1853-1857.

第12章 无人驾驶定位技术

无人驾驶过程中，首先要回答的问题就是 Where am I（我在哪里）。完成这个过程需要的技术就是无人驾驶定位技术。RTK（Real - Time Kinematic）载波相位差分技术是常用的定位技术，定位过程中需要结合 GPS 和 IMU 的输入进行综合考量。在当前自动驾驶汽车传感器类型和数量日益增长的大背景下，使用多传感器融合定位的方法逐渐主流化。

本章主要讲解无人驾驶中常用的定位技术以及在百度 Apollo 系统中定位模型的具体实现。内容分为两部分：第一部分着重介绍无人驾驶中常用的 RTK 定位技术以及在百度 Apollo 系统中定位模型的具体实现；第二部分着重介绍无人驾驶中常用的多传感器融合定位技术以及在百度 Apollo 系统中定位模型的具体实现。

12.1 RTK 定位技术

首先我们来介绍 RTK-GPS 定位法的概念和基本原理。RTK 载波相位差分技术是实时处理两个测量站载波相位观测量的差分方法，将基准站采集的载波相位发给用户接收机，进行求差解算坐标。这是一种新的常用的 GPS 测量方法，以前的静态、快速静态、动态测量都需要事后进行解算才能获得厘米级的精度，而 RTK 是能够在野外实时得到厘米级定位精度的测量方法，它采用了载波相位动态实时差分方法，是 GPS 应用的重大里程碑，它的出现为工程放样、地形测图等各种控制测量带来了新曙光，极大地提高了野外作业效率。高精度的 GPS 测量必须采用载波相位观测值，RTK 定位技术就是基于载波相位观测值的实时动态定位技术，它能够实时地提供测站点在指定坐标系中的三维定位结果，并达到厘米级精度。在 RTK 作业模式下，基准站通过数据链将其观测值和测站坐标信息一起传送给流动站。流动站不仅通过数据链接收来自基准站的数据，还要采集 GPS 观测数据，并在系统内组成差分观测值进行实时处理，同时给出厘米级定位结果，历时不足一秒钟。流动站可处于静止状态，也可以处于运动状态；可以在固定点上先进行初始化后再进入动态作业，也可在动态条件下直接开机，并在动态环境下完成整周模糊度的搜索求解。在整周未知数解固定后，即可进行每个历元的实时处理，只要能保持4颗以上卫星相位观测值的跟踪和必要的几何图形，流动站就可以随时给出厘米级定位结果。

另一个核心硬件输入是惯性测量单元（Inertial Measurement Unit，IMU），它由3个单轴的加速度计和3个单轴的陀螺仪组成，加速度计检测物体在载体坐标系统独立三轴的加速度信号，而陀螺仪检测载体相对于导航坐标系的角速度信号，对这些信号进行处理之后，便可解算出物体的姿态。值得注意的是，IMU 提供的是一个相对的定位信息，它的作用是测量相对于起点物体所运动的路线，所以它并不能提供所在的具体位置的信息。IMU 常常和 GPS 一起使用，当在某些 GPS 信号微弱的地方时，IMU 就可以发挥作用，可以让汽车继续获得绝对位置的信息，不至于“迷路”。

接下来我们分析一下 Apollo 中 RTK 定位的实现过程的核心方法 ComposeLocalizationMsg，代码位于 modules/localization/rtk/rtk_localization.cc 中。

```
  //实施运动融合的核心算法
  void RTKLocalization::ComposeLocalizationMsg(
      const localization::Gps &gps_msg, const localization::CorrectedImu
&imu_msg,
      LocalizationEstimate *localization) {
    //清除定位消息对象的内存，保证定位消息对象最终输出正常
  localization->Clear();
  //构建定位消息对象的消息头
    FillLocalizationMsgHeader(localization);
  //通过 gps 时钟信息设定定位对象的信息时间
    localization->set_measurement_time(gps_msg.header().timestamp_sec());

    //结合 gps 和 imu 信息完成定位信息（mutable_pose 包含位置、方向、速度、加速度等）的构建
    auto mutable_pose = localization→mutable_pose();
  //按照 gps 输入信息首先构建定位信息
    if (gps_msg.has_localization()) {
      const auto &pose = gps_msg.localization();

      if (pose.has_position()) {
        // 根据 gps 输入构建位置信息
        mutable_pose->mutable_position()->set_x(pose.position().x() -
                                                map_offset_[0]);
        mutable_pose->mutable_position()->set_y(pose.position().y() -
                                                map_offset_[1]);
        mutable_pose->mutable_position()->set_z(pose.position().z() -
                                                map_offset_[2]);
      }

      // 根据 gps 输入构建方向信息
      if (pose.has_orientation()) {
```

```
      mutable_pose->mutable_orientation()->CopyFrom(pose.orientation());
      double heading = common::math::QuaternionToHeading(
          pose.orientation().qw(), pose.orientation().qx(),
          pose.orientation().qy(), pose.orientation().qz());
      mutable_pose->set_heading(heading);
    }
    // 根据 gps 输入构建速度信息
    if (pose.has_linear_velocity()) {
mutable_pose->mutable_linear_velocity()->CopyFrom(pose.linear_velocity());
    }
  }
  //按照 imu 输入信息首先构建定位信息
   if (imu_msg.has_imu()) {
    const auto &imu = imu_msg.imu();
    // 按照 imu 输入信息首先构建加速度信息
    if (imu.has_linear_acceleration()) {
      if (localization->pose().has_orientation()) {
        // imu 变量使用的核心步骤是坐标转化，将车辆坐标映射成地图坐标
        Vector3d orig(imu.linear_acceleration().x(),
                      imu.linear_acceleration().y(),
                      imu.linear_acceleration().z());
        Vector3d vec = common::math::QuaternionRotate(
            localization->pose().orientation(), orig);
        mutable_pose->mutable_linear_acceleration()->set_x(vec[0]);
        mutable_pose->mutable_linear_acceleration()->set_y(vec[1]);
        mutable_pose->mutable_linear_acceleration()->set_z(vec[2]);

        //加速度信息存储
        mutable_pose->mutable_linear_acceleration_vrf()->CopyFrom(
            imu.linear_acceleration());
      } else {
        AERROR << "[PrepareLocalizationMsg]: "
               << "fail to convert linear_acceleration";
      }
    }

    // 按照 imu 输入信息首先构建角速度信息
    if (imu.has_angular_velocity()) {
      if (localization->pose().has_orientation()) {
        // imu 变量使用的核心步骤是坐标转化，将车辆坐标映射成地图坐标
        Vector3d orig(imu.angular_velocity().x(), imu.angular_velocity().y(),
                      imu.angular_velocity().z());
```

```
        Vector3d vec = common::math::QuaternionRotate(
            localization->pose().orientation(), orig);
        mutable_pose->mutable_angular_velocity()->set_x(vec[0]);
        mutable_pose->mutable_angular_velocity()->set_y(vec[1]);
        mutable_pose->mutable_angular_velocity()->set_z(vec[2]);

         //角速度信息存储
        mutable_pose->mutable_angular_velocity_vrf()->CopyFrom(
            imu.angular_velocity());
      } else {
        AERROR << "[PrepareLocalizationMsg]: "
               << "fail to convert angular_velocity";
      }
    }

    // 按照 imu 输入信息首先构建欧拉角信息
    if (imu.has_euler_angles()) {
      mutable_pose->mutable_euler_angles()->CopyFrom(imu.euler_angles());
    }
  }
  return;
}
```

12.2 多传感器融合定位技术

多传感器融合算法针对车辆上所有类型的传感器组件，包括之前使用的惯性测量单元（IMU）、全球定位系统（GPS）的输入以及摄像头、激光雷达等其他传感器的输入。

12.2.1 激光雷达简介

作为自动驾驶的传感器中价格最高的硬件系统，激光雷达有其自身独特的特性。激光雷达物体识别最大的优点是可以完全排除光线的干扰，无论是白天还是黑夜，无论是树影斑驳的林荫道还是光线急剧变化的隧道出口，都没有问题。其次，激光雷达可以轻易获得深度信息，而对摄像头系统来说这非常困难。再次，激光雷达的有效距离远在摄像头之上，更远的有效距离等于加大了安全冗余。最后，激光雷达的 3D 云点与摄像头的 2D 图像，两者在进行深度学习目标识别时，2D 图像容易发生透视变形（Perspective Distortion）。简单地说，透视变形指的是一个物体及其周围区域与标准镜头中看到的相比完全不同，由于远近特征的相对比例变化，发生了弯曲或变形。这是透镜的固有特性（凸透镜汇聚光线，凹透镜发散光线），所以无法消除，只能改善。而 3D 就不会有这个问题，所以 3D 图像的深度学习使用的神经网络可以更加简单一点。另外，激光雷达也可以识别颜色和车道线。实际上激光雷达与摄像头没有本质区别，其最大的区别除了激光雷达是主动发射激光，是主动传感器外，只是光电接收二极管不同。摄像头可以做到的，激光雷达都能够做到，只是目前激光雷达的点云密集度还不能和 300 万像素级的摄像头比。无人车领域对目标图像的识别不仅仅是识别，还包括分割和追踪。分割就是用物体框框出目标，对于 2D 图像来说，只能用 2D 框分割出目标，而激光雷达图像则可以做到 3D 框，安全性更高。追踪则是预测出车辆或行人可能的运动轨迹。

12.2.2 扩展卡尔曼滤波原理

在多传感器融合算法上，目前多采用扩展卡尔曼滤波（Extended Kalman Filter，EKF）算法。卡尔曼滤波算法能够在线性高斯模型的条件下，对目标的状态做出最优的估计，得到较好的跟踪效果。对非线性滤波问题常用的处理方法是利用线性化技巧将其转化为一个近似的线性滤波问题。因此，可以利用非线性函数的局部性特性将非线性模型局部化，再利用卡尔曼滤波算法完成滤波跟踪。扩展卡尔曼滤波就是基于这样的思想，将系统的非线性函数进行一阶 Taylor 展开，得到线性化的系统方程，从而完成对目标的滤波估计等处理。

非线性系统离散动态方程可以表示为：

$$X(k+1)=f[k,X(k)]+G(k)W(k) \tag{1}$$

$$Z(k)=h[k,X(k)]+V(k) \tag{2}$$

这里为了便于数学处理，假定没有控制量的输入，并假定过程噪声是均值为零的高斯白噪声，且噪声分布矩阵$G(k)$是已知的。其中，观测噪声$V(k)$是加和均值为零的高斯白噪声。假定过程噪声和观测噪声序列是彼此独立的，并且有初始状态估计$\hat{X}(0|0)$和协方差矩阵$P(0|0)$。和线性系统的情况一样，我们可以得到扩展卡尔曼滤波算法如下：

$$\hat{X}(k|k+1)=f(\hat{X}(k|k)) \tag{3}$$

$$P(k+1|k)=\Phi(k+1|k)P(k|k)\Phi^{T}(k+1|k)+Q(k+1) \tag{4}$$

$$K(k+1)=P(k+1|k)H^{T}(k+1)[H(k+1)P(k+1|k)H^{T}(k+1)+R(k+1)]^{-1} \tag{5}$$

$$\hat{X}(K+1|k+1)=\hat{X}(K+1|k)+K(k+1)[Z(k+1)-h(\hat{X}(k+1|k))] \tag{6}$$

$$P(k+1)=[I-K(k+1)H(k+1)]P(k+1|k) \tag{7}$$

这里需要重点说明的是，状态转移$\Phi(k+1|k)$和量测矩阵$H(k+1)$是由f和 h 的雅克比矩阵代替的。其雅克比矩阵的求法如下：

假如状态变量有 n 维，即 X=[x_1 x_2…x_n]，则使用状态方程对各维求偏导：

$$\Phi(k+1)=\frac{\partial f}{\partial x}=\frac{\partial f}{\partial x_1}+\frac{\partial f}{\partial x_2}+\frac{\partial f}{\partial x_3}+\ldots+\frac{\partial f}{\partial x_n} \tag{8}$$

$$H(k+1)=\frac{\partial h}{\partial X}=\frac{\partial h}{\partial x_1}+\frac{\partial h}{\partial x_2}+\frac{\partial h}{\partial x_3}+\ldots+\frac{\partial h}{\partial x_n} \tag{9}$$

12.2.3 百度 Apollo 相关源码分析

接下来我们分析 Apollo 中的多传感器融合定位的核心组件 MSFLocalizationComponent，代码位于 modules/localization/msf/msf_localization_component.cc 中。

```
namespace apollo {
namespace localization {

using apollo::common::time::Clock;

MSFLocalizationComponent::MSFLocalizationComponent() {}
//多传感器融合定位组件初始化方法
bool MSFLocalizationComponent::Init() {
//设定组件时钟
  Clock::SetMode(Clock::CYBER);
//设定定位组件消息发布对象
  publisher_.reset(new LocalizationMsgPublisher(this->node_));
//初始化融合定位组件配置参数
  if (!InitConfig()) {
    AERROR << "Init Config failed.";
    return false;
  }
//初始化融合定位组件 IO 部分
  if (!InitIO()) {
    AERROR << "Init IO failed.";
    return false;
  }

  return true;
}
//初始化融合定位组件配置参数具体实现
bool MSFLocalizationComponent::InitConfig() {
  lidar_topic_ = FLAGS_lidar_topic;
  bestgnsspos_topic_ = FLAGS_gnss_best_pose_topic;
  gnss_heading_topic_ = FLAGS_heading_topic;
//定位组件消息发布对象配置初始化
  if (!publisher_->InitConfig()) {
    AERROR << "Init publisher config failed.";
    return false;
  }
//定位组件初始化以及初始化结果检测
  if (!localization_.Init().ok()) {
    AERROR << "Init class MSFLocalization failed.";
```

```
      return false;
    }
    return true;
  }
  //初始化融合定位组件 IO 部分
  bool MSFLocalizationComponent::InitIO() {
    cyber::ReaderConfig reader_config;
    reader_config.channel_name = lidar_topic_;
    reader_config.pending_queue_size = 1;
  //初始化激光雷达通信模块
    std::function<void(const std::shared_ptr<drivers::PointCloud>&)>
        lidar_register_call = std::bind(&MSFLocalization::OnPointCloud,
                                        &localization_, std::placeholders::_1);
  //创建激光雷达点云读对象，使用 Cyber 的 API，具体过程可参考 10.3 节的内容
    lidar_listener_ = this->node_->CreateReader<drivers::PointCloud>(
        reader_config, lidar_register_call);
  //初始化 GPS 通信模块
    std::function<void(const std::shared_ptr<drivers::gnss::GnssBestPose>&)>
        bestgnsspos_register_call =
            std::bind(&MSFLocalization::OnGnssBestPose, &localization_,
                      std::placeholders::_1);
   / /创建 GPS 定位信息读对象，使用 Cyber 的 API，具体过程可参考 10.3 节的内容
  bestgnsspos_listener_ =
        this->node_->CreateReader<drivers::gnss::GnssBestPose>(
            bestgnsspos_topic_, bestgnsspos_register_call);

    std::function<void(const std::shared_ptr<drivers::gnss::Heading>&)>
        gnss_heading_call = std::bind(&MSFLocalization::OnGnssHeading,
                                      &localization_, std::placeholders::_1);
   / /创建 GPS 定位头对象的读对象，使用 Cyber 的 API，具体过程可参考 10.3 节的内容
    gnss_heading_listener_ =
this->node_->CreateReader<drivers::gnss::Heading>(
        gnss_heading_topic_, gnss_heading_call);

    // 初始化写对象，publisher 就是对 writer 对象的二次封装
    if (!publisher_->InitIO()) {
      AERROR << "Init publisher io failed.";
      return false;
    }
   //设定写对象
    localization_.SetPublisher(publisher_);
```

```
  return true;
}
//融合定位组件的 Proc 方法，此方法为通信过程中的核心调用
bool MSFLocalizationComponent::Proc(
    const std::shared_ptr<drivers::gnss::Imu>& imu_msg) {
  localization_.OnRawImu(imu_msg);
  return true;
}
//为组件定制的写对象 LocalizationMsgPublisher
LocalizationMsgPublisher::LocalizationMsgPublisher(
    const std::shared_ptr<cyber::Node>& node)
    : node_(node), tf2_broadcaster_(node) {}
//写对象 LocalizationMsgPublisher 初始化配置的方法，主要是设定与其他需要通信模块的通信
id 和 topic
bool LocalizationMsgPublisher::InitConfig() {
  localization_topic_ = FLAGS_localization_topic;
  broadcast_tf_frame_id_ = FLAGS_broadcast_tf_frame_id;
  broadcast_tf_child_frame_id_ = FLAGS_broadcast_tf_child_frame_id;
  lidar_local_topic_ = FLAGS_localization_lidar_topic;
  gnss_local_topic_ = FLAGS_localization_gnss_topic;
  localization_status_topic_ = FLAGS_localization_msf_status;

  return true;
}

//写对象 LocalizationMsgPublisher 初始化写组件的方法,主要调用 Cyber 中的 CreateWriter API
LocalizationMsgPublisher::InitIO() {
//初始化 localization 写组件
  localization_talker_ =
      node_->CreateWriter<LocalizationEstimate>(localization_topic_);
//初始化激光雷达写组件
  lidar_local_talker_ =
      node_->CreateWriter<LocalizationEstimate>(lidar_local_topic_);

  //初始化 GPS 写组件
gnss_local_talker_ =
      node_->CreateWriter<LocalizationEstimate>(gnss_local_topic_);
 //初始化与定位状态通信的写组件
  localization_status_talker_ =
      node_->CreateWriter<LocalizationStatus>(localization_status_topic_);
  return true;
}
//通过 LocalizationMsgPublisher 进行定位信息的转换
```

```
void LocalizationMsgPublisher::PublishPoseBroadcastTF(
    const LocalizationEstimate& localization) {
  // 声明带有转化标记的定位信息
  apollo::transform::TransformStamped tf2_msg;
//信息头填充
  auto mutable_head = tf2_msg.mutable_header();
  mutable_head->set_timestamp_sec(localization.measurement_time());
  mutable_head->set_frame_id(broadcast_tf_frame_id_);
  tf2_msg.set_child_frame_id(broadcast_tf_child_frame_id_);
//3 维定位信息填充
  auto mutable_translation =
tf2_msg.mutable_transform()->mutable_translation();
  mutable_translation->set_x(localization.pose().position().x());
  mutable_translation->set_y(localization.pose().position().y());
  mutable_translation->set_z(localization.pose().position().z());
//方向信息填充
  auto mutable_rotation = tf2_msg.mutable_transform()->mutable_rotation();
  mutable_rotation->set_qx(localization.pose().orientation().qx());
  mutable_rotation->set_qy(localization.pose().orientation().qy());
  mutable_rotation->set_qz(localization.pose().orientation().qz());
  mutable_rotation->set_qw(localization.pose().orientation().qw());
//发布信息
  tf2_broadcaster_.SendTransform(tf2_msg);
  return;
}
//发布定位信息
void LocalizationMsgPublisher::PublishPoseBroadcastTopic(
    const LocalizationEstimate& localization) {
  localization_talker_->Write(localization);
  return;
}
//发布定位信息给 GPS 单元
void LocalizationMsgPublisher::PublishLocalizationMsfGnss(
    const LocalizationEstimate& localization) {
  gnss_local_talker_->Write(localization);
  return;
}
//发布定位信息给激光雷达单元
void LocalizationMsgPublisher::PublishLocalizationMsfLidar(
    const LocalizationEstimate& localization) {
  lidar_local_talker_->Write(localization);
  return;
}
```

```
//发布定位状态信息
void LocalizationMsgPublisher::PublishLocalizationStatus(
    const LocalizationStatus& localization_status) {
  localization_status_talker_->Write(localization_status);
}

}  // namespace localization
}  // namespace apollo
```

msf_localization.cc

```
namespace apollo {
namespace localization {

using apollo::common::Status;

MSFLocalization::MSFLocalization()
: monitor_logger_(
apollo::common::monitor::MonitorMessageItem::LOCALIZATION),
localization_state_(msf::LocalizationMeasureState::OK),
pcd_msg_index_(-1) {}

Status MSFLocalization::Init() {
InitParams();

return localization_integ_.Init(localization_param_);
}

void MSFLocalization::InitParams() {
// 完成不同传感器模块的集成工作
localization_param_.is_ins_can_self_align = FLAGS_integ_ins_can_self_align;
localization_param_.is_sins_align_with_vel =
FLAGS_integ_sins_align_with_vel;
localization_param_.is_sins_state_check = FLAGS_integ_sins_state_check;
localization_param_.sins_state_span_time = FLAGS_integ_sins_state_span_time;
localization_param_.sins_state_pos_std = FLAGS_integ_sins_state_pos_std;
localization_param_.vel_threshold_get_yaw = FLAGS_vel_threshold_get_yaw;
localization_param_.is_trans_gpstime_to_utctime =
FLAGS_trans_gpstime_to_utctime;
localization_param_.gnss_mode = FLAGS_gnss_mode;
localization_param_.is_using_raw_gnsspos = true;
```

```
    // 全球定位模块
    localization_param_.enable_ins_aid_rtk = FLAGS_enable_ins_aid_rtk;

    //激光雷达模块
    localization_param_.map_path = FLAGS_map_dir + "/" + FLAGS_local_map_name;
    localization_param_.lidar_extrinsic_file = FLAGS_lidar_extrinsics_file;
    localization_param_.lidar_height_file = FLAGS_lidar_height_file;
    localization_param_.lidar_height_default = FLAGS_lidar_height_default;
    localization_param_.localization_mode = FLAGS_lidar_localization_mode;
    localization_param_.lidar_yaw_align_mode = FLAGS_lidar_yaw_align_mode;
    localization_param_.lidar_filter_size = FLAGS_lidar_filter_size;
    localization_param_.map_coverage_theshold =
FLAGS_lidar_map_coverage_theshold;
    localization_param_.imu_lidar_max_delay_time =
FLAGS_lidar_imu_max_delay_time;
    localization_param_.if_use_avx = FLAGS_if_use_avx;

    AINFO << "map: " << localization_param_.map_path;
    AINFO << "lidar_extrin: " << localization_param_.lidar_extrinsic_file;
    AINFO << "lidar_height: " << localization_param_.lidar_height_file;

    localization_param_.utm_zone_id = FLAGS_local_utm_zone_id;
    // 尝试从 local_map 文件夹加载区域 ID

    if (FLAGS_if_utm_zone_id_from_folder) {
    bool success = LoadZoneIdFromFolder(localization_param_.map_path,
    &localization_param_.utm_zone_id);
    if (!success) {
    AWARN << "Can't load utm zone id from map folder, use default value.";
    }
    }
    AINFO << "utm zone id: " << localization_param_.utm_zone_id;

    //车辆外部惯性测量单元
    imu_vehicle_quat_.x() = FLAGS_imu_vehicle_qx;
    imu_vehicle_quat_.y() = FLAGS_imu_vehicle_qy;
    imu_vehicle_quat_.z() = FLAGS_imu_vehicle_qz;
    imu_vehicle_quat_.w() = FLAGS_imu_vehicle_qw;
    // 尝试从文件中载入 IMU 的动量
    if (FLAGS_if_vehicle_imu_from_file) {
    double qx = 0.0;
    double qy = 0.0;
    double qz = 0.0;
```

```
double qw = 0.0;

AINFO << "Vehile imu file: " << FLAGS_vehicle_imu_file;
if (LoadImuVehicleExtrinsic(FLAGS_vehicle_imu_file, &qx, &qy, &qz, &qw)) {
imu_vehicle_quat_.x() = qx;
imu_vehicle_quat_.y() = qy;
imu_vehicle_quat_.z() = qz;
imu_vehicle_quat_.w() = qw;
} else {
AWARN << "Can't load imu vehicle quat from file, use default value.";
}
}
AINFO << "imu_vehicle_quat: " << imu_vehicle_quat_.x() << " "
<< imu_vehicle_quat_.y() << " " << imu_vehicle_quat_.z() << " "
<< imu_vehicle_quat_.w();

// 需要初始化的其他参数
localization_param_.enable_lidar_localization =
FLAGS_enable_lidar_localization;

if (!FLAGS_if_imuant_from_file) {
localization_param_.imu_to_ant_offset.offset_x = FLAGS_imu_to_ant_offset_x;
localization_param_.imu_to_ant_offset.offset_y = FLAGS_imu_to_ant_offset_y;
localization_param_.imu_to_ant_offset.offset_z = FLAGS_imu_to_ant_offset_z;
localization_param_.imu_to_ant_offset.uncertainty_x =
FLAGS_imu_to_ant_offset_ux;
localization_param_.imu_to_ant_offset.uncertainty_y =
FLAGS_imu_to_ant_offset_uy;
localization_param_.imu_to_ant_offset.uncertainty_z =
FLAGS_imu_to_ant_offset_uz;
} else {
double offset_x = 0.0;
double offset_y = 0.0;
double offset_z = 0.0;
double uncertainty_x = 0.0;
double uncertainty_y = 0.0;
double uncertainty_z = 0.0;
AINFO << "Ant imu lever arm file: " << FLAGS_ant_imu_leverarm_file;
CHECK(LoadGnssAntennaExtrinsic(FLAGS_ant_imu_leverarm_file, &offset_x,
&offset_y, &offset_z, &uncertainty_x,
&uncertainty_y, &uncertainty_z));
localization_param_.ant_imu_leverarm_file = FLAGS_ant_imu_leverarm_file;
```

```
localization_param_.imu_to_ant_offset.offset_x = offset_x;
localization_param_.imu_to_ant_offset.offset_y = offset_y;
localization_param_.imu_to_ant_offset.offset_z = offset_z;
localization_param_.imu_to_ant_offset.uncertainty_x = uncertainty_x;
localization_param_.imu_to_ant_offset.uncertainty_y = uncertainty_y;
localization_param_.imu_to_ant_offset.uncertainty_z = uncertainty_z;

AINFO << localization_param_.imu_to_ant_offset.offset_x << " "
<< localization_param_.imu_to_ant_offset.offset_y << " "
<< localization_param_.imu_to_ant_offset.offset_z << " "
<< localization_param_.imu_to_ant_offset.uncertainty_x << " "
<< localization_param_.imu_to_ant_offset.uncertainty_y << " "
<< localization_param_.imu_to_ant_offset.uncertainty_z;
}

localization_param_.imu_delay_time_threshold_1 =
FLAGS_imu_delay_time_threshold_1;
localization_param_.imu_delay_time_threshold_2 =
FLAGS_imu_delay_time_threshold_2;
localization_param_.imu_delay_time_threshold_3 =
FLAGS_imu_delay_time_threshold_3;

localization_param_.imu_missing_time_threshold_1 =
FLAGS_imu_missing_time_threshold_1;
localization_param_.imu_missing_time_threshold_2 =
FLAGS_imu_missing_time_threshold_2;
localization_param_.imu_missing_time_threshold_3 =
FLAGS_imu_missing_time_threshold_3;

localization_param_.bestgnsspose_loss_time_threshold =
FLAGS_bestgnsspose_loss_time_threshold;
localization_param_.lidar_loss_time_threshold =
FLAGS_lidar_loss_time_threshold;

localization_param_.localization_std_x_threshold_1 =
FLAGS_localization_std_x_threshold_1;
localization_param_.localization_std_y_threshold_1 =
FLAGS_localization_std_y_threshold_1;

localization_param_.localization_std_x_threshold_2 =
FLAGS_localization_std_x_threshold_2;
localization_param_.localization_std_y_threshold_2 =
FLAGS_localization_std_y_threshold_2;
```

```
}
//激光雷达点云数据处理
void MSFLocalization::OnPointCloud(
const std::shared_ptr<drivers::PointCloud> &message) {
++pcd_msg_index_;
if (pcd_msg_index_ % FLAGS_point_cloud_step != 0) {
return;
}
//处理点云信息
localization_integ_.PcdProcess(*message);
//根据点云处理结果得到最近的定位对象
const auto &result = localization_integ_.GetLastestLidarLocalization();

if (result.state() == msf::LocalizationMeasureState::OK ||
result.state() == msf::LocalizationMeasureState::VALID) {
//发布点云数据转化后的定位对象
publisher_->PublishLocalizationMsfLidar(result.localization());
}

return;
}
//处理惯性测量单元的原始数据
void MSFLocalization::OnRawImu(
const std::shared_ptr<drivers::gnss::Imu> &imu_msg) {
if (FLAGS_imu_coord_rfu) {
localization_integ_.RawImuProcessRfu(*imu_msg);
} else {
localization_integ_.RawImuProcessFlu(*imu_msg);
}
//根据惯性单元处理结果得到最近的定位对象
const auto &result = localization_integ_.GetLastestIntegLocalization();

// 合并定位状态信息
LocalizationStatus status;
apollo::common::Header *status_headerpb = status.mutable_header();
status_headerpb->set_timestamp_sec(
result.localization().header().timestamp_sec());
status.set_fusion_status(
static_cast<MeasureState>(result.integ_status().integ_state));
status.set_state_message(result.integ_status().state_message);
status.set_measurement_time(result.localization().measurement_time());
publisher_->PublishLocalizationStatus(status);
```

```
if (result.state() == msf::LocalizationMeasureState::OK ||
result.state() == msf::LocalizationMeasureState::VALID) {
// 计算 orientation_vehicle_world
LocalizationEstimate local_result = result.localization();
CompensateImuVehicleExtrinsic(&local_result);
//发布得到的定位对象
publisher_->PublishPoseBroadcastTF(local_result);
publisher_->PublishPoseBroadcastTopic(local_result);
}

localization_state_ = result.state();

return;
}
//处理全球导航卫星系统的最优定位数据信息
void MSFLocalization::OnGnssBestPose(
const std::shared_ptr<drivers::gnss::GnssBestPose> &bestgnsspos_msg) {
if ((localization_state_ == msf::LocalizationMeasureState::OK ||
localization_state_ == msf::LocalizationMeasureState::VALID) &&
FLAGS_gnss_only_init) {
return;
}
//处理全球导航卫星系统的数据信息
localization_integ_.GnssBestPoseProcess(*bestgnsspos_msg);
//根据处理结果得到最近的定位对象
const auto &result = localization_integ_.GetLastestGnssLocalization();
//发布得到的定位对象
if (result.state() == msf::LocalizationMeasureState::OK ||
result.state() == msf::LocalizationMeasureState::VALID) {
publisher_->PublishLocalizationMsfGnss(result.localization());
}

return;
}
//处理全球导航卫星系统的原始观测数据信息
void MSFLocalization::OnGnssRtkObs(
const std::shared_ptr<drivers::gnss::EpochObservation> &raw_obs_msg) {
if ((localization_state_ == msf::LocalizationMeasureState::OK ||
localization_state_ == msf::LocalizationMeasureState::VALID) &&
FLAGS_gnss_only_init) {
return;
}
//处理全球导航卫星系统的数据信息
```

```
localization_integ_.RawObservationProcess(*raw_obs_msg);
//根据处理结果得到最近的定位对象
const auto &result = localization_integ_.GetLastestGnssLocalization();
//发布得到的定位对象
if (result.state() == msf::LocalizationMeasureState::OK ||
result.state() == msf::LocalizationMeasureState::VALID) {
publisher_->PublishLocalizationMsfGnss(result.localization());
}

return;
}

//处理全球导航卫星系统的星历数据信息
void MSFLocalization::OnGnssRtkEph(
const std::shared_ptr<drivers::gnss::GnssEphemeris> &gnss_orbit_msg) {
if ((localization_state_ == msf::LocalizationMeasureState::OK ||
localization_state_ == msf::LocalizationMeasureState::VALID) &&
FLAGS_gnss_only_init) {
return;
}
//处理全球导航卫星系统的星历数据信息
localization_integ_.RawEphemerisProcess(*gnss_orbit_msg);
return;
}
//处理全球导航卫星系统的头部数据信息
void MSFLocalization::OnGnssHeading(
const std::shared_ptr<drivers::gnss::Heading> &gnss_heading_msg) {
if ((localization_state_ == msf::LocalizationMeasureState::OK ||
localization_state_ == msf::LocalizationMeasureState::VALID) &&
FLAGS_gnss_only_init) {
return;
}
//处理全球导航卫星系统的头部数据信息
localization_integ_.GnssHeadingProcess(*gnss_heading_msg);
return;
}
//绑定 MSFLocalization 对象的 LocalizationMsgPublisher
void MSFLocalization::SetPublisher(
const std::shared_ptr<LocalizationMsgPublisher> &publisher) {
publisher_ = publisher;
}
//补充惯性测量单元数据
void MSFLocalization::CompensateImuVehicleExtrinsic(
```

```
LocalizationEstimate *local_result) {
CHECK_NOTNULL(local_result);
// 将定位信息映射至车辆坐标系中
apollo::localization::Pose *posepb_loc = local_result->mutable_pose();
const apollo::common::Quaternion &orientation = posepb_loc->orientation();
const Eigen::Quaternion<double> quaternion(
orientation.qw(), orientation.qx(), orientation.qy(), orientation.qz());
Eigen::Quaternion<double> quat_vehicle_world = quaternion * imu_vehicle_quat_;

// 根据车辆旋转程度设置车头方向
posepb_loc->set_heading(common::math::QuaternionToHeading(
quat_vehicle_world.w(), quat_vehicle_world.x(), quat_vehicle_world.y(),
quat_vehicle_world.z()));

// 根据车辆旋转程度设置欧拉角度
apollo::common::Point3D *eulerangles = posepb_loc->mutable_euler_angles();
common::math::EulerAnglesZXYd euler_angle(
quat_vehicle_world.w(), quat_vehicle_world.x(), quat_vehicle_world.y(),
quat_vehicle_world.z());
eulerangles->set_x(euler_angle.pitch());
eulerangles->set_y(euler_angle.roll());
eulerangles->set_z(euler_angle.yaw());
}
//载入 Gnss 天线信息，用来进行定位信息校准
bool MSFLocalization::LoadGnssAntennaExtrinsic(
const std::string &file_path, double *offset_x, double *offset_y,
double *offset_z, double *uncertainty_x, double *uncertainty_y,
double *uncertainty_z) {
YAML::Node config = YAML::LoadFile(file_path);
if (config["leverarm"]) {
if (config["leverarm"]["primary"]["offset"]) {
*offset_x = config["leverarm"]["primary"]["offset"]["x"].as<double>();
*offset_y = config["leverarm"]["primary"]["offset"]["y"].as<double>();
*offset_z = config["leverarm"]["primary"]["offset"]["z"].as<double>();

if (config["leverarm"]["primary"]["uncertainty"]) {
*uncertainty_x =
config["leverarm"]["primary"]["uncertainty"]["x"].as<double>();
*uncertainty_y =
config["leverarm"]["primary"]["uncertainty"]["y"].as<double>();
*uncertainty_z =
config["leverarm"]["primary"]["uncertainty"]["z"].as<double>();
}
```

```
return true;
}
}
return false;
}
//载入惯性测量单元测量信息，用来进行定位信息校准
bool MSFLocalization::LoadImuVehicleExtrinsic(const std::string &file_path,
double *quat_qx, double *quat_qy,
double *quat_qz,
double *quat_qw) {
if (!cyber::common::PathExists(file_path)) {
return false;
}
YAML::Node config = YAML::LoadFile(file_path);
if (config["transform"]) {
if (config["transform"]["translation"]) {
if (config["transform"]["rotation"]) {
*quat_qx = config["transform"]["rotation"]["x"].as<double>();
*quat_qy = config["transform"]["rotation"]["y"].as<double>();
*quat_qz = config["transform"]["rotation"]["z"].as<double>();
*quat_qw = config["transform"]["rotation"]["w"].as<double>();
return true;
}
}
}
return false;
}
//从文件夹中载入区域编号信息
bool MSFLocalization::LoadZoneIdFromFolder(const std::string &folder_path,
int *zone_id) {
std::string map_zone_id_folder;
if (cyber::common::DirectoryExists(folder_path + "/map/000/north")) {
map_zone_id_folder = folder_path + "/map/000/north";
} else if (cyber::common::DirectoryExists(folder_path + "/map/000/south")) {
map_zone_id_folder = folder_path + "/map/000/south";
} else {
return false;
}

auto folder_list = cyber::common::ListSubPaths(map_zone_id_folder);
for (auto itr = folder_list.begin(); itr != folder_list.end(); ++itr) {
*zone_id = std::stoi(*itr);
return true;
```

```
}
return false;
}

} // namespace localization
} // namespace apollo
```

参考文献

[1] RTCM Recommended Standards for Differential GNSS (Global Navigation Satellite Systems) Service version 2.3, August 20, 2001.

[2] RTCM Standard 10403.2, Differential GNSS (Global Navigation Satellite Systems) Services - version 3, February 1, 2013.

[3] Dissanayake G, Durrant-Whyte H, Bailey T. A computationally efficient solution to the simultaneous localisation and map building (SLAM) problem[C]// IEEE International Conference on Robotics & Automation. 2000.

[4] RTCM Standard 10403.2, Differential GNSS (Global Navigation Satellite Systems) Services - version 3, with amendment 1/2, november 7, 2013.

[5] Proposal of new RTCM SSR Messages (ssr_1_gal_qzss_sbas_dbs_v05) 2014/04/17.

[6] Montenbruck O, Ebinuma T, Lightsey E G, et al. A real-time kinematic GPS sensor for spacecraft relative navigation[J]. Aerospace Science and Technology, 2002, 6(6):435-449.

第 13 章

无人驾驶预测技术

在无人驾驶系统中，对地图、定位以及传感器的原始数据进行综合处理的技术就是无人驾驶预测技术。无人驾驶感知子系统需要检测到所有的障碍物，并且按照一定的规则对障碍物进行分类。通过不同的障碍物类型最终会决定程序的控制策略。比如我们可能会选择绕过前面的车辆，而对于墙壁必须执行停车策略。

本章主要讲解无人驾驶预测技术的核心概念，结合 Apollo 系统中的具体实现分析机器学习技术如何应用到无人驾驶预测核心过程中。本章内容分为四部分：第一部分着重介绍无人驾驶预测技术的核心概念，通过代码分析说明主要业务流程；第二部分详细分析成本评估器在百度 Apollo 预测模块中的模型、业务流程及其对外接口；第三部分详细分析多层感知器评估器在百度 Apollo 预测模块中的模型、业务流程及其对外接口；第四部分详细分析循环神经网络评估器在百度 Apollo 预测模块中的模型、业务流程及其对外接口。

13.1 预测模块简介

预测模块主要用于接收感知模块的障碍物消息，包括障碍物的基本感知信息，如位置、朝向角、速度、加速度等，以及通过无人驾驶车辆的定位消息来预测障碍物未来一段时间的带有置信度的轨迹，此预测数据将为接下来的路径规划提供依据，以便做出合理舒适的规划路径。其算法输入部分包括来自感知模块的障碍物数据和来自定位模块的自车定位数据，算法输出部分是每个障碍物的预测轨迹。

预测部分的主要逻辑在预测业务类 Prediction 中，我们直接看 prediction.cc，在其 Init()入口函数中加载了 prediction_conf_配置文件及 adapter_conf_配置文件。然后使用这两个配置文件分别实例化 4 个单体类的对像，代码如下：

```
  // Initialization of all managers
```

```
AdapterManager::instance()->Init(adapter_conf_);
ContainerManager::instance()->Init(adapter_conf_);
EvaluatorManager::instance()->Init(prediction_conf_);
PredictorManager::instance()->Init(prediction_conf_);
```

此处 instance() 是一个静态成员函数，后续都可以通过此函数来获取此处初始化后的类实例。之后分别设置了接收到感知消息时的回调函数 onPerception()和接收到自定位消息时的回调函数 onLocalization()。此模块执行后，每当接收到感知消息和自定位消息都会回调相应的函数。此处也可以看到本模块的 3 个主要功能：

- Container，存储订阅获取的消息信息，目前支持感知障碍物信息、车辆定位信息、车辆规划信息。
- Evaluator，分别预测障碍物的路径和速度，同时使用_prediction/data_中的模型来给一个路径打一个分数，表示障碍物走此路径的可能性。同时开源了 3 种 Evaluator: Cost Evaluator 用回归损失函数来计算概率; MLP Evaluator 用 MLP（多层感知器）模型来计算概率; RNN Evaluator 用 RNN（循环神经网络）模型来计算概率。
- Predictor，生成所有障碍物的移动轨迹。目前支持的预测器有 5 个：Lane Sequence 表示沿车道中心线走；Free Movement 表示自由移动；Regional Movement 表示在一个可能的区域内移动；Single Lane 表示在高速场景下沿车道行驶，没在车道上的障碍物就忽略了；Move Sequence 表示障碍物按照它的运动模式沿车道行驶，比如匀速、减速。

下面着重分析业务类 Prediction 中的 3 个回调函数：自定位回调（onLocalization()）、感知回调（Prediction::RunOnce()）及规划回调（Prediction::OnPlanning()）来顺带分析这 3 个主要的功能。接收到自定位消息后回调函数 onLocalization()，其主要涉及 Container 功能，源码如下：

```
void Prediction::OnLocalization(const LocalizationEstimate& localization) {
  PoseContainer* pose_container = dynamic_cast<PoseContainer*>(

ContainerManager::instance()->GetContainer(AdapterConfig::LOCALIZATION));
  CHECK_NOTNULL(pose_container);
  pose_container->Insert(localization);

  ADEBUG << "Received a localization message ["
         << localization.ShortDebugString() << "].";
}
```

首先获取自定位信息的容器指针 pose_container，然后把获取到的自定位信息插入容器指针中。

接收感知消息的回调函数 Prediction::RunOnce，其主要涉及 3 个主要功能模块中的 Evaluator 和 Predictor，源码如下：

```
void Prediction::RunOnce(const PerceptionObstacles& perception_obstacles) {
  if (FLAGS_prediction_test_mode && FLAGS_prediction_test_duration > 0.0 &&
      (Clock::NowInSeconds() - start_time_ > FLAGS_prediction_test_duration))
```

```
{
    AINFO << "Prediction finished running in test mode";
    ros::shutdown();
  }

  // Update relative map if needed
  AdapterManager::Observe();
  if (FLAGS_use_navigation_mode && !PredictionMap::Ready()) {
    AERROR << "Relative map is empty.";
    return;
  }

  double start_timestamp = Clock::NowInSeconds();

  // Insert obstacle
  ObstaclesContainer* obstacles_container =
dynamic_cast<ObstaclesContainer*>(
      ContainerManager::instance()->GetContainer(
          AdapterConfig::PERCEPTION_OBSTACLES));
  CHECK_NOTNULL(obstacles_container);
  obstacles_container->Insert(perception_obstacles);

  ADEBUG << "Received a perception message ["
         << perception_obstacles.ShortDebugString() << "].";

  // Update ADC status
  PoseContainer* pose_container = dynamic_cast<PoseContainer*>(

ContainerManager::instance()->GetContainer(AdapterConfig::LOCALIZATION));
  ADCTrajectoryContainer* adc_container =
dynamic_cast<ADCTrajectoryContainer*>(
      ContainerManager::instance()->GetContainer(
          AdapterConfig::PLANNING_TRAJECTORY));
  CHECK_NOTNULL(pose_container);
  CHECK_NOTNULL(adc_container);

  PerceptionObstacle* adc = pose_container->ToPerceptionObstacle();
  if (adc != nullptr) {
    obstacles_container->InsertPerceptionObstacle(*adc, adc->timestamp());
    double x = adc->position().x();
    double y = adc->position().y();
    ADEBUG << "Get ADC position [" << std::fixed << std::setprecision(6) << x
           << ", " << std::fixed << std::setprecision(6) << y << "].";
```

```
    Vec2d adc_position(x, y);
    adc_container->SetPosition(adc_position);
  }

  // Make evaluations
  EvaluatorManager::instance()->Run(perception_obstacles);

  // No prediction for offline mode
  if (FLAGS_prediction_offline_mode) {
    return;
  }

  // Make predictions
  PredictorManager::instance()->Run(perception_obstacles);

  auto prediction_obstacles =
      PredictorManager::instance()->prediction_obstacles();
  prediction_obstacles.set_start_timestamp(start_timestamp);
  prediction_obstacles.set_end_timestamp(Clock::NowInSeconds());
  prediction_obstacles.mutable_header()->set_lidar_timestamp(
      perception_obstacles.header().lidar_timestamp());
  prediction_obstacles.mutable_header()->set_camera_timestamp(
      perception_obstacles.header().camera_timestamp());
  prediction_obstacles.mutable_header()->set_radar_timestamp(
      perception_obstacles.header().radar_timestamp());

  if (FLAGS_prediction_test_mode) {
    for (auto const& prediction_obstacle :
         prediction_obstacles.prediction_obstacle()) {
      for (auto const& trajectory : prediction_obstacle.trajectory()) {
        for (auto const& trajectory_point : trajectory.trajectory_point()) {
          if (!ValidationChecker::ValidTrajectoryPoint(trajectory_point)) {
            AERROR << "Invalid trajectory point ["
                   << trajectory_point.ShortDebugString() << "]";
            return;
          }
        }
      }
    }
  }

  Publish(&prediction_obstacles);
}
```

同样是先获取感知障碍物的容器类指针 obstacles_container，检查是否获取成功，然后把此次回调接收到的感知障碍物消息插入窗口中，然后把 pose_container 中的行车信息转化成感知障碍物的格式插入障碍物的容器类中。这里也获取了自车轨迹容器指针 adc_container 以把自车的位置信息（一个二维的坐标点）更新到 adc_container 中，此时感知的容器中就同时有了过去一段时间各个时刻的障碍物信息及自车的自定位信息。然后通过函数 EvaluatorManager::instance()->Run(perception_obstacles);实现 Evaluator 的功能，通过 PredictorManager::instance()->Run(perception_obstacles);实现 Predictor 的功能。接下就从 PredictorManager 中获取预测的障碍物轨迹，对预测的障碍物消息加上 Header 信息。最后利用 Publish(&prediction_obstacles);发布预测的障碍物轨迹消息即可。

接收到自车轨迹消息后的回调函数为 Prediction::OnPlanning()，类似于获取自定位信息，只是把收到的消息放入容器中。

```
void Prediction::OnPlanning(const planning::ADCTrajectory& adc_trajectory) {
  ADCTrajectoryContainer* adc_trajectory_container =
      dynamic_cast<ADCTrajectoryContainer*>(
          ContainerManager::instance()->GetContainer(
              AdapterConfig::PLANNING_TRAJECTORY));
  CHECK_NOTNULL(adc_trajectory_container);
  adc_trajectory_container->Insert(adc_trajectory);

  ADEBUG << "Received a planning message [" << adc_trajectory.ShortDebugString()
         << "].";
}
```

首先获取存放规划轨迹的容器指针，然后把收到的最新轨迹插入此容器中。

所有的容器都继承 Container 中的 container.h 类，adc_trajectory_container 虽说是容器，但里面存放轨迹的只有一个轨迹的成员，也就是说这里只存放了一条最新的轨迹。

13.2　成本评估器：由一组成本函数计算概率

成本评估器中的预测概率是由一组成本函数计算的，换句话说，就是使用简单的逻辑回归模型进行预测。核心代码位于 modules/prediction/evaluator/vehicle/cost_evaluator.cc 文件中。接下来，我们使用注释的方式来详细分析代码的逻辑流程。

百度 Apollo 相关源码分析：

```
namespace apollo {
namespace prediction {

//载入在数学工具中的 Sigmoid 函数
using apollo::prediction::math_util::Sigmoid;
//核心方法进行障碍物的评估，输入是障碍物的对象
void CostEvaluator::Evaluate(Obstacle* obstacle_ptr) {
```

```
    //检测障碍物是否存在，非空检测是优秀的编程习惯
   CHECK_NOTNULL(obstacle_ptr);
    int id = obstacle_ptr->id();
    if (!obstacle_ptr->latest_feature().IsInitialized()) {
      AERROR << "Obstacle [" << id << "] has no latest feature.";
      return;
    }
  //得到障碍物对象中最近的属性数据，对属性数据进行非空检测（包含属性中是否含有车道线信息）
    Feature* latest_feature_ptr = obstacle_ptr->mutable_latest_feature();
    CHECK_NOTNULL(latest_feature_ptr);
    if (!latest_feature_ptr->has_lane() ||
        !latest_feature_ptr->lane().has_lane_graph()) {
      ADEBUG << "Obstacle [" << id << "] has no lane graph.";
      return;
    }
  //通过最近的属性信息得到障碍物的长度
    double obstacle_length = 0.0;
    if (latest_feature_ptr->has_length()) {
      obstacle_length = latest_feature_ptr->length();
    }
  //通过最近的属性信息得到障碍物的宽度
    double obstacle_width = 0.0;
    if (latest_feature_ptr->has_width()) {
      obstacle_width = latest_feature_ptr->width();
    }
  //得到障碍物对象属性信息中的车道线集合信息
    LaneGraph* lane_graph_ptr =
        latest_feature_ptr->mutable_lane()->mutable_lane_graph();
    CHECK_NOTNULL(lane_graph_ptr);
    if (lane_graph_ptr->lane_sequence_size() == 0) {
      AERROR << "Obstacle [" << id << "] has no lane sequences.";
      return;
    }
  //遍历车道线集合信息
    for (int i = 0; i < lane_graph_ptr->lane_sequence_size(); ++i) {
      LaneSequence* lane_sequence_ptr =
lane_graph_ptr->mutable_lane_sequence(i);
      CHECK_NOTNULL(lane_sequence_ptr);
  //判断每条车道线能够通过障碍物的概率，核心方法 ComputeProbability
      double probability =
          ComputeProbability(obstacle_length, obstacle_width,
*lane_sequence_ptr);
      lane_sequence_ptr->set_probability(probability);
```

```
    }
  }
  //判断每条车道线能够通过障碍物的概率，核心方法 ComputeProbability
  double CostEvaluator::ComputeProbability(const double obstacle_length,
                                           const double obstacle_width,
                                           const LaneSequence& lane_sequence) {
  //前方横向距离成本计算
    double front_lateral_distance_cost =
        FrontLateralDistanceCost(obstacle_length, obstacle_width,
lane_sequence);
  //对前方横向距离成本计算的结果使用 sigmoid 函数进行逻辑回归
    return Sigmoid(front_lateral_distance_cost);
  }
  //前方横向距离成本计算方法
  double CostEvaluator::FrontLateralDistanceCost(
      const double obstacle_length, const double obstacle_width,
      const LaneSequence& lane_sequence) {
  //车道序列参数非空判断
    if (lane_sequence.lane_segment_size() == 0 ||
        lane_sequence.lane_segment(0).lane_point_size() == 0) {
      AWARN << "Empty lane sequence.";
      return 0.0;
    }
    const LanePoint& lane_point = lane_sequence.lane_segment(0).lane_point(0);
    double lane_l = -lane_point.relative_l();
  //三角函数计算跨越距离
    double distance = std::abs(lane_l - obstacle_length / 2.0 *
                                        std::sin(lane_point.angle_diff()));
    double half_lane_width = lane_point.width() / 2.0;
    //计算前方横向距离成本
  return half_lane_width - distance;
  }

  }  // namespace prediction
  }  // namespace apollo
```

13.3 MLP 评估器：用 MLP 模型计算概率

MLP 评估器中的预测概率是由 MLP 模型函数计算的，换句话说，就是使用全连接神经网络模型进行预测。核心代码位于 modules/prediction/evaluator/vehicle/cruise_mlp_evaluator.cc 文件中。接下来，我们使用注释的方式来详细分析代码的逻辑流程。

百度 Apollo 相关源码分析：

```
#include "modules/prediction/evaluator/vehicle/cruise_mlp_evaluator.h"

#include <omp.h>

#include <limits>
#include <utility>

#include "cyber/common/file.h"
#include "modules/prediction/common/feature_output.h"
#include "modules/prediction/common/prediction_gflags.h"
#include "modules/prediction/common/prediction_system_gflags.h"
#include "modules/prediction/common/prediction_util.h"
#include "modules/prediction/container/container_manager.h"
#include "modules/prediction/container/obstacles/obstacles_container.h"

namespace apollo {
namespace prediction {

using apollo::common::adapter::AdapterConfig;
using apollo::cyber::common::GetProtoFromFile;
using apollo::prediction::math_util::Sigmoid;

// 用于计算向量平均值的辅助函数
double ComputeMean(const std::vector<double>& nums, size_t start, size_t end) {
  int count = 0;
  double sum = 0.0;
// 遍历向量中的元素进行求和
  for (size_t i = start; i <= end && i < nums.size(); i++) {
    sum += nums[i];
    ++count;
  }
//计算向量平均值
  return (count == 0) ? 0.0 : sum / count;
}

CruiseMLPEvaluator::CruiseMLPEvaluator() : device_(torch::kCPU) {
  LoadModels();
}

void CruiseMLPEvaluator::Clear() {}
//核心方法进行障碍物的评估，输入是障碍物的对象
void CruiseMLPEvaluator::Evaluate(Obstacle* obstacle_ptr) {
```

```
    //输入非空等脏检测
    omp_set_num_threads(1);
    Clear();
    CHECK_NOTNULL(obstacle_ptr);
    int id = obstacle_ptr->id();
    if (!obstacle_ptr->latest_feature().IsInitialized()) {
      AERROR << "Obstacle [" << id << "] has no latest feature.";
      return;
    }
  //得到障碍物对象中最近的属性数据，对属性数据进行非空检测（包含属性中是否含有车道线信息）
    Feature* latest_feature_ptr = obstacle_ptr->mutable_latest_feature();
    CHECK_NOTNULL(latest_feature_ptr);
    if (!latest_feature_ptr->has_lane() ||
        !latest_feature_ptr->lane().has_lane_graph()) {
      ADEBUG << "Obstacle [" << id << "] has no lane graph.";
      return;
    }
    LaneGraph* lane_graph_ptr =
        latest_feature_ptr->mutable_lane()->mutable_lane_graph();
    CHECK_NOTNULL(lane_graph_ptr);
    if (lane_graph_ptr->lane_sequence_size() == 0) {
      AERROR << "Obstacle [" << id << "] has no lane sequences.";
      return;
    }

    ADEBUG << "There are " << lane_graph_ptr->lane_sequence_size()
          << " lane sequences with probabilities:";
    // 对于每个可能的车道序列，提取需要输入训练模型中的特征，然后计算障碍物进入该车道序列的可
能性
    for (int i = 0; i < lane_graph_ptr->lane_sequence_size(); ++i) {
      LaneSequence* lane_sequence_ptr =
lane_graph_ptr->mutable_lane_sequence(i);
      CHECK_NOTNULL(lane_sequence_ptr);
      std::vector<double> feature_values;
  //提取特征输入值
      ExtractFeatureValues(obstacle_ptr, lane_sequence_ptr, &feature_values);
      if (feature_values.size() !=
          OBSTACLE_FEATURE_SIZE + SINGLE_LANE_FEATURE_SIZE * LANE_POINTS_SIZE) {
        lane_sequence_ptr->set_probability(0.0);
        ADEBUG << "Skip lane sequence due to incorrect feature size";
        continue;
      }
```

```
    // 将功能插入数据以便学习

    if (FLAGS_prediction_offline_mode == 2) {
      std::vector<double> interaction_feature_values;
      SetInteractionFeatureValues(obstacle_ptr, lane_sequence_ptr,
                                  &interaction_feature_values);
      if (interaction_feature_values.size() != INTERACTION_FEATURE_SIZE) {
        ADEBUG << "Obstacle [" << id << "] has fewer than "
               << "expected lane feature_values"
               << interaction_feature_values.size() << ".";
        return;
      }
      ADEBUG << "Interaction feature size = "
             << interaction_feature_values.size();
      feature_values.insert(feature_values.end(),
                            interaction_feature_values.begin(),
                            interaction_feature_values.end());
      FeatureOutput::InsertDataForLearning(*latest_feature_ptr,
feature_values,
                                           "lane_scanning", lane_sequence_ptr);
      ADEBUG << "Save extracted features for learning locally.";
      return; // 跳过离线模式下进行概率计算的过程
    }

    //多层感知器模型处理前面的特征值集合得到能否通过的预测结果
   std::vector<torch::jit::IValue> torch_inputs;
    int input_dim = static_cast<int>(
      OBSTACLE_FEATURE_SIZE + SINGLE_LANE_FEATURE_SIZE * LANE_POINTS_SIZE);
    torch::Tensor torch_input = torch::zeros({1, input_dim});
    for (size_t i = 0; i < feature_values.size(); ++i) {
      torch_input[0][i] = static_cast<float>(feature_values[i]);
    }
    torch_inputs.push_back(torch_input.to(device_));
    if (lane_sequence_ptr->vehicle_on_lane()) {
      ModelInference(torch_inputs, torch_go_model_ptr_, lane_sequence_ptr);
    } else {
      ModelInference(torch_inputs, torch_cutin_model_ptr_,
lane_sequence_ptr);
    }
  }
}
//从车道线对象和障碍物对象中提取特征值形成特征向量输入
void CruiseMLPEvaluator::ExtractFeatureValues(
```

```
    Obstacle* obstacle_ptr, LaneSequence* lane_sequence_ptr,
    std::vector<double>* feature_values) {
  // 函数相关参数的脏检查
  CHECK_NOTNULL(obstacle_ptr);
  CHECK_NOTNULL(lane_sequence_ptr);
  int id = obstacle_ptr->id();

  //提取障碍物对象中多层感知器输入需要的特征值并存储
  std::vector<double> obstacle_feature_values;
  SetObstacleFeatureValues(obstacle_ptr, &obstacle_feature_values);
  if (obstacle_feature_values.size() != OBSTACLE_FEATURE_SIZE) {
    ADEBUG << "Obstacle [" << id << "] has fewer than "
           << "expected obstacle feature_values "
           << obstacle_feature_values.size() << ".";
    return;
  }
  ADEBUG << "Obstacle feature size = " << obstacle_feature_values.size();
  feature_values->insert(feature_values->end(),
obstacle_feature_values.begin(),
                        obstacle_feature_values.end());

 //提取车道对象中多层感知器输入需要的特征值并存储
  std::vector<double> lane_feature_values;
  SetLaneFeatureValues(obstacle_ptr, lane_sequence_ptr,
&lane_feature_values);
  if (lane_feature_values.size() !=
      SINGLE_LANE_FEATURE_SIZE * LANE_POINTS_SIZE) {
    ADEBUG << "Obstacle [" << id << "] has fewer than "
           << "expected lane feature_values" << lane_feature_values.size()
           << ".";
    return;
  }
  ADEBUG << "Lane feature size = " << lane_feature_values.size();
  feature_values->insert(feature_values->end(), lane_feature_values.begin(),
                        lane_feature_values.end());
 }
 //提取障碍物对象中多层感知器输入需要的特征值
void CruiseMLPEvaluator::SetObstacleFeatureValues(
    const Obstacle* obstacle_ptr, std::vector<double>* feature_values) {
  // 函数相关参数的脏检查和使用变量的声明过程
  CHECK_NOTNULL(obstacle_ptr);
  feature_values->clear();
  feature_values->reserve(OBSTACLE_FEATURE_SIZE);
```

```
    std::vector<double> thetas;
    std::vector<double> lane_ls;
    std::vector<double> dist_lbs;
    std::vector<double> dist_rbs;
    std::vector<int> lane_types;
    std::vector<double> speeds;
    std::vector<double> timestamps;

    std::vector<double> has_history(FLAGS_cruise_historical_frame_length,
1.0);
    std::vector<std::pair<double, double>> pos_history(
        FLAGS_cruise_historical_frame_length, std::make_pair(0.0, 0.0));
    std::vector<std::pair<double, double>> vel_history(
        FLAGS_cruise_historical_frame_length, std::make_pair(0.0, 0.0));
    std::vector<std::pair<double, double>> acc_history(
        FLAGS_cruise_historical_frame_length, std::make_pair(0.0, 0.0));
    std::vector<double>
vel_heading_history(FLAGS_cruise_historical_frame_length,
                                              0.0);
    std::vector<double> vel_heading_changing_rate_history(
        FLAGS_cruise_historical_frame_length, 0.0);

    // 获取障碍物的当前位置以建立相对坐标系
   const Feature& obs_curr_feature = obstacle_ptr->latest_feature();
    double obs_curr_heading = obs_curr_feature.velocity_heading();
    std::pair<double, double> obs_curr_pos = std::make_pair(
        obs_curr_feature.position().x(), obs_curr_feature.position().y());
    double obs_feature_history_start_time =
        obstacle_ptr->timestamp() - FLAGS_prediction_trajectory_time_length;
    int count = 0;
    // int num_available_history_frames = 0;
    double prev_timestamp = obs_curr_feature.timestamp();

    // 从最新的时间戳开始并向后移动
    ADEBUG << "Obstacle has " << obstacle_ptr->history_size()
           << " history timestamps.";
    for (std::size_t i = 0; i < obstacle_ptr->history_size(); ++i) {
      const Feature& feature = obstacle_ptr->feature(i);
      if (!feature.IsInitialized()) {
        continue;
      }
      if (feature.timestamp() < obs_feature_history_start_time) {
        break;
```

```
    }
    if (!feature.has_lane()) {
      ADEBUG << "Feature has no lane. Quit.";
    }
    //在相对坐标系中提取 23 个特征值
    if (feature.has_lane() && feature.lane().has_lane_feature()) {
      thetas.push_back(feature.lane().lane_feature().angle_diff());
      lane_ls.push_back(feature.lane().lane_feature().lane_l());

dist_lbs.push_back(feature.lane().lane_feature().dist_to_left_boundary());
      dist_rbs.push_back(
          feature.lane().lane_feature().dist_to_right_boundary());
      lane_types.push_back(feature.lane().lane_feature().lane_turn_type());
      timestamps.push_back(feature.timestamp());
      speeds.push_back(feature.speed());
      ++count;
    } else {
      ADEBUG << "Feature has no lane_feature!!!";
      ADEBUG << feature.lane().current_lane_feature_size();
    }

    // 这些是基于相对坐标系的新特征值

    if (i >= FLAGS_cruise_historical_frame_length) {
      continue;
    }
    if (i != 0 && has_history[i - 1] == 0.0) {
      has_history[i] = 0.0;
      continue;
    }
    if (feature.has_position()) {
      pos_history[i] = WorldCoordToObjCoord(
          std::make_pair(feature.position().x(), feature.position().y()),
          obs_curr_pos, obs_curr_heading);
    } else {
      has_history[i] = 0.0;
    }
    if (feature.has_velocity()) {
      auto vel_end = WorldCoordToObjCoord(
          std::make_pair(feature.velocity().x(), feature.velocity().y()),
          obs_curr_pos, obs_curr_heading);
      auto vel_begin = WorldCoordToObjCoord(std::make_pair(0.0, 0.0),
                                            obs_curr_pos, obs_curr_heading);
```

```
      vel_history[i] = std::make_pair(vel_end.first - vel_begin.first,
                                      vel_end.second - vel_begin.second);
    } else {
      has_history[i] = 0.0;
    }
    if (feature.has_acceleration()) {
      auto acc_end =
          WorldCoordToObjCoord(std::make_pair(feature.acceleration().x(),
                                              feature.acceleration().y()),
                               obs_curr_pos, obs_curr_heading);
      auto acc_begin = WorldCoordToObjCoord(std::make_pair(0.0, 0.0),
                                            obs_curr_pos, obs_curr_heading);
      acc_history[i] = std::make_pair(acc_end.first - acc_begin.first,
                                      acc_end.second - acc_begin.second);
    } else {
      has_history[i] = 0.0;
    }
    if (feature.has_velocity_heading()) {
      vel_heading_history[i] =
          WorldAngleToObjAngle(feature.velocity_heading(), obs_curr_heading);
      if (i != 0) {
        vel_heading_changing_rate_history[i] =
            (vel_heading_history[i - 1] - vel_heading_history[i]) /
            (FLAGS_double_precision + feature.timestamp() - prev_timestamp);
        prev_timestamp = feature.timestamp();
      }
    } else {
      has_history[i] = 0.0;
    }
  }
  if (count <= 0) {
    ADEBUG << "There is no feature with lane info. Quit.";
    return;
  }

  // 下面的部分是设定 23 个特征值的过程
  ///////////////////////////////////////////////////////////////
//特征值提取计算过程
  int curr_size = 5;
  int hist_size = static_cast<int>(obstacle_ptr->history_size());
  double theta_mean = ComputeMean(thetas, 0, hist_size - 1);
  double theta_filtered = ComputeMean(thetas, 0, curr_size - 1);
  double lane_l_mean = ComputeMean(lane_ls, 0, hist_size - 1);
```

```
double lane_l_filtered = ComputeMean(lane_ls, 0, curr_size - 1);
double speed_mean = ComputeMean(speeds, 0, hist_size - 1);

double time_diff = timestamps.front() - timestamps.back();
double dist_lb_rate = (timestamps.size() > 1)
                      ? (dist_lbs.front() - dist_lbs.back()) / time_diff
                      : 0.0;
double dist_rb_rate = (timestamps.size() > 1)
                      ? (dist_rbs.front() - dist_rbs.back()) / time_diff
                      : 0.0;

double delta_t = 0.0;
if (timestamps.size() > 1) {
  delta_t = (timestamps.front() - timestamps.back()) /
           static_cast<double>(timestamps.size() - 1);
}
double angle_curr = ComputeMean(thetas, 0, curr_size - 1);
double angle_prev = ComputeMean(thetas, curr_size, 2 * curr_size - 1);
double angle_diff =
    (hist_size >= 2 * curr_size) ? angle_curr - angle_prev : 0.0;

double lane_l_curr = ComputeMean(lane_ls, 0, curr_size - 1);
double lane_l_prev = ComputeMean(lane_ls, curr_size, 2 * curr_size - 1);
double lane_l_diff =
    (hist_size >= 2 * curr_size) ? lane_l_curr - lane_l_prev : 0.0;

double angle_diff_rate = 0.0;
double lane_l_diff_rate = 0.0;
if (delta_t > std::numeric_limits<double>::epsilon()) {
  angle_diff_rate = angle_diff / (delta_t * curr_size);
  lane_l_diff_rate = lane_l_diff / (delta_t * curr_size);
}

double acc = 0.0;
double jerk = 0.0;
if (static_cast<int>(speeds.size()) >= 3 * curr_size &&
    delta_t > std::numeric_limits<double>::epsilon()) {
  double speed_1st_recent = ComputeMean(speeds, 0, curr_size - 1);
  double speed_2nd_recent = ComputeMean(speeds, curr_size, 2 * curr_size - 1);
  double speed_3rd_recent =
      ComputeMean(speeds, 2 * curr_size, 3 * curr_size - 1);
  acc = (speed_1st_recent - speed_2nd_recent) / (curr_size * delta_t);
  jerk = (speed_1st_recent - 2 * speed_2nd_recent + speed_3rd_recent) /
```

```
        (curr_size * curr_size * delta_t * delta_t);
  }

  double dist_lb_rate_curr = 0.0;
  if (hist_size >= 2 * curr_size &&
      delta_t > std::numeric_limits<double>::epsilon()) {
    double dist_lb_curr = ComputeMean(dist_lbs, 0, curr_size - 1);
    double dist_lb_prev = ComputeMean(dist_lbs, curr_size, 2 * curr_size - 1);
    dist_lb_rate_curr = (dist_lb_curr - dist_lb_prev) / (curr_size * delta_t);
  }

  double dist_rb_rate_curr = 0.0;
  if (hist_size >= 2 * curr_size &&
      delta_t > std::numeric_limits<double>::epsilon()) {
    double dist_rb_curr = ComputeMean(dist_rbs, 0, curr_size - 1);
    double dist_rb_prev = ComputeMean(dist_rbs, curr_size, 2 * curr_size - 1);
    dist_rb_rate_curr = (dist_rb_curr - dist_rb_prev) / (curr_size * delta_t);
  }
//特征值存储到 feature_values 对象中的过程
  feature_values->push_back(theta_filtered);
  feature_values->push_back(theta_mean);
  feature_values->push_back(theta_filtered - theta_mean);
  feature_values->push_back(angle_diff);
  feature_values->push_back(angle_diff_rate);

  feature_values->push_back(lane_l_filtered);
  feature_values->push_back(lane_l_mean);
  feature_values->push_back(lane_l_filtered - lane_l_mean);
  feature_values->push_back(lane_l_diff);
  feature_values->push_back(lane_l_diff_rate);

  feature_values->push_back(speed_mean);
  feature_values->push_back(acc);
  feature_values->push_back(jerk);

  feature_values->push_back(dist_lbs.front());
  feature_values->push_back(dist_lb_rate);
  feature_values->push_back(dist_lb_rate_curr);

  feature_values->push_back(dist_rbs.front());
  feature_values->push_back(dist_rb_rate);
  feature_values->push_back(dist_rb_rate_curr);
```

```
    feature_values->push_back(lane_types.front() == 0 ? 1.0 : 0.0);
    feature_values->push_back(lane_types.front() == 1 ? 1.0 : 0.0);
    feature_values->push_back(lane_types.front() == 2 ? 1.0 : 0.0);
    feature_values->push_back(lane_types.front() == 3 ? 1.0 : 0.0);

    for (std::size_t i = 0; i < FLAGS_cruise_historical_frame_length; i++) {
      feature_values->push_back(has_history[i]);
      feature_values->push_back(pos_history[i].first);
      feature_values->push_back(pos_history[i].second);
      feature_values->push_back(vel_history[i].first);
      feature_values->push_back(vel_history[i].second);
      feature_values->push_back(acc_history[i].first);
      feature_values->push_back(acc_history[i].second);
      feature_values->push_back(vel_heading_history[i]);
      feature_values->push_back(vel_heading_changing_rate_history[i]);
    }
  }

  //提取障碍物对象和车道线对象协同计算产生的多层感知器输入需要的特征值
  void CruiseMLPEvaluator::SetInteractionFeatureValues(
      Obstacle* obstacle_ptr, LaneSequence* lane_sequence_ptr,
      std::vector<double>* feature_values) {
    // 向前或后退需要的变量：relative_s、relative_l、speed、length
    feature_values->clear();
    // 首先需要初始化在车辆前后邻近车辆的障碍物对象
    NearbyObstacle forward_obstacle;
    NearbyObstacle backward_obstacle;
    forward_obstacle.set_s(FLAGS_default_s_if_no_obstacle_in_lane_sequence);
    forward_obstacle.set_l(FLAGS_default_l_if_no_obstacle_in_lane_sequence);

backward_obstacle.set_s(-FLAGS_default_s_if_no_obstacle_in_lane_sequence);
    backward_obstacle.set_l(FLAGS_default_l_if_no_obstacle_in_lane_sequence);

    for (const auto& nearby_obstacle : lane_sequence_ptr->nearby_obstacle()) {
      if (nearby_obstacle.s() < 0.0) {
        if (nearby_obstacle.s() > backward_obstacle.s()) {
          backward_obstacle.set_id(nearby_obstacle.id());
          backward_obstacle.set_s(nearby_obstacle.s());
          backward_obstacle.set_l(nearby_obstacle.l());
        }
      } else {
        if (nearby_obstacle.s() < forward_obstacle.s()) {
          forward_obstacle.set_id(nearby_obstacle.id());
```

```
        forward_obstacle.set_s(nearby_obstacle.s());
        forward_obstacle.set_l(nearby_obstacle.l());
      }
    }
  }

  auto obstacles_container =
      ContainerManager::Instance()->GetContainer<ObstaclesContainer>(
          AdapterConfig::PERCEPTION_OBSTACLES);
  // 设置前向障碍物的特征值
  feature_values->push_back(forward_obstacle.s());
  feature_values->push_back(forward_obstacle.l());
  if (!forward_obstacle.has_id()) {  // no forward obstacle
    feature_values->push_back(0.0);
    feature_values->push_back(0.0);
  } else {
    Obstacle* forward_obs_ptr =
        obstacles_container->GetObstacle(forward_obstacle.id());
    const Feature& feature = forward_obs_ptr->latest_feature();
    feature_values->push_back(feature.length());
    feature_values->push_back(feature.speed());
  }

  //设置后向障碍物的特征值
  feature_values->push_back(backward_obstacle.s());
  feature_values->push_back(backward_obstacle.l());
  if (!backward_obstacle.has_id()) {  // no forward obstacle
    feature_values->push_back(0.0);
    feature_values->push_back(0.0);
  } else {
    Obstacle* backward_obs_ptr =
        obstacles_container->GetObstacle(backward_obstacle.id());
    const Feature& feature = backward_obs_ptr->latest_feature();
    feature_values->push_back(feature.length());
    feature_values->push_back(feature.speed());
  }
}
//提取车道线对象中多层感知器输入需要的特征值
void CruiseMLPEvaluator::SetLaneFeatureValues(
    const Obstacle* obstacle_ptr, const LaneSequence* lane_sequence_ptr,
    std::vector<double>* feature_values) {
  // 函数相关参数的脏检查
  feature_values->clear();
```

```
    feature_values->reserve(SINGLE_LANE_FEATURE_SIZE * LANE_POINTS_SIZE);
    const Feature& feature = obstacle_ptr->latest_feature();
    if (!feature.IsInitialized()) {
      ADEBUG << "Obstacle [" << obstacle_ptr->id() << "] has no latest feature.";
      return;
    } else if (!feature.has_position()) {
      ADEBUG << "Obstacle [" << obstacle_ptr->id() << "] has no position.";
      return;
    }

    double heading = feature.velocity_heading();
    for (int i = 0; i < lane_sequence_ptr->lane_segment_size(); ++i) {
      if (feature_values->size() >= SINGLE_LANE_FEATURE_SIZE * LANE_POINTS_SIZE)
{
        break;
      }
      const LaneSegment& lane_segment = lane_sequence_ptr->lane_segment(i);
      for (int j = 0; j < lane_segment.lane_point_size(); ++j) {
        if (feature_values->size() >=
            SINGLE_LANE_FEATURE_SIZE * LANE_POINTS_SIZE) {
          break;
        }
        const LanePoint& lane_point = lane_segment.lane_point(j);
        if (!lane_point.has_position()) {
          AERROR << "Lane point has no position.";
          continue;
        }

        std::pair<double, double> relative_s_l = WorldCoordToObjCoord(
            std::make_pair(lane_point.position().x(),
lane_point.position().y()),
            std::make_pair(feature.position().x(), feature.position().y()),
            heading);
        double relative_ang = WorldAngleToObjAngle(lane_point.heading(),
heading);

        feature_values->push_back(relative_s_l.second);
        feature_values->push_back(relative_s_l.first);
        feature_values->push_back(relative_ang);
        feature_values->push_back(lane_point.kappa());
      }
    }
```

```
  // 如果车道点不够，就应用线性外推法
  std::size_t size = feature_values->size();
  while (size >= 10 && size < SINGLE_LANE_FEATURE_SIZE * LANE_POINTS_SIZE) {
    double relative_l_new = 2 * feature_values->operator[](size - 5) -
                            feature_values->operator[](size - 10);
    double relative_s_new = 2 * feature_values->operator[](size - 4) -
                            feature_values->operator[](size - 9);
    double relative_ang_new = feature_values->operator[](size - 3);

    feature_values->push_back(relative_l_new);
    feature_values->push_back(relative_s_new);
    feature_values->push_back(relative_ang_new);
    feature_values->push_back(0.0);

    size = feature_values->size();
  }
}
//载入多层感知器模型到多层感知器评估器中
void CruiseMLPEvaluator::LoadModels() {
  torch::set_num_threads(1);
  torch_go_model_ptr_ =
      torch::jit::load(FLAGS_torch_vehicle_cruise_go_file, device_);
  torch_cutin_model_ptr_ =
      torch::jit::load(FLAGS_torch_vehicle_cruise_cutin_file, device_);
}
//使用多层感知器模型和我们的输入进行预测的过程
void CruiseMLPEvaluator::ModelInference(
    const std::vector<torch::jit::IValue>& torch_inputs,
    std::shared_ptr<torch::jit::script::Module> torch_model_ptr,
    LaneSequence* lane_sequence_ptr) {
  auto torch_output_tuple = torch_model_ptr->forward(torch_inputs).toTuple();
  auto probability_tensor = torch_output_tuple->elements()[0].toTensor();
  auto finish_time_tensor = torch_output_tuple->elements()[1].toTensor();
  lane_sequence_ptr->set_probability(Sigmoid(
      static_cast<double>(probability_tensor.accessor<float, 2>()[0][0])));
  lane_sequence_ptr->set_time_to_lane_center(
      static_cast<double>(finish_time_tensor.accessor<float, 2>()[0][0]));
}

}  // namespace prediction
}  // namespace Apollo
```

13.4 RNN 评估器：用 RNN 模型计算概率

RNN 评估器中的预测概率是由 RNN 模型函数计算的，换句话说，就是使用循环神经网络模型进行预测。核心代码位于 modules/prediction/evaluator/vehicle/rnn_evaluator.cc 文件中。接下来，我们使用注释的方式来详细分析代码的逻辑流程。

百度 Apollo 相关源码分析：

```
#include "modules/prediction/evaluator/vehicle/rnn_evaluator.h"

#include <memory>
#include <utility>

#include "cyber/common/file.h"
#include "modules/prediction/common/prediction_gflags.h"
#include "modules/prediction/common/prediction_map.h"

namespace apollo {
namespace prediction {

using apollo::hdmap::LaneInfo;
//循环神经网络评估器构造方法
RNNEvaluator::RNNEvaluator() { LoadModel(FLAGS_evaluator_vehicle_rnn_file); }
//使用循环神经网络评估器评估障碍物能够通过
void RNNEvaluator::Evaluate(Obstacle* obstacle_ptr) {
 //检测障碍物是否存在，非空检测是优秀的编程习惯
 Clear();
  CHECK_NOTNULL(obstacle_ptr);

  int id = obstacle_ptr->id();
  if (!obstacle_ptr->latest_feature().IsInitialized()) {
    ADEBUG << "Obstacle [" << id << "] has no latest feature.";
    return;
  }
//得到障碍物对象中最近的属性数据，对属性数据进行非空检测（包含属性中是否含有车道线信息）

  Feature* latest_feature_ptr = obstacle_ptr->mutable_latest_feature();
  CHECK_NOTNULL(latest_feature_ptr);
  if (!latest_feature_ptr->has_lane() ||
      !latest_feature_ptr->lane().has_lane_graph()) {
    ADEBUG << "Obstacle [" << id << "] has no lane graph.";
    return;
  }
```

```
  LaneGraph* lane_graph_ptr =
      latest_feature_ptr->mutable_lane()->mutable_lane_graph();
  CHECK_NOTNULL(lane_graph_ptr);
  if (lane_graph_ptr->lane_sequence_size() == 0) {
    ADEBUG << "Obstacle [" << id << "] has no lane sequences.";
    return;
  }
//分别构建障碍物和车道线特征向量矩阵
  Eigen::MatrixXf obstacle_feature_mat;
  std::unordered_map<int, Eigen::MatrixXf> lane_feature_mats;
  if (ExtractFeatureValues(obstacle_ptr, &obstacle_feature_mat,
                          &lane_feature_mats) != 0) {
    ADEBUG << "Fail to extract feature from obstacle";
    return;
  }
  if (obstacle_feature_mat.rows() != 1 ||
      obstacle_feature_mat.size() != DIM_OBSTACLE_FEATURE) {
    ADEBUG << "Dim of obstacle feature is wrong!";
    return;
  }

//使用循环神经网络算法预测车辆能够通过障碍物
 Eigen::MatrixXf pred_mat;
  std::vector<Eigen::MatrixXf> states;
//循环神经网络模块脏检查和模块状态初始化
  if (!obstacle_ptr->RNNEnabled()) {
    obstacle_ptr->InitRNNStates();
  }
  obstacle_ptr->GetRNNStates(&states);
//遍历车道线集合信息
  for (int i = 0; i < lane_graph_ptr->lane_sequence_size(); ++i) {
    LaneSequence* lane_sequence_ptr =
lane_graph_ptr->mutable_lane_sequence(i);
    int seq_id = lane_sequence_ptr->lane_sequence_id();
    if (lane_feature_mats.find(seq_id) == lane_feature_mats.end()) {
      ADEBUG << "Fail to access seq-" << seq_id << " feature!";
      continue;
    }
    const Eigen::MatrixXf& lane_feature_mat = lane_feature_mats[seq_id];
    if (lane_feature_mat.cols() != DIM_LANE_POINT_FEATURE) {
      ADEBUG << "Lane feature dim of seq-" << seq_id << " is wrong!";
      continue;
```

```
    }
  model_ptr_->SetState(states);
  //判断每条车道线能够通过障碍物的概率，Run 为循环神经网络评估的核心过程
    model_ptr_->Run({obstacle_feature_mat, lane_feature_mat}, &pred_mat);
    double probability = pred_mat(0, 0);
    ADEBUG << "Probability = " << probability;
    double acceleration = pred_mat(0, 1);
    if (std::isnan(probability) || std::isinf(probability)) {
      ADEBUG << "Fail to compute probability.";
      continue;
    }
    if (std::isnan(acceleration) || std::isinf(acceleration)) {
      ADEBUG << "Fail to compute acceleration.";
      continue;
    }
    lane_sequence_ptr->set_probability(probability);
    lane_sequence_ptr->set_acceleration(acceleration);
  }
  model_ptr_->State(&states);
  obstacle_ptr->SetRNNStates(states);
}

void RNNEvaluator::Clear() {}
//从文件中载入循环神经网络算法模型
void RNNEvaluator::LoadModel(const std::string& model_file) {
  NetParameter net_parameter;
  CHECK(cyber::common::GetProtoFromFile(model_file, &net_parameter))
      << "Unable to load model file: " << model_file << ".";

  ADEBUG << "Succeeded in loading the model file: " << model_file << ".";
  model_ptr_ = network::RnnModel::Instance();
  model_ptr_->LoadModel(net_parameter);
}
//从车道线对象和障碍物对象中提取特征值形成特征向量输入

int RNNEvaluator::ExtractFeatureValues(
    Obstacle* obstacle, Eigen::MatrixXf* const obstacle_feature_mat,
    std::unordered_map<int, Eigen::MatrixXf>* const lane_feature_mats) {
  std::vector<float> obstacle_features;
  std::vector<float> lane_features;
//从障碍物对象中提取特征值形成特征向量输入
  if (SetupObstacleFeature(obstacle, &obstacle_features) != 0) {
    ADEBUG << "Reset rnn state";
```

```
      obstacle->InitRNNStates();
    }
    if (static_cast<int>(obstacle_features.size()) != DIM_OBSTACLE_FEATURE) {
      AWARN << "Obstacle feature size: " << obstacle_features.size();
      return -1;
    }
    obstacle_feature_mat->resize(1, obstacle_features.size());
    for (size_t i = 0; i < obstacle_features.size(); ++i) {
      (*obstacle_feature_mat)(0, i) = obstacle_features[i];
    }
  //从车道线对象中提取特征值形成特征向量输入
    Feature* feature = obstacle->mutable_latest_feature();
    if (!feature->has_lane() || !feature->lane().has_lane_graph()) {
      AWARN << "Fail to access lane graph!";
      return -1;
    }
    int routes = feature->lane().lane_graph().lane_sequence_size();
  //遍历可能的所有车道线
    for (int i = 0; i < routes; ++i) {
      LaneSequence lane_seq = feature->lane().lane_graph().lane_sequence(i);
      int seq_id = lane_seq.lane_sequence_id();
      if (SetupLaneFeature(*feature, lane_seq, &lane_features) != 0) {
        ADEBUG << "Fail to setup lane sequence feature!";
        continue;
      }
      int dim = DIM_LANE_POINT_FEATURE;
      Eigen::MatrixXf mat(lane_features.size() / dim, dim);
      for (int j = 0; j < static_cast<int>(lane_features.size()); ++j) {
        mat(j / dim, j % dim) = lane_features[j];
      }
      (*lane_feature_mats)[seq_id] = std::move(mat);
    }
    return 0;
  }
  //从障碍物对象中提取特征值形成特征向量输入
  int RNNEvaluator::SetupObstacleFeature(
      Obstacle* obstacle, std::vector<float>* const feature_values) {
    feature_values->clear();
    feature_values->reserve(DIM_OBSTACLE_FEATURE);
  //函数使用常量声明
    float heading = 0.0f;
    float speed = 0.0f;
    float lane_l = 0.0f;
```

```
float theta = 0.0f;
float dist_lb = 1.0f;
float dist_rb = 1.0f;
if (obstacle->history_size() < 1) {
  AWARN << "Size of feature less than 1!";
  return -1;
}
//按照时间序列取得多个障碍物特征值集合
bool success_setup = false;
int ret = 0;
//设定障碍物最多取出 3 个
size_t num = obstacle->history_size() > 3 ? 3 : obstacle->history_size();
//遍历障碍物集合取出各个障碍物特征值
for (size_t i = 0; i < num; ++i) {
  Feature* fea = obstacle->mutable_feature(i);
  if (fea == nullptr) {
    ADEBUG << "Fail to get " << i << "-th feature from obstacle";
    continue;
}
//取得与每个障碍物相关的车道线对象
  if (!fea->has_lane() || !fea->lane().has_lane_feature() ||
      !fea->lane().lane_feature().has_lane_id()) {
    ADEBUG << "Fail to access lane feature from " << i << "-the feature";
    continue;
}
//车道线对象相关的特征属性：lane_l、theta、dist_lb、dist_rb
  LaneFeature* p_lane_fea = fea->mutable_lane()->mutable_lane_feature();
  lane_l = static_cast<float>(p_lane_fea->lane_l());
  theta = static_cast<float>(p_lane_fea->angle_diff());
  dist_lb = static_cast<float>(p_lane_fea->dist_to_left_boundary());
  dist_rb = static_cast<float>(p_lane_fea->dist_to_right_boundary());
//取得车速和车头朝向
  if (!fea->has_speed() || !fea->velocity_heading()) {
    ADEBUG << "Fail to access speed and velocity heading from " << i
        << "-the feature";
    continue;
  }
  speed = static_cast<float>(fea->speed());
  heading = static_cast<float>(fea->velocity_heading());
  success_setup = true;
  ADEBUG << "Success to setup obstacle feature!";
//输出取得的特征向量集合
  Feature* fea_pre = nullptr;
```

```
    if (i + 1 < obstacle->history_size()) {
      fea_pre = obstacle->mutable_feature(i + 1);
    }
    if (fea_pre != nullptr) {
      if (fea_pre->lane().has_lane_feature() &&
          fea_pre->lane().lane_feature().has_lane_id()) {
        std::string lane_id_pre = fea_pre->lane().lane_feature().lane_id();
        if (lane_id_pre != p_lane_fea->lane_id() &&
            IsCutinInHistory(p_lane_fea->lane_id(), lane_id_pre)) {
          ADEBUG << "Obstacle [" << fea->id() << "] cut in from " << lane_id_pre
                 << " to " << p_lane_fea->lane_id() << ", reset";
          ret = -1;
        }
      }
    }
    break;
  }

  if (!success_setup) {
    return -1;
  }
  feature_values->push_back(heading);
  feature_values->push_back(speed);
  feature_values->push_back(lane_l);
  feature_values->push_back(dist_lb);
  feature_values->push_back(dist_rb);
  feature_values->push_back(theta);
  return ret;
}
//从车道线对象中提取特征值形成特征向量输入
int RNNEvaluator::SetupLaneFeature(const Feature& feature,
                                   const LaneSequence& lane_sequence,
                                   std::vector<float>* const feature_values) {
//函数需要的脏检查
  CHECK_LE(LENGTH_LANE_POINT_SEQUENCE, FLAGS_max_num_lane_point);
  feature_values->clear();
  feature_values->reserve(static_cast<size_t>(DIM_LANE_POINT_FEATURE) *
                          static_cast<size_t>(LENGTH_LANE_POINT_SEQUENCE));
  LanePoint* p_lane_point = nullptr;
  int counter = 0;
//遍历车道线对象集合
  for (int seg_i = 0; seg_i < lane_sequence.lane_segment_size(); ++seg_i) {
    if (counter > LENGTH_LANE_POINT_SEQUENCE) {
```

```
        break;
      }
    LaneSegment lane_seg = lane_sequence.lane_segment(seg_i);
    //遍历每个车道线中的车道线道路点集合
      for (int pt_i = 0; pt_i < lane_seg.lane_point_size(); ++pt_i) {
        p_lane_point = lane_seg.mutable_lane_point(pt_i);
        if (p_lane_point->has_relative_s() &&
            p_lane_point->relative_s() < FLAGS_rnn_min_lane_relatice_s) {
          continue;
        }
        if (!feature.has_position() || !p_lane_point->has_position()) {
          ADEBUG << "Feature or lane_point has no position!";
          continue;
        }
    //车道线相关特征值计算：diff_x、diff_y、angle
        float diff_x = static_cast<float>(p_lane_point->position().x() -
                                          feature.position().x());
        float diff_y = static_cast<float>(p_lane_point->position().y() -
                                          feature.position().y());
        float angle = std::atan2(diff_y, diff_x);
    //存储计算出的车道线特征值

feature_values->push_back(static_cast<float>(p_lane_point->heading()));

feature_values->push_back(static_cast<float>(p_lane_point->angle_diff()));
        feature_values->push_back(static_cast<float>(
            p_lane_point->relative_l() -
feature.lane().lane_feature().lane_l()));
        feature_values->push_back(angle);
        ++counter;
        if (counter > LENGTH_LANE_POINT_SEQUENCE) {
          ADEBUG << "Full the lane point sequence";
          break;
        }
      }
    }
    ADEBUG << "Lane sequence feature size: " << counter;
    if (counter < DIM_LANE_POINT_FEATURE) {
      AWARN << "Fail to setup lane feature!";
      return -1;
    }
    return 0;
  }
```

```
//检测新车道线对象是否能够连接现有车道线
bool RNNEvaluator::IsCutinInHistory(const std::string& curr_lane_id,
                                    const std::string& prev_lane_id) {
  std::shared_ptr<const LaneInfo> curr_lane_info =
      PredictionMap::LaneById(curr_lane_id);
  std::shared_ptr<const LaneInfo> prev_lane_info =
      PredictionMap::LaneById(prev_lane_id);
//通过高精度地图封装来判断两个车道之间是否存在继承关系
  if (!PredictionMap::IsSuccessorLane(curr_lane_info, prev_lane_info)) {
    return true;
  }
  return false;
}

}  // namespace prediction
}  // namespace apollo
```

参考文献

[1] Lee G H. Motion Estimation for Self-Driving Cars with a Generalized Camera[C]// Computer Vision & Pattern Recognition. IEEE, 2013.

[2] Redmon J, Farhadi A. [IEEE 2017 IEEE Conference on Computer Vision and Pattern Recognition (CVPR) - Honolulu, HI (2017.7.21-2017.7.26)] 2017 IEEE Conference on Computer Vision and Pattern Recognition (CVPR) - YOLO9000: Better, Faster, Stronger[J]. 2017:6517-6525.

[3] Qin S J, Badgwell T A. An Overview of Nonlinear Model Predictive Control Applications[M]// Nonlinear Model Predictive Control. 2000.

[4] Cheang T K, Chong Y S, Tay Y H. Segmentation-free Vehicle License Plate Recognition using ConvNet-RNN[J]. 2017.

[5] Palaiahnakote S, Dongqi T, Maryam A, et al. CNN-RNN based method for license plate recognition[J]. CAAI Transactions on Intelligence Technology, 2018.

第 14 章

无人驾驶规划策略

无人驾驶中重要的问题是怎么开车，这个问题是由无人驾驶规划模块来完成的。“凡事预则立，不预则废”这句古语很形象地诠释了无人驾驶规划模块的意义。具体来说，我们进行驾驶规划时需要分析清楚起点和终点之间的路径是怎样的，这个路径是否能够躲避开静态或动态的障碍物，以及计算出过程中的速度和速度变化方式等问题。

本章主要讲解无人驾驶预测技术的核心概念，结合 Apollo 系统中的具体实现分析机器学习技术如何应用到无人驾驶规划核心过程中。本章内容分为四部分：第一部分着重介绍无人驾驶规划技术的核心概念，通过代码分析说明主要业务流程；第二部分详细分析路径规划在百度 Apollo 规划模块中的具体模型、业务流程及其对外接口；第三部分详细分析障碍物规划在百度 Apollo 规划模块中的具体模型、业务流程及其对外接口；第四部分详细分析速度规划在百度 Apollo 规划模块中的具体模型、业务流程及其对外接口。

14.1 规划模块简介

14.1.1 规划业务流程分析

在规划中，我们通过结合高精度地图，进行定位和预测来构建车辆轨迹。规划过程主要分为两步：第一步是路线规划，地图路线导航，从 A 前往 B，Apollo 中通过路线规划模块处理该任务；第二步是轨迹规划，目标是生成免碰撞和舒适的可执行轨迹，该轨迹由一系列点定义，每个点都有一个关联速度和一个指示何时应抵达那个点的时间戳。

路线规划有 3 个输入：

- 地图，例如百度 Apollo 框架提供的地图数据包括公路网和实时交通信息。
- 我们当前在地图上的位置（A）。
- 我们在地图上的目的地（B）。

接下来将输入的地图数据转化为“图”（Graph），利用 A*算法生成轨迹。轨迹生成的目标是生成由一系列路径点定义的轨迹，为每个路径点分配一个时间戳和速度，我们让一条曲线与这些路径点拟合，生成轨迹的几何表征，然后多条轨迹选择最佳的一条。由于移动的障碍物可能会暂时阻挡部分路段，轨迹中的每个路点都有时间戳，因此可以将时间戳与预测模块的输出相结合，以确保在计划通过时，轨迹上的每个路径点均未被占用。这些时间戳创建了一个三维轨迹，每个路径点由空间中的两个维度以及时间上的第三个维度来定义。我们还为每个路径点指定了一个速度，速度用于确保车辆按时到达每个路径点，并且设置了多种约束条件，例如轨迹应能免于碰撞， 要让乘客感到舒适，路径点之间的过渡以及速度的任何变化都必须平滑，可行的实际路径的验证和对现实交通规则匹配。

14.1.2 Frenet 坐标系

空间曲线的曲率、挠率和 Frenet 公式是空间曲线基本理论的一部分，它是以空间曲线的密切平面和基本三棱形的知识作为基础的。空间曲线的曲率、挠率和 Frenet 公式在空间曲线的基本理论中占有重要位置。曲线的曲率和挠率完全决定了曲线的形状，当曲线的曲率和挠率之间满足多种不同的关系时，就会得到不同类型的曲线。在 Apollo 系统中，Frenet 坐标以路中心线为 y 轴，距中心线平移的距离为 x 轴。

14.1.3 路径-速度解耦

将路段分割成单元格，然后对这些单元格中的点进行随机采样，通过从每个单元格中取一个点并将点连接，就创建了候选路径。通过重复此过程，可以构建多个候选路径，使用成本函数对这些路径进行评估，并选择成本最低的路径。成本函数可能需要考虑的因素有：与车道中心的偏离、与障碍物的距离、速度和曲率的变化、对车辆的压力或任何其他因素。

然后选择与该路径关联的速度曲线，一个被称为“ST 图”的工具可以帮助我们设计和选择速度曲线。在 ST 图中，s 表示车辆的纵向位移 t 表示时间。ST 图上的曲线是对车辆运动的描述，因为它说明了车辆在不同时间的位置，由于速度是位置变化的速率，因此可以通过查看曲线的斜率从 ST 图上推断速度。

为得到最佳速度曲线，需要将 ST 图离散为多个单元格，单元格之间的速度有所变化，但在每个单元格内的速度保持不变，该方法可简化速度曲线的构建，并维持曲线的近似度。在 ST 图中，可以将障碍物绘制为在特定时间段内阻挡道路的某些部分的矩形。例如，假设使用预测模块预测车辆，将在 t_0 到 t_1 的时间段内驶入车道，由于该车将在此期间占据位置 s_0 到 s_1。因此，我们在 ST 图上绘制了一个矩形，它将在时间段 t_0 到 t_1 期间阻挡位置 s_0 到 s_1。为避免碰撞，速度曲线不得与此矩形相交，既然有了一幅各种单元格被阻挡的 ST 图，我们便可以使用优化引擎为该图选择最佳的速度曲线，优化算法通过复杂的数学运算来搜索，在实际应用中需要考虑各种限制条件，从而降低解决方案成本，这些限制可能包括：交通规则限制、距离限制（如与障碍物的距离）、汽车的物理限制（如加速度限制）。为了将离散解决方案转换为平滑轨迹，可使用“二次规划”技术，二次规划将平滑的非线性曲线与这些分段式线性段拟合。

14.1.4　三维轨迹生成 Lattice

一旦路径和速度曲线就绪，通过增加时间的维度便可以用其构建三维轨迹。接下来通过分离轨迹的纵向和横向分量将三维问题分解成两个单独的二维问题。其中，一个二维轨迹是具有时间戳的纵向轨迹，我们称之为 ST 轨迹；另一个二维轨迹是相对于纵向轨迹的横向偏移，我们称之为 SL 轨迹。然后将它们合并为生成纵向和横向二维轨迹，我们先将初始车辆状态投射到 ST 坐标系和 SL 坐标系中，通过对预选模式中的多个候选最终状态进行采样来选择最终车辆状态，对于每个候选的最终状态，我们构建了一组轨迹，将车辆从其初始状态转换为最终状态，使用成本函数对这些轨迹进行评估，并选择成本最低的轨迹。

14.1.5　车辆状态

车辆状态主要包括以下 3 种。

- 巡航：车辆将在完成规划步骤后定速行驶。
- 跟随：尝试在时间 t 出现在某辆车后面 跟随车辆。
- 停止：速度和加速度会被修正为 0。

一旦同时拥有了 ST 和 SL 轨迹，就需要将它们重新转换为笛卡尔坐标系，然后可以将它们相结合，构建由二维路径点和一维时间戳组成的三维轨迹。ST 轨迹是随时间变化的纵向位移，SL 轨迹是纵向轨迹上每个点的横向偏移，由于两个轨迹都有纵坐标 S，因此可以通过将其 S 值进行匹配来合并其轨迹。

14.2　路径规划

在 Apollo 系统中，路径规划算法集中在 NaviPathDecider 对象中，对应源代码文件为 modules/planning/navi/decider/navi_path_decider.cpp。下面我们以注释的方式来了解一下代码的具体执行流程。

百度 Apollo 相关源码分析：

```
/**
 * @file
 * @brief This file provides the implementation of the class "NaviPathDecider".
```

```
 */

#include "modules/planning/navi/decider/navi_path_decider.h"

#include <algorithm>
#include <utility>

#include "cyber/common/log.h"
#include "modules/common/configs/vehicle_config_helper.h"
#include "modules/common/math/vec2d.h"
#include "modules/planning/common/planning_gflags.h"
#include "modules/planning/proto/sl_boundary.pb.h"

namespace apollo {
namespace planning {

using apollo::common::Status;
using apollo::common::math::Box2d;
using apollo::common::math::Vec2d;

NaviPathDecider::NaviPathDecider() : NaviTask("NaviPathDecider") {
  // TODO(all): 添加定制开发的其他相关模块的初始化操作
}
//根据配置对象初始化导航路径规划器
bool NaviPathDecider::Init(const PlanningConfig& config) {
  move_dest_lane_config_talbe_.clear();
  max_speed_levels_.clear();
//根据 PlanningConfig 对象实例化 PlannerNaviConfig 对象
  PlannerNaviConfig planner_navi_conf =
      config.navigation_planning_config().planner_navi_config();
//提取目的地车道线配置表
  config_ = planner_navi_conf.navi_path_decider_config();
  auto move_dest_lane_config_talbe = config_.move_dest_lane_config_talbe();
//遍历配置表中的每个对象
  for (const auto& item : move_dest_lane_config_talbe.lateral_shift()) {
//确定最高时速
    double max_speed_level = item.max_speed();
    double max_move_dest_lane_shift_y = item.max_move_dest_lane_shift_y();
    if (move_dest_lane_config_talbe_.find(max_speed_level) ==
        move_dest_lane_config_talbe_.end()) {
      move_dest_lane_config_talbe_.emplace(
          std::make_pair(max_speed_level, max_move_dest_lane_shift_y));
      max_speed_levels_.push_back(max_speed_level);
```

```
    }
  }
  AINFO << "Maximum speeds and move to dest lane config: ";
  for (const auto& data : move_dest_lane_config_talbe_) {
    auto max_speed = data.first;
    auto max_move_dest_lane_shift_y = data.second;
    AINFO << "[max_speed : " << max_speed
          << " ,max move dest lane shift y : " << max_move_dest_lane_shift_y
          << "]";
  }
//确定其他相关参数
  max_keep_lane_distance_ = config_.max_keep_lane_distance();
  max_keep_lane_shift_y_ = config_.max_keep_lane_shift_y();
  min_keep_lane_offset_ = config_.min_keep_lane_offset();
  keep_lane_shift_compensation_ = config_.keep_lane_shift_compensation();
  start_plan_point_from_ = config_.start_plan_point_from();
  move_dest_lane_compensation_ = config_.move_dest_lane_compensation();
//根据配置对象初始化导航路径规划器
  is_init_ = obstacle_decider_.Init(config);
  return is_init_;
}
//外部调用的路径规划接口方法
Status NaviPathDecider::Execute(Frame* frame,
                                ReferenceLineInfo* const reference_line_info) {
  NaviTask::Execute(frame, reference_line_info);
  vehicle_state_ = frame->vehicle_state();
  cur_reference_line_lane_id_ = reference_line_info->Lanes().Id();
//利用车道线和障碍物相关信息进行路径规划
  auto ret = Process(reference_line_info->reference_line(),
                     frame->PlanningStartPoint(), frame->obstacles(),
                     reference_line_info->path_decision(),
                     reference_line_info->mutable_path_data());
  RecordDebugInfo(reference_line_info->path_data());
//规划失败，提示系统不能继续自动驾驶过程
  if (ret != Status::OK()) {
    reference_line_info->SetDrivable(false);
    AERROR << "参考线 " << reference_line_info->Lanes().Id()
           << " is not drivable after " << Name();
  }
  return ret;
}
//利用车道线和障碍物相关信息进行路径规划核心方法
apollo::common::Status NaviPathDecider::Process(
```

```
    const ReferenceLine& reference_line,
    const common::TrajectoryPoint& init_point,
    const std::vector<const Obstacle*>& obstacles,
PathDecision* const path_decision, PathData* const path_data) {
//输出对象非空检测
  CHECK_NOTNULL(path_decision);
  CHECK_NOTNULL(path_data);
//通过车辆状态初始化路径规划起始点信息
  start_plan_point_.set_x(vehicle_state_.x());
  start_plan_point_.set_y(vehicle_state_.y());
  start_plan_point_.set_theta(vehicle_state_.heading());
  start_plan_v_ = vehicle_state_.linear_velocity();
  start_plan_a_ = vehicle_state_.linear_acceleration();
  if (start_plan_point_from_ == 1) {
    // 从路径规划时间表开始重置路径规划起始点
    start_plan_point_.set_x(init_point.path_point().x());
    start_plan_point_.set_y(init_point.path_point().y());
    start_plan_point_.set_theta(init_point.path_point().theta());
    start_plan_v_ = init_point.v();
    start_plan_a_ = init_point.a();
  }

  //从参考线截取路径点
  std::vector<apollo::common::PathPoint> path_points;
  if (!GetBasicPathData(reference_line, &path_points)) {
    AERROR << "Get path points from 参考线 failed";
    return Status(apollo::common::ErrorCode::PLANNING_ERROR,
              "NaviPathDecider GetBasicPathData");
  }

  // 根据起始平面点和参考线的位置，将从参考线截获的轨迹沿其轴线移动到目的地车道线
  double dest_ref_line_y = path_points[0].y();

  ADEBUG << "in current plan cycle, adc to ref line distance : "
         << dest_ref_line_y << "lane id : " << cur_reference_line_lane_id_;
  MoveToDestLane(dest_ref_line_y, &path_points);

  KeepLane(dest_ref_line_y, &path_points);

  DiscretizedPath discretized_path(path_points);
  path_data->SetReferenceLine(&(reference_line_info_->reference_line()));
  if (!path_data->SetDiscretizedPath(discretized_path)) {
    AERROR << "Set path data failed.";
```

```
    return Status(apollo::common::ErrorCode::PLANNING_ERROR,
                  "NaviPathDecider SetDiscretizedPath");
  }

  return Status::OK();
}
// 根据起始平面点和参考线的位置，将从参考线截获的轨迹沿其轴线移动到目的地车道线
void NaviPathDecider::MoveToDestLane(
    const double dest_ref_line_y,
    std::vector<common::PathPoint>* const path_points) {
  double dest_lateral_distance = std::fabs(dest_ref_line_y);
  if (dest_lateral_distance < max_keep_lane_distance_) {
    return;
  }

  //计算横向位移范围和 theta 变化率
  double max_shift_y = CalculateDistanceToDestLane();

  double actual_start_point_y = std::copysign(max_shift_y, dest_ref_line_y);
  // 横向移动路径指向 y 坐标最大值
  double lateral_shift_value = -dest_ref_line_y + actual_start_point_y;
  //方向盘对左侧比对右侧更敏感，需要右侧的补偿值
  lateral_shift_value =
      lateral_shift_value > 0.0
          ? (lateral_shift_value - move_dest_lane_compensation_)
          : lateral_shift_value;
  ADEBUG << "in current plan cycle move to dest lane, adc shift to dest "
          "参考线 : "
         << lateral_shift_value;
  std::transform(path_points->begin(), path_points->end(),
path_points->begin(),
                 [lateral_shift_value](common::PathPoint& old_path_point) {
                   common::PathPoint new_path_point = old_path_point;
                   double new_path_point_y =
                       old_path_point.y() + lateral_shift_value;
                   new_path_point.set_y(new_path_point_y);
                   return new_path_point;
                 });

  return;
}
//根据起始平面点和参考线的位置进行车道线保持操作
void NaviPathDecider::KeepLane(
```

```
      const double dest_ref_line_y,
      std::vector<common::PathPoint>* const path_points) {
    double dest_lateral_distance = std::fabs(dest_ref_line_y);
    if (dest_lateral_distance <= max_keep_lane_distance_) {
  //取得进行实际目的路径点 y 值路径规划的输入值（参考线、障碍物对象、路径决策结果）
      auto& reference_line = reference_line_info_->reference_line();
      auto obstacles = frame_->obstacles();
  auto* path_decision = reference_line_info_->path_decision();
  //计算实际目的路径点 y 值路径规划的输入值
      double actual_dest_point_y =
          NudgeProcess(reference_line, *path_points, obstacles, *path_decision,
                 vehicle_state_);

      double actual_dest_lateral_distance = std::fabs(actual_dest_point_y);
  double actual_shift_y = 0.0;
  //计算车辆侧向移动值
      if (actual_dest_lateral_distance > min_keep_lane_offset_) {
        double lateral_shift_value = 0.0;
        lateral_shift_value =
            (actual_dest_lateral_distance < max_keep_lane_shift_y_ +
                                      min_keep_lane_offset_ -
                                      keep_lane_shift_compensation_)
                ? (actual_dest_lateral_distance - min_keep_lane_offset_ +
                  keep_lane_shift_compensation_)
                : max_keep_lane_shift_y_;
        actual_shift_y = std::copysign(lateral_shift_value,
actual_dest_point_y);
      }

      ADEBUG << "in current plan cycle keep lane, actual dest : "
             << actual_dest_point_y << " adc shift to dest : " << actual_shift_y;
  //封装车道保持结果并输出，std::transform 在指定的范围内应用于给定的操作，并将结果存储在
指定的另一个范围内。要使用 std::transform 函数需要包含<algorithm>头文件
      std::transform(
          path_points->begin(), path_points->end(), path_points->begin(),
          [actual_shift_y](common::PathPoint& old_path_point) {
            common::PathPoint new_path_point = old_path_point;
            double new_path_point_y = old_path_point.y() + actual_shift_y;
            new_path_point.set_y(new_path_point_y);
            return new_path_point;
          });
    }
```

```
    return;
  }
  //调试信息记录函数，记录的内容是路径信息对象
  void NaviPathDecider::RecordDebugInfo(const PathData& path_data) {
    const auto& path_points = path_data.discretized_path();
    auto* ptr_optimized_path = reference_line_info_->mutable_debug()
                                   ->mutable_planning_data()
                                   ->add_path();
    ptr_optimized_path->set_name(Name());
    ptr_optimized_path->mutable_path_point()->CopyFrom(
        {path_points.begin(), path_points.end()});
  }
  //获得基本的路径信息
  bool NaviPathDecider::GetBasicPathData(
      const ReferenceLine& reference_line,
      std::vector<common::PathPoint>* const path_points) {
    CHECK_NOTNULL(path_points);

    double min_path_len = config_.min_path_length();
    //得到最短的路径规划长度，使用公式： s = v0 * t + 1 / 2.0 * a * t^2
    double path_len = start_plan_v_ * config_.min_look_forward_time() +
                      start_plan_a_ * pow(0.1, 2) / 2.0;
    path_len = std::max(path_len, min_path_len);

    const double reference_line_len = reference_line.Length();
    if (reference_line_len < path_len) {
      AERROR << "参考线 is too short to generate path trajectory( s = "
             << reference_line_len << ").";
      return false;
    }

    // 获取参考线上的起始计划点的映射 s 并获取参考线的长度
    auto start_plan_point_project = reference_line.GetReferencePoint(
        start_plan_point_.x(), start_plan_point_.y());
    common::SLPoint sl_point;
    if (!reference_line.XYToSL(start_plan_point_project.ToPathPoint(0.0),
                               &sl_point)) {
      AERROR << "Failed to get start plan point s from reference "
                "line.";
      return false;
    }
    auto start_plan_point_project_s = sl_point.has_s() ? sl_point.s() : 0.0;
```

```
  //从参考线获取基本路径点
  ADEBUG << "Basic path data len ; " << reference_line_len;
  constexpr double KDenseSampleUnit = 0.50;
  constexpr double KSparseSmapleUnit = 2.0;
  for (double s = start_plan_point_project_s; s < reference_line_len;
       s += ((s < path_len) ? KDenseSampleUnit : KSparseSmapleUnit)) {
    const auto& ref_point = reference_line.GetReferencePoint(s);
    auto path_point = ref_point.ToPathPoint(s - start_plan_point_project_s);
    path_points->emplace_back(path_point);
  }

  if (path_points->empty()) {
    AERROR << "path poins is empty.";
    return false;
  }

  return true;
}
//路径规划车辆变线操作前是否安全的判断
bool NaviPathDecider::IsSafeChangeLane(const ReferenceLine& reference_line,
                                       const PathDecision& path_decision) {
//路径规划车辆变线操作前是否安全的判断使用变量准备
  const auto& adc_param =
      common::VehicleConfigHelper::GetConfig().vehicle_param();

  Vec2d adc_position(start_plan_point_.x(), start_plan_point_.y());
  Vec2d vec_to_center(
      (adc_param.front_edge_to_center() - adc_param.back_edge_to_center()) /
          2.0,
      (adc_param.left_edge_to_center() - adc_param.right_edge_to_center()) /
          2.0);
  Vec2d adc_center(adc_position +
                   vec_to_center.rotate(start_plan_point_.theta()));
  Box2d adc_box(adc_center, start_plan_point_.theta(), adc_param.length(),
                adc_param.width());
  SLBoundary adc_sl_boundary;
//计算 ADC 边界是否存在
  if (!reference_line.GetSLBoundary(adc_box, &adc_sl_boundary)) {
    AERROR << "Failed to get ADC boundary from box: " << adc_box.DebugString();
    return false;
  }
//遍历路径中的所有障碍物，计算发生碰撞的可能性
  for (const auto* obstacle : path_decision.obstacles().Items()) {
```

```
    const auto& sl_boundary = obstacle->PerceptionSLBoundary();
    constexpr double kLateralShift = 6.0;
    if (sl_boundary.start_l() < -kLateralShift ||
        sl_boundary.end_l() > kLateralShift) {
      continue;
    }
  //安全常量声明
    constexpr double kSafeTime = 3.0;
    constexpr double kForwardMinSafeDistance = 6.0;
    constexpr double kBackwardMinSafeDistance = 8.0;
  //车辆安全距离计算
    const double kForwardSafeDistance = std::max(
        kForwardMinSafeDistance,
        ((vehicle_state_.linear_velocity() - obstacle->speed()) * kSafeTime));
    const double kBackwardSafeDistance = std::max(
        kBackwardMinSafeDistance,
        ((obstacle->speed() - vehicle_state_.linear_velocity()) * kSafeTime));
  //ADC 边界和安全距离进行比较
    if (sl_boundary.end_s() >
            adc_sl_boundary.start_s() - kBackwardSafeDistance &&
        sl_boundary.start_s() <
            adc_sl_boundary.end_s() + kForwardSafeDistance) {
      return false;
    }
  }

  return true;
}
//通过车道线、路径点、障碍物信息进行路径规划中的车辆侧向微调
double NaviPathDecider::NudgeProcess(
    const ReferenceLine& reference_line,
    const std::vector<common::PathPoint>& path_data_points,
    const std::vector<const Obstacle*>& obstacles,
    const PathDecision& path_decision,
    const common::VehicleState& vehicle_state) {
  double nudge_position_y = 0.0;

  if (!FLAGS_enable_nudge_decision) {
    nudge_position_y = path_data_points[0].y();
    return nudge_position_y;
  }

  //获得车辆侧向微调的距离
```

```
    int lane_obstacles_num = 0;
    constexpr double KNudgeEpsilon = 1e-6;
    double nudge_distance = obstacle_decider_.GetNudgeDistance(
        obstacles, reference_line, path_decision, path_data_points,
vehicle_state,
        &lane_obstacles_num);
    // 调整路径规划的起始点
    if (std::fabs(nudge_distance) > KNudgeEpsilon) {
      ADEBUG << "need latteral nudge distance : " << nudge_distance;
      nudge_position_y = nudge_distance;
      last_lane_id_to_nudge_flag_[cur_reference_line_lane_id_] = true;
    } else {
      //一旦当前车道有障碍物，保持最后一个微调路径方向的路径在这种情况下无微调距离
      bool last_plan_has_nudge = false;
      if (last_lane_id_to_nudge_flag_.find(cur_reference_line_lane_id_) !=
          last_lane_id_to_nudge_flag_.end()) {
        last_plan_has_nudge =
            last_lane_id_to_nudge_flag_[cur_reference_line_lane_id_];
      }

      if (last_plan_has_nudge && lane_obstacles_num != 0) {
        ADEBUG << "Keepping last nudge path direction";
        nudge_position_y = vehicle_state_.y();
      } else {
        // 这种情况下不需要进行微调
        last_lane_id_to_nudge_flag_[cur_reference_line_lane_id_] = false;
        nudge_position_y = path_data_points[0].y();
      }
    }

    return nudge_position_y;
  }
  //计算到规划目的车道的距离
  double NaviPathDecider::CalculateDistanceToDestLane() {
    // 根据车辆的当前状态匹配配置文件中的适当横向换档参数
    double move_distance = 0.0;
    double max_adc_speed =
        start_plan_v_ + start_plan_a_ * 1.0 / FLAGS_planning_loop_rate;
    auto max_speed_level_itr = std::upper_bound(
        max_speed_levels_.begin(), max_speed_levels_.end(), max_adc_speed);
    if (max_speed_level_itr != max_speed_levels_.end()) {
      auto max_speed_level = *max_speed_level_itr;
      move_distance = move_dest_lane_config_talbe_[max_speed_level];
```

```
  }

  return move_distance;
}

}  // namespace planning
}  // namespace apollo
```

14.3　障碍物规划

在 Apollo 系统中，障碍物规划算法集中在 NaviObstacleDecider 对象中，对应源代码文件为 modules/planning/navi/decider/navi_obstacle_decider.cpp。下面以注释的方式来了解一下代码的具体执行流程。

百度 Apollo 相关源码分析：

```
/**
 * @file
 * @brief This file provides the implementation of the class
 * "NaviObstacleDecider".
 */
#include "modules/planning/navi/decider/navi_obstacle_decider.h"

#include <algorithm>
#include <limits>
#include <utility>

#include "cyber/common/log.h"
#include "modules/common/math/line_segment2d.h"
#include "modules/common/math/linear_interpolation.h"
#include "modules/common/math/path_matcher.h"
#include "modules/planning/common/planning_gflags.h"

namespace apollo {
namespace planning {

using apollo::common::PathPoint;
using apollo::common::math::PathMatcher;
using apollo::common::math::Vec2d;
using apollo::common::util::MakePathPoint;

namespace {
```

```
constexpr double kEpislon = 1e-6;
} // namespace

NaviObstacleDecider::NaviObstacleDecider() : NaviTask("NaviObstacleDecider") {}
//根据配置对象初始化导航障碍物规划器
bool NaviObstacleDecider::Init(const PlanningConfig& config) {
//根据 PlanningConfig 对象实例化 PlannerNaviConfig 对象
  PlannerNaviConfig planner_navi_conf =
      config.navigation_planning_config().planner_navi_config();
  config_ = planner_navi_conf.navi_obstacle_decider_config();
  return true;
}
//利用路径数据点集合和参考线得到最小的车道线宽度
double NaviObstacleDecider::GetMinLaneWidth(
    const std::vector<common::PathPoint>& path_data_points,
    const ReferenceLine& reference_line) {
  double min_lane_width = std::numeric_limits<double>::max();
  double lane_left_width = 0.0;
  double lane_right_width = 0.0;
//遍历整个路径数据点集合
  for (const auto& path_data_point : path_data_points) {
//使用参考线对象得到车道线宽度，宽度存在于 lane_left_width 和 lane_right_width 两个变量中
bool ret = reference_line.GetLaneWidth(
        path_data_point.s(), &lane_left_width, &lane_right_width);
    if (ret) {
      double lane_width = lane_left_width + lane_right_width;
//将这个路径点的车道宽度与之前的路径点的宽度中的最小值进行比较，如果当前路径点宽度小，就替换最小宽度
      if (lane_width < min_lane_width) {
        min_lane_width = lane_width;
      }
    }
  }
  return min_lane_width;
}
//添加障碍物偏移方向
void NaviObstacleDecider::AddObstacleOffsetDirection(
    const common::PathPoint& projection_point,
    const std::vector<common::PathPoint>& path_data_points,
    const Obstacle* current_obstacle, const double proj_len, double* dist) {
  Vec2d p1(0.0, 0.0);
  Vec2d p2(0.0, 0.0);
```

```
    p1.set_x(projection_point.x());
    p1.set_y(projection_point.y());
    if ((proj_len + 1) > path_data_points.back().s()) {
      p2.set_x(path_data_points.back().x());
      p2.set_y(path_data_points.back().y());
    } else {
  //MatchToPath 方法中使用了线性近似的插值运算
      auto point = PathMatcher::MatchToPath(path_data_points, (proj_len + 1));
      p2.set_x(point.x());
      p2.set_y(point.y());
    }
    auto d = ((current_obstacle->Perception().position().x() - p1.x()) *
              (p2.y() - p1.y())) -
             ((current_obstacle->Perception().position().y() - p1.y()) *
              (p2.x() - p1.x()));
    if (d > 0) {
      *dist = *dist * -1;
    }
  }
  //判断是否需要进行障碍物对象的过滤操作
  bool NaviObstacleDecider::IsNeedFilterObstacle(
      const Obstacle* current_obstacle, const PathPoint&
vehicle_projection_point,
      const std::vector<common::PathPoint>& path_data_points,
      const common::VehicleState& vehicle_state,
      PathPoint* projection_point_ptr) {
    bool is_filter = true;
  //通过障碍物位置点和路径点集合
    *projection_point_ptr = PathMatcher::MatchToPath(
        path_data_points, current_obstacle->Perception().position().x(),
        current_obstacle->Perception().position().y());
    ADEBUG << "obstacle distance : " << projection_point_ptr->s()
           << "vehicle distance : " << vehicle_projection_point.s();
  //判断路径点的投影和车辆点之间的距离是否大于根据车速进行的线性车速预测数据，如果大于，就应
该进行过滤
    if ((projection_point_ptr->s() - vehicle_projection_point.s()) >
        (config_.judge_dis_coeff() * vehicle_state.linear_velocity() +
         config_.basis_dis_value())) {
      return is_filter;
    }
  //对车辆前后边缘位置进行计算
    double vehicle_frontedge_position =
        vehicle_projection_point.s() + VehicleParam().length();
```

```
  double vehicle_backedge_position = vehicle_projection_point.s();
//对障碍物前后位置进行计算
  double obstacle_start_position =
      projection_point_ptr->s() - current_obstacle->Perception().length() /
2.0;
  double obstacle_end_position =
      projection_point_ptr->s() + current_obstacle->Perception().length() /
2.0;
//后向车辆和障碍物的距离和安全距离进行比较
  if ((vehicle_backedge_position - obstacle_end_position) >
      config_.safe_distance()) {
    return is_filter;
  }
//前向车辆和障碍物的距离和安全距离进行比较
  if ((obstacle_start_position - vehicle_frontedge_position) >
      config_.safe_distance()) {
    if (!current_obstacle->IsStatic()) {
      if (current_obstacle->Perception().velocity().x() > 0.0) {
        return is_filter;
      }
    }
  }
//不满足上述过滤条件的情况，返回 False，表示不需要过滤
  is_filter = false;
  return is_filter;
}
//根据路径规划结果对障碍物信息进行处理
void NaviObstacleDecider::ProcessObstacle(
    const std::vector<const Obstacle*>& obstacles,
    const std::vector<common::PathPoint>& path_data_points,
    const PathDecision& path_decision, const double min_lane_width,
const common::VehicleState& vehicle_state) {
//定义计算距离的回调接口
  auto func_distance = [](const PathPoint& point, const double x,
                          const double y) {
    double dx = point.x() - x;
    double dy = point.y() - y;
    return sqrt(dx * dx + dy * dy);
  };
//初始化函数中需要的路径点和车辆投影点
  PathPoint projection_point = MakePathPoint(0.0, 0.0, 0.0, 0.0, 0.0, 0.0, 0.0);
  PathPoint point = MakePathPoint(0.0, 0.0, 0.0, 0.0, 0.0, 0.0, 0.0);
  PathPoint vehicle_projection_point =
```

```
    PathMatcher::MatchToPath(path_data_points, 0, 0);
//遍历整个障碍物集合
  for (const auto& current_obstacle : obstacles) {
//通过 IsNeedFilterObstacle 确定当前障碍物对象否需要分析
    bool is_continue = IsNeedFilterObstacle(
        current_obstacle, vehicle_projection_point, path_data_points,
        vehicle_state, &projection_point);
    if (is_continue) {
      continue;//不需要分析的障碍物对象当前循环直接跳过
}
//计算障碍物和投影点之间的距离
    auto dist = func_distance(projection_point,
                              current_obstacle->Perception().position().x(),
                              current_obstacle->Perception().position().y());
//判断障碍物和投影点之间的距离是否能够和车辆干涉
    if (dist <
        (config_.max_nudge_distance()+current_obstacle->Perception().width()+
         VehicleParam().left_edge_to_center())) {
      auto proj_len = projection_point.s();
      if (std::fabs(proj_len) <= kEpislon ||
          proj_len >= path_data_points.back().s()) {
        continue;
      }
//添加障碍物偏移方向
      AddObstacleOffsetDirection(projection_point, path_data_points,
                                 current_obstacle, proj_len, &dist);
//存储最终障碍物规划结果
      obstacle_lat_dist_.emplace(std::pair<double, double>(
          current_obstacle->Perception().width(), dist));
    }
  }
}
//获取障碍物实际偏移距离
double NaviObstacleDecider::GetObstacleActualOffsetDistance(
    std::map<double, double>::iterator iter, const double right_nudge_lane,
    const double left_nudge_lane, int* lane_obstacles_num) {
  auto obs_width = iter->first;
  auto lat_dist = iter->second;
  ADEBUG << "get obstacle width : " << obs_width
         << "get lattitude distance : " << lat_dist;
//障碍物实际偏移距离
  auto actual_dist = std::fabs(lat_dist) - obs_width / 2.0 -
```

```
                    VehicleParam().left_edge_to_center();
//确定车道线上相关的障碍物数量
  if (last_nudge_dist_ > 0) {
    if (lat_dist < 0) {
      if (actual_dist < std::fabs(right_nudge_lane) + config_.safe_distance())
{
        *lane_obstacles_num = *lane_obstacles_num + 1;
      }
    }
  } else if (last_nudge_dist_ < 0) {
    if (lat_dist > 0) {
      if (actual_dist < std::fabs(left_nudge_lane) + config_.safe_distance())
{
        *lane_obstacles_num = *lane_obstacles_num + 1;
      }
    }
  }

  if ((last_lane_obstacles_num_ != 0) && (*lane_obstacles_num == 0) &&
      (!is_obstacle_stable_)) {
    is_obstacle_stable_ = true;
    statist_count_ = 0;
    ADEBUG << "begin keep obstacles";
  }

  if (is_obstacle_stable_) {
    ++statist_count_;
    if (statist_count_ > config_.cycles_number()) {
      is_obstacle_stable_ = false;
    } else {
      *lane_obstacles_num = last_lane_obstacles_num_;
    }
    ADEBUG << "statist_count_ : " << statist_count_;
  }
  last_lane_obstacles_num_ = *lane_obstacles_num;
  ADEBUG << "last_nudge_dist : " << last_nudge_dist_
         << "lat_dist : " << lat_dist << "actual_dist : " << actual_dist;
  return actual_dist;
}
//记录上次的微调距离
void NaviObstacleDecider::RecordLastNudgeDistance(const double nudge_dist) {
  double tolerance = config_.nudge_allow_tolerance();
```

```
  if (std::fabs(nudge_dist) > tolerance) {
    if (std::fabs(nudge_dist) > std::fabs(last_nudge_dist_)) {
      last_nudge_dist_ = nudge_dist;
    }
    no_nudge_num_ = 0;
  } else {
    ++no_nudge_num_;
  }

  if (no_nudge_num_ >= config_.cycles_number()) {
    last_nudge_dist_ = 0.0;
  }
}
//微调距离通过车辆状态进行优化
void NaviObstacleDecider::SmoothNudgeDistance(
const common::VehicleState& vehicle_state, double* nudge_dist) {
//函数脏检查
  CHECK_NOTNULL(nudge_dist);
//通过车速来进行是否优化的判断
  if (vehicle_state.linear_velocity() < config_.max_allow_nudge_speed()) {
    ++limit_speed_num_;
  } else {
    limit_speed_num_ = 0;
  }
  if (limit_speed_num_ < config_.cycles_number()) {
    *nudge_dist = 0;
  }
//通过微调系统配置容忍界限来判断
  if (std::fabs(*nudge_dist) > config_.nudge_allow_tolerance()) {
    ++eliminate_clutter_num_;
  } else {
    eliminate_clutter_num_ = 0;
  }
  if (eliminate_clutter_num_ < config_.cycles_number()) {
    *nudge_dist = 0;
  }
  ADEBUG << "eliminate_clutter_num_: " << eliminate_clutter_num_;
}

//根据障碍物、参考线、路径点和车辆状态来计算微调距离
double NaviObstacleDecider::GetNudgeDistance(
    const std::vector<const Obstacle*>& obstacles,
    const ReferenceLine& reference_line, const PathDecision& path_decision,
```

```
      const std::vector<common::PathPoint>& path_data_points,
      const common::VehicleState& vehicle_state, int* lane_obstacles_num) {
    CHECK_NOTNULL(lane_obstacles_num);

    // 计算车道上左右可移动的距离
    double left_nudge_lane = 0.0;
    double right_nudge_lane = 0.0;
    double routing_y = path_data_points[0].y();
    double min_lane_width = GetMinLaneWidth(path_data_points, reference_line);

    ADEBUG << "get min_lane_width: " << min_lane_width;
    if (routing_y <= 0.0) {
      left_nudge_lane = min_lane_width / 2.0 - std::fabs(routing_y) -
                        VehicleParam().left_edge_to_center();
      right_nudge_lane = -1.0 * (min_lane_width / 2.0 + std::fabs(routing_y) -
                                 VehicleParam().right_edge_to_center());
    } else {
      left_nudge_lane = min_lane_width / 2.0 + std::fabs(routing_y) -
                        VehicleParam().left_edge_to_center();
      right_nudge_lane = -1.0 * (min_lane_width / 2.0 - std::fabs(routing_y) -
                                 VehicleParam().right_edge_to_center());
    }

    // 根据障碍物的位置计算左右可微调整的距离
    double left_nudge_obstacle = 0.0;
    double right_nudge_obstacle = 0.0;

    // 计算当前车道障碍物的数量
    obstacle_lat_dist_.clear();
    ProcessObstacle(obstacles, path_data_points, path_decision, min_lane_width,
                    vehicle_state);
    for (auto iter = obstacle_lat_dist_.begin(); iter !=
obstacle_lat_dist_.end();
         ++iter) {
      auto actual_dist = GetObstacleActualOffsetDistance(
          iter, right_nudge_lane, left_nudge_lane, lane_obstacles_num);
      auto lat_dist = iter->second;
      if (actual_dist > config_.min_nudge_distance() &&
          actual_dist < config_.max_nudge_distance()) {
        auto need_nudge_dist = config_.max_nudge_distance() - actual_dist;
        if (lat_dist >= 0.0) {
          right_nudge_obstacle =
              -1.0 * std::max(std::fabs(right_nudge_obstacle), need_nudge_dist);
```

```
      } else {
        left_nudge_obstacle =
            std::max(std::fabs(left_nudge_obstacle), need_nudge_dist);
      }
    }
  }
  ADEBUG << "get left_nudge_lane: " << left_nudge_lane
         << "get right_nudge_lane : " << right_nudge_lane
         << "get left_nudge_obstacle: " << left_nudge_obstacle
         << "get right_nudge_obstacle : " << right_nudge_obstacle;
  //获取适当的微调距离值
  double nudge_dist = 0.0;
  if (std::fabs(left_nudge_obstacle) > kEpislon &&
      std::fabs(right_nudge_obstacle) <= kEpislon) {
    nudge_dist = std::min(left_nudge_lane, left_nudge_obstacle);
  } else if (std::fabs(right_nudge_obstacle) > kEpislon &&
             std::fabs(left_nudge_obstacle) <= kEpislon) {
    nudge_dist = std::max(right_nudge_lane, right_nudge_obstacle);
  }

  ADEBUG << "get nudge distance : " << nudge_dist
         << "get lane_obstacles_num : " << *lane_obstacles_num;
  RecordLastNudgeDistance(nudge_dist);
  SmoothNudgeDistance(vehicle_state, &nudge_dist);
  KeepNudgePosition(nudge_dist, lane_obstacles_num);
  ADEBUG << "last nudge distance : " << nudge_dist;
  return nudge_dist;
}
//保持微调位置
void NaviObstacleDecider::KeepNudgePosition(const double nudge_dist,
                                            int* lane_obstacles_num) {
//通过 nudge_dist 的数据判断是否需要保持微调位置
if (std::fabs(nudge_dist) > config_.nudge_allow_tolerance() &&
      std::fabs(last_nudge_dist_) < config_.nudge_allow_tolerance() &&
      !keep_nudge_flag_) {
    cycles_count_ = 0;
    keep_nudge_flag_ = true;
  }
//微调位置保持过程

  if (keep_nudge_flag_) {
    ++cycles_count_;;
    if (cycles_count_ > config_.max_keep_nudge_cycles()) {
```

```
      *lane_obstacles_num = 0;
      keep_nudge_flag_ = false;
    } else {
      *lane_obstacles_num = 1;
    }
  }
  ADEBUG << "get lane_obstacles_num : " << *lane_obstacles_num;
}
//获取危险障碍物的信息
void NaviObstacleDecider::GetUnsafeObstaclesInfo(
    const std::vector<common::PathPoint>& path_data_points,
    const std::vector<const Obstacle*>& obstacles) {
  // 找到参考线的起点
  double reference_line_y = path_data_points[0].y();

  // 根据参考线的位置判断安全行车范围
  double unsafe_refline_pos_y = 0.0;
  double unsafe_car_pos_y = 0.0;
  std::pair<double, double> unsafe_range;
  if (reference_line_y < 0.0) {
    unsafe_refline_pos_y = reference_line_y -
                           VehicleParam().right_edge_to_center() -
                           config_.speed_decider_detect_range();
    unsafe_car_pos_y = VehicleParam().right_edge_to_center() +
                       config_.speed_decider_detect_range();
    unsafe_range = std::make_pair(unsafe_refline_pos_y, unsafe_car_pos_y);
  } else {
    unsafe_refline_pos_y = reference_line_y +
                           VehicleParam().left_edge_to_center() +
                           config_.speed_decider_detect_range();
    unsafe_car_pos_y = -1.0 * (VehicleParam().left_edge_to_center() +
                               config_.speed_decider_detect_range());
    unsafe_range = std::make_pair(unsafe_car_pos_y, unsafe_refline_pos_y);
  }
  // 读取障碍物编号列表
  unsafe_obstacle_info_.clear();
  PathPoint vehicle_projection_point =
      PathMatcher::MatchToPath(path_data_points, 0, 0);
  for (const auto& iter : obstacles) {
    double obstacle_y = iter->Perception().position().y();
    if ((obstacle_y > unsafe_range.first &&
         obstacle_y < unsafe_range.second) ||
        (iter->Perception().velocity().y() >
```

```
        config_.lateral_velocity_value()) ||
       (iter->Perception().velocity().y() <
        -1.0 * config_.lateral_velocity_value())) {
     auto projection_point = PathMatcher::MatchToPath(
         path_data_points, iter->Perception().position().x(),
         iter->Perception().position().y());
     if (vehicle_projection_point.s() >= projection_point.s()) {
       continue;
     }
     auto front_distance =
         (projection_point.s() - iter->Perception().length() / 2.0) -
         vehicle_projection_point.s();
     auto ref_theta = projection_point.theta();
     auto project_velocity =
         iter->Perception().velocity().x() * std::cos(ref_theta) +
         iter->Perception().velocity().y() * std::sin(ref_theta);
     ADEBUG << "Lateral speed : " << iter->Perception().velocity().y();
     unsafe_obstacle_info_.emplace_back(iter->Id(), front_distance,
                                        project_velocity);
   }
  }
 }
}  // namespace planning
}  // namespace apollo
```

14.4　速度规划

在 Apollo 系统中，速度规划算法集中在 NaviSpeedDecider 对象中，对应源代码文件为 modules/planning/navi/decider/navi_speed_decider.cpp。下面以注释的方式来了解一下代码的具体执行流程。

百度 Apollo 相关源码分析：

```
namespace apollo {
namespace planning {

using apollo::common::ErrorCode;
using apollo::common::PathPoint;
using apollo::common::Status;
using apollo::common::VehicleConfigHelper;
using apollo::common::math::Clamp;
//速度规划算法中使用的常量
namespace {
```

```
constexpr double kTsGraphSStep = 0.4;
constexpr size_t kFallbackSpeedPointNum = 4;
constexpr double kSpeedPointSLimit = 200.0;
constexpr double kSpeedPointTimeLimit = 50.0;
constexpr double kZeroSpeedEpsilon = 1.0e-3;
constexpr double kZeroAccelEpsilon = 1.0e-3;
constexpr double kDecelCompensationLimit = 2.0;
constexpr double kKappaAdjustRatio = 20.0;
}  // namespace

NaviSpeedDecider::NaviSpeedDecider() : NaviTask("NaviSpeedDecider") {}

bool NaviSpeedDecider::Init(const PlanningConfig& planning_config) {
  CHECK_GT(FLAGS_planning_upper_speed_limit, 0.0);
//初始化 NavigationPlanningConfig 对象
  NavigationPlanningConfig config =
      planning_config.navigation_planning_config();
//初始化方法配置对象中相关参数脏检查
  CHECK(config.has_planner_navi_config());
  CHECK(config.planner_navi_config().has_navi_speed_decider_config());
  CHECK(config.planner_navi_config()
            .navi_speed_decider_config()
            .has_preferred_accel());
  CHECK(config.planner_navi_config()
            .navi_speed_decider_config()
            .has_preferred_decel());
  CHECK(

config.planner_navi_config().navi_speed_decider_config().has_max_accel());
  CHECK(

config.planner_navi_config().navi_speed_decider_config().has_max_decel());
  CHECK(config.planner_navi_config()
            .navi_speed_decider_config()
            .has_preferred_jerk());
  CHECK(config.planner_navi_config()
            .navi_speed_decider_config()
            .has_obstacle_buffer());
  CHECK(config.planner_navi_config()
            .navi_speed_decider_config()
            .has_safe_distance_base());
  CHECK(config.planner_navi_config()
            .navi_speed_decider_config()
```

```
          .has_safe_distance_ratio());
  CHECK(config.planner_navi_config()
          .navi_speed_decider_config()
          .has_following_accel_ratio());
  CHECK(config.planner_navi_config()
          .navi_speed_decider_config()
          .has_soft_centric_accel_limit());
  CHECK(config.planner_navi_config()
          .navi_speed_decider_config()
          .has_hard_centric_accel_limit());
  CHECK(config.planner_navi_config()
          .navi_speed_decider_config()
          .has_hard_speed_limit());
  CHECK(config.planner_navi_config()
          .navi_speed_decider_config()
          .has_hard_accel_limit());
  CHECK(config.planner_navi_config()
          .navi_speed_decider_config()
          .has_enable_safe_path());
  CHECK(config.planner_navi_config()
          .navi_speed_decider_config()
          .has_enable_planning_start_point());
  CHECK(config.planner_navi_config()
          .navi_speed_decider_config()
          .has_enable_accel_auto_compensation());
  CHECK(config.planner_navi_config()
          .navi_speed_decider_config()
          .has_kappa_preview());
  CHECK(config.planner_navi_config()
          .navi_speed_decider_config()
          .has_kappa_threshold());
 //通过系统设置初始化相关的变量初始值
  max_speed_ = FLAGS_planning_upper_speed_limit;
  preferred_accel_ = std::abs(config.planner_navi_config()
                              .navi_speed_decider_config()
                              .preferred_accel());
  preferred_decel_ = std::abs(config.planner_navi_config()
                              .navi_speed_decider_config()
                              .preferred_decel());
  preferred_jerk_ = std::abs(config.planner_navi_config()
                             .navi_speed_decider_config()
                             .preferred_jerk());
  max_accel_ = std::abs(
```

```
        config.planner_navi_config().navi_speed_decider_config().max_accel());
    max_decel_ = std::abs(
        config.planner_navi_config().navi_speed_decider_config().max_decel());
    preferred_accel_ = std::min(max_accel_, preferred_accel_);
    preferred_decel_ = std::min(max_decel_, preferred_accel_);

    obstacle_buffer_ = std::abs(config.planner_navi_config()
                              .navi_speed_decider_config()
                              .obstacle_buffer());
    safe_distance_base_ = std::abs(config.planner_navi_config()
                                 .navi_speed_decider_config()
                                 .safe_distance_base());
    safe_distance_ratio_ = std::abs(config.planner_navi_config()
                                  .navi_speed_decider_config()
                                  .safe_distance_ratio());
    following_accel_ratio_ = std::abs(config.planner_navi_config()
                                    .navi_speed_decider_config()
                                    .following_accel_ratio());
    soft_centric_accel_limit_ = std::abs(config.planner_navi_config()
                                       .navi_speed_decider_config()
                                       .soft_centric_accel_limit());
    hard_centric_accel_limit_ = std::abs(config.planner_navi_config()
                                       .navi_speed_decider_config()
                                       .hard_centric_accel_limit());
    soft_centric_accel_limit_ =
        std::min(hard_centric_accel_limit_, soft_centric_accel_limit_);
    hard_speed_limit_ = std::abs(config.planner_navi_config()
                               .navi_speed_decider_config()
                               .hard_speed_limit());
    hard_accel_limit_ = std::abs(config.planner_navi_config()
                               .navi_speed_decider_config()
                               .hard_accel_limit());
    enable_safe_path_ = config.planner_navi_config()
                        .navi_speed_decider_config()
                        .enable_safe_path();
    enable_planning_start_point_ = config.planner_navi_config()
                                   .navi_speed_decider_config()
                                   .enable_planning_start_point();
    enable_accel_auto_compensation_ = config.planner_navi_config()
                                      .navi_speed_decider_config()
                                      .enable_accel_auto_compensation();
    kappa_preview_ =
```

```
config.planner_navi_config().navi_speed_decider_config().kappa_preview();
    kappa_threshold_ = config.planner_navi_config()
                           .navi_speed_decider_config()
                           .kappa_threshold();
  //使用配置文件进行实际初始化过程，调用 init 方法
    return obstacle_decider_.Init(planning_config);
  }
  //外部调用的路径规划接口方法
  Status NaviSpeedDecider::Execute(Frame* frame,
                                   ReferenceLineInfo* reference_line_info) {
    NaviTask::Execute(frame, reference_line_info);

    // 获得车辆定速巡航速度
    const auto& planning_target = reference_line_info_->planning_target();
    preferred_speed_ = planning_target.has_cruise_speed()
                           ? std::abs(planning_target.cruise_speed())
                           : 0.0;
    preferred_speed_ = std::min(max_speed_, preferred_speed_);

    // 车辆预期状态
    const auto& planning_start_point = frame->PlanningStartPoint();
    auto expected_v =
       planning_start_point.has_v() ? planning_start_point.v() : 0.0;
    auto expected_a =
       planning_start_point.has_a() ? planning_start_point.a() : 0.0;

    // 车辆当前状态
    const auto& vehicle_state = frame->vehicle_state();
    auto current_v = vehicle_state.has_linear_velocity()
                         ? vehicle_state.linear_velocity()
                         : 0.0;
    auto current_a = vehicle_state.has_linear_acceleration()
                         ? vehicle_state.linear_acceleration()
                         : 0.0;

    // 获取起始点信息
    double start_v;
    double start_a;
    double start_da;

    if (enable_planning_start_point_) {
      start_v = std::max(0.0, expected_v);
      start_a = expected_a;
```

```
      start_da = 0.0;
    } else {
      start_v = std::max(0.0, current_v);
      start_a = current_a;
      start_da = 0.0;
    }

    // 加速度自动补偿
    if (enable_accel_auto_compensation_) {
      if (prev_v_ > 0.0 && current_v > 0.0 && prev_v_ > expected_v) {
        auto raw_ratio = (prev_v_ - expected_v) / (prev_v_ - current_v);
        raw_ratio = Clamp(raw_ratio, 0.0, kDecelCompensationLimit);
        decel_compensation_ratio_ = (decel_compensation_ratio_ + raw_ratio) /
2.0;
        decel_compensation_ratio_ =
            Clamp(decel_compensation_ratio_, 1.0, kDecelCompensationLimit);
        ADEBUG << "change decel_compensation_ratio: " <<
decel_compensation_ratio_
               << " raw: " << raw_ratio;
      }
      prev_v_ = current_v;
    }

    //获取离散化路径信息
    auto& discretized_path =
reference_line_info_->path_data().discretized_path();
  //进行速度规划过程
    auto ret = MakeSpeedDecision(
        start_v, start_a, start_da, discretized_path, frame_->obstacles(),
        [&](const std::string& id) { return frame_->Find(id); },
        reference_line_info_->mutable_speed_data());
    RecordDebugInfo(reference_line_info->speed_data());
  //速度规划失败处理逻辑
    if (ret != Status::OK()) {
      reference_line_info->SetDrivable(false);
      AERROR << "参考线 " << reference_line_info->Lanes().Id()
             << " is not drivable after " << Name();
    }

    return ret;
  }
  //进行速度规划过程
  Status NaviSpeedDecider::MakeSpeedDecision(
```

```
    double start_v, double start_a, double start_da,
    const std::vector<PathPoint>& path_points,
    const std::vector<const Obstacle*>& obstacles,
    const std::function<const Obstacle*(const std::string&)>& find_obstacle,
SpeedData* const speed_data) {
//函数脏检测
  CHECK_NOTNULL(speed_data);
  CHECK_GE(path_points.size(), 2);
//确定速度规划的距离
  auto start_s = path_points.front().has_s() ? path_points.front().s() : 0.0;
  auto end_s = path_points.back().has_s() ? path_points.back().s() : start_s;
  auto planning_length = end_s - start_s;

  ADEBUG << "start to make speed decision,  start_v: " << start_v
         << " start_a: " << start_a << " start_da: " << start_da
         << " start_s: " << start_s << " planning_length: " << planning_length;
//起始速度太快的情况下，停止速度规划
  if (start_v > max_speed_) {
    AERROR << "exceeding maximum allowable speed.";
    return Status(ErrorCode::PLANNING_ERROR,
                  "exceeding maximum allowable speed.");
  }
//确定起始速度
  start_v = std::max(0.0, start_v);

  auto s_step = planning_length > kTsGraphSStep
                    ? kTsGraphSStep
                    : planning_length / kFallbackSpeedPointNum;

  //初始化 t-s 图
  ts_graph_.Reset(s_step, planning_length, start_v, start_a, start_da);

  // 添加 t-s 约束
  auto ret = AddPerceptionRangeConstraints();
  if (ret != Status::OK()) {
    AERROR << "Add t-s constraints base on range of perception failed";
    return ret;
  }

  ret = AddObstaclesConstraints(start_v, planning_length, path_points,
                                obstacles, find_obstacle);
  if (ret != Status::OK()) {
    AERROR << "Add t-s constraints base on obstacles failed";
```

```
    return ret;
  }

  ret = AddCentricAccelerationConstraints(path_points);
  if (ret != Status::OK()) {
    AERROR << "Add t-s constraints base on centric acceleration failed";
    return ret;
  }

  ret = AddConfiguredConstraints();
  if (ret != Status::OK()) {
    AERROR << "Add t-s constraints base on configs failed";
    return ret;
  }

  //创建速度点集合
  std::vector<NaviSpeedTsPoint> ts_points;
  if (ts_graph_.Solve(&ts_points) != Status::OK()) {
    AERROR << "Solve speed points failed";
    speed_data->clear();
    speed_data->AppendSpeedPoint(0.0 + start_s, 0.0, 0.0, -max_decel_, 0.0);
    speed_data->AppendSpeedPoint(0.0 + start_s, 1.0, 0.0, -max_decel_, 0.0);
    return Status::OK();
  }

  speed_data->clear();
  for (auto& ts_point : ts_points) {
    if (ts_point.s > kSpeedPointSLimit || ts_point.t > kSpeedPointTimeLimit)
      break;

    if (ts_point.v > hard_speed_limit_) {
      AERROR << "The v: " << ts_point.v << " of point with s: " << ts_point.s
             << " and t: " << ts_point.t << "is greater than hard_speed_limit "
             << hard_speed_limit_;
      ts_point.v = hard_speed_limit_;
    }

    if (std::abs(ts_point.v) < kZeroSpeedEpsilon) ts_point.v = 0.0;

    if (ts_point.a > hard_accel_limit_) {
      AERROR << "The a: " << ts_point.a << " of point with s: " << ts_point.s
             << " and t: " << ts_point.t << "is greater than hard_accel_limit "
             << hard_accel_limit_;
```

```
      ts_point.a = hard_accel_limit_;
    }

    if (std::abs(ts_point.a) < kZeroAccelEpsilon) ts_point.a = 0.0;

    // 进行加速度调节
    if (enable_accel_auto_compensation_) {
      if (ts_point.a > 0)
        ts_point.a *= accel_compensation_ratio_;
      else
        ts_point.a *= decel_compensation_ratio_;
    }

    speed_data->AppendSpeedPoint(ts_point.s + start_s, ts_point.t, ts_point.v,
                                 ts_point.a, ts_point.da);
  }

  if (speed_data->size() == 1) {
    const auto& prev = speed_data->back();
    speed_data->AppendSpeedPoint(prev.s(), prev.t() + 1.0, 0.0, 0.0, 0.0);
  }

  return Status::OK();
}
//设置障碍物约束条件
Status NaviSpeedDecider::AddObstaclesConstraints(
    double vehicle_speed, double path_length,
    const std::vector<PathPoint>& path_points,
    const std::vector<const Obstacle*>& obstacles,
    const std::function<const Obstacle*(const std::string&)>& find_obstacle) {
  const auto& vehicle_config = VehicleConfigHelper::Instance()->GetConfig();
  auto front_edge_to_center =
      vehicle_config.vehicle_param().front_edge_to_center();
  auto get_obstacle_distance = [&](double d) -> double {
    return std::max(0.0, d - front_edge_to_center - obstacle_buffer_);
  };
  auto get_safe_distance = [&](double v) -> double {
    return safe_distance_ratio_ * v + safe_distance_base_ +
           front_edge_to_center + obstacle_buffer_;
  };

  // 增加感知障碍物信息
  obstacle_decider_.GetUnsafeObstaclesInfo(path_points, obstacles);
```

```
  for (const auto& info : obstacle_decider_.UnsafeObstacles()) {
    const auto& id = std::get<0>(info);
    const auto* obstacle = find_obstacle(id);
    if (obstacle != nullptr) {
      auto s = std::get<1>(info);
      auto obstacle_distance = get_obstacle_distance(s);
      auto obstacle_speed = std::max(std::get<2>(info), 0.0);
      auto safe_distance = get_safe_distance(obstacle_speed);
      AINFO << "obstacle with id: " << id << " s: " << s
            << " distance: " << obstacle_distance
            << " speed: " << obstacle_speed
            << " safe_distance: " << safe_distance;

      ts_graph_.UpdateObstacleConstraints(obstacle_distance, safe_distance,
                                          following_accel_ratio_,
                                          obstacle_speed, preferred_speed_);
    }
  }

  //在规划路径终点设定最后一个障碍物
  if (enable_safe_path_) {
    auto obstacle_distance = get_obstacle_distance(path_length);
    auto safe_distance = get_safe_distance(0.0);
    ts_graph_.UpdateObstacleConstraints(obstacle_distance, safe_distance,
                                        following_accel_ratio_, 0.0,
                                        preferred_speed_);
  }

  return Status::OK();
}
//添加中心加速约束
Status NaviSpeedDecider::AddCentricAccelerationConstraints(
const std::vector<PathPoint>& path_points) {
//如果规划路径点小于两个，就无法进行中心加速约束计算
  if (path_points.size() < 2) {
    AERROR << "Too few path points";
    return Status(ErrorCode::PLANNING_ERROR, "too few path points.");
  }
//中心加速约束计算中使用的常量初始化
  double max_kappa = 0.0;
  double max_kappa_v = std::numeric_limits<double>::max();
  double preffered_kappa_v = std::numeric_limits<double>::max();
  double max_kappa_s = 0.0;
```

```
  const auto bs = path_points[0].has_s() ? path_points[0].s() : 0.0;

  struct CLimit {
   double s;
   double v_max;
   double v_preffered;
  };
  std::vector<CLimit> c_limits;
  c_limits.resize(path_points.size() - 1);
 //遍历所有路径点
  for (size_t i = 1; i < path_points.size(); i++) {
   const auto& prev = path_points[i - 1];
   const auto& cur = path_points[i];
   auto start_s = prev.has_s() ? prev.s() - bs : 0.0;
   start_s = std::max(0.0, start_s);
   auto end_s = cur.has_s() ? cur.s() - bs : 0.0;
   end_s = std::max(0.0, end_s);
   auto start_k = prev.has_kappa() ? prev.kappa() : 0.0;
   auto end_k = cur.has_kappa() ? cur.kappa() : 0.0;
   auto kappa = std::abs((start_k + end_k) / 2.0);
   if (std::abs(kappa) < kappa_threshold_) kappa /= kKappaAdjustRatio;

   auto v_preffered = std::min(std::sqrt(soft_centric_accel_limit_ / kappa),
                     std::numeric_limits<double>::max());
   auto v_max = std::min(std::sqrt(hard_centric_accel_limit_ / kappa),
                  std::numeric_limits<double>::max());

   c_limits[i - 1].s = end_s;
   c_limits[i - 1].v_max = v_max;
   c_limits[i - 1].v_preffered = v_preffered;

   if (kappa > max_kappa) {
    max_kappa = kappa;
    max_kappa_v = v_max;
    preffered_kappa_v = v_preffered;
    max_kappa_s = start_s;
   }
  }

  //用 kappa 增加中心加速度的速度限制
  for (size_t i = 0; i < c_limits.size(); i++) {
   for (size_t j = i; j - i < (size_t)(kappa_preview_ / kTsGraphSStep) &&
            j < c_limits.size();
```

```
        j++)
      c_limits[i].v_preffered =
          std::min(c_limits[j].v_preffered, c_limits[i].v_preffered);
  }

  double start_s = 0.0;
  for (size_t i = 0; i < c_limits.size(); i++) {
   auto end_s = c_limits[i].s;
   auto v_max = c_limits[i].v_max;
   auto v_preffered = c_limits[i].v_preffered;

   NaviSpeedTsConstraints constraints;
   constraints.v_max = v_max;
   constraints.v_preffered = v_preffered;
   ts_graph_.UpdateRangeConstraints(start_s, end_s, constraints);

   start_s = end_s;
  }

  AINFO << "add speed limit for centric acceleration with kappa: " << max_kappa
      << " v_max: " << max_kappa_v << " v_preffered: " << preffered_kappa_v
      << " s: " << max_kappa_s;

  return Status::OK();
}
//添加配置文件相关约束
Status NaviSpeedDecider::AddConfiguredConstraints() {
  NaviSpeedTsConstraints constraints;
  constraints.v_max = max_speed_;
  constraints.v_preffered = preferred_speed_;
  constraints.a_max = max_accel_;
  constraints.a_preffered = preferred_accel_;
  constraints.b_max = max_decel_;
  constraints.b_preffered = preferred_decel_;
  constraints.da_preffered = preferred_jerk_;
  ts_graph_.UpdateConstraints(constraints);

  return Status::OK();
}
//记录调试的速度信息
void NaviSpeedDecider::RecordDebugInfo(const SpeedData& speed_data) {
  auto* debug = reference_line_info_->mutable_debug();
  auto ptr_speed_plan = debug->mutable_planning_data()->add_speed_plan();
```

```
  ptr_speed_plan->set_name(Name());
  ptr_speed_plan->mutable_speed_point()->CopyFrom(
      {speed_data.begin(), speed_data.end()});
}

}  // namespace planning
}  // namespace apollo
```

```
/**
 * @file
 * @brief This file provides the implementation of the class "NaviSpeedTsGraph".
 */
//速度规划核心类 NaviSpeedTsGraph
#include "modules/planning/navi/decider/navi_speed_ts_graph.h"

#include <algorithm>

#include "cyber/common/log.h"
#include "modules/common/math/math_utils.h"

namespace apollo {
namespace planning {

using apollo::common::ErrorCode;
using apollo::common::Status;
using apollo::common::math::Clamp;
//定义速度规划中所需要的常量
namespace {
constexpr double kDoubleEpsilon = 1.0e-6;
constexpr double kDefaultSStep = 1.0;
constexpr double kDefaultSMax = 2.0;
constexpr double kDefaultSafeDistanceRatio = 1.0;
constexpr double kDefaultSafeDistanceBase = 2.0;
constexpr double kSafeDistanceSmooth = 3.0;
constexpr double kFollowSpeedSmooth = 0.25;
constexpr double kInfinityValue = 1.0e8;
}  // namespace
//检测速度规划中约束的赋值是否正确
static void CheckConstraints(const NaviSpeedTsConstraints& constraints) {
  CHECK_GE(constraints.t_min, 0.0);
```

```
  CHECK_GE(constraints.v_max, 0.0);
  CHECK_GE(constraints.v_max, constraints.v_preffered);
  CHECK_GE(constraints.a_max, 0.0);
  CHECK_GE(constraints.a_max, constraints.a_preffered);
  CHECK_GE(constraints.b_max, 0.0);
  CHECK_GE(constraints.b_max, constraints.b_preffered);
  CHECK_GE(constraints.da_max, 0.0);
  CHECK_GE(constraints.da_max, constraints.da_preffered);
}
//通过比较默认约束值和设定约束值的关系确定在速度规划中使用的约束值
static void CombineConstraints(const NaviSpeedTsConstraints& constraints,
                               NaviSpeedTsConstraints* dst) {
  dst->t_min = std::max(constraints.t_min, dst->t_min);
  dst->v_max = std::min(constraints.v_max, dst->v_max);
  dst->v_preffered = std::min(constraints.v_preffered, dst->v_preffered);
  dst->a_max = std::min(constraints.a_max, dst->a_max);
  dst->a_preffered = std::min(constraints.a_preffered, dst->a_preffered);
  dst->b_max = std::min(constraints.b_max, dst->b_max);
  dst->b_preffered = std::min(constraints.b_preffered, dst->b_preffered);
  dst->da_max = std::min(constraints.da_max, dst->da_max);
  dst->da_preffered = std::min(constraints.da_preffered, dst->da_preffered);
}
//NaviSpeedTsGraph 对象构造方法
NaviSpeedTsGraph::NaviSpeedTsGraph() {
  Reset(kDefaultSStep, kDefaultSMax, 0.0, 0.0, 0.0);
}
//重置 NaviSpeedTsGraph 对象中的基础变量
void NaviSpeedTsGraph::Reset(double s_step, double s_max, double start_v,
                             double start_a, double start_da) {
  CHECK_GT(s_step, 0.0);
  CHECK_GE(s_max, s_step);
  CHECK_GE(start_v, 0.0);

  s_step_ = s_step;
  start_v_ = start_v;
  start_a_ = start_a;
  start_da_ = start_da;

  auto point_num = (size_t)((s_max + s_step_) / s_step_);
  constraints_.clear();
  constraints_.resize(point_num);
}
//更新约束值，调用约束值检测和确定约束值的方法
```

```
void NaviSpeedTsGraph::UpdateConstraints(
    const NaviSpeedTsConstraints& constraints) {
  CheckConstraints(constraints);

  for (auto& pc : constraints_) CombineConstraints(constraints, &pc);
}
//更新一段范围的约束值
void NaviSpeedTsGraph::UpdateRangeConstraints(
    double start_s, double end_s, const NaviSpeedTsConstraints& constraints) {
//函数脏检测
CHECK_GE(start_s, 0.0);
  CHECK_GE(end_s, start_s);
  CheckConstraints(constraints);
//确认更新范围的起止点
  auto start_idx = (size_t)(std::floor(start_s / s_step_));
  auto end_idx = (size_t)(std::ceil(end_s / s_step_));
//更新约束值
  if (start_idx == end_idx) {
    CombineConstraints(constraints, &constraints_[start_idx]);
  } else {
    for (size_t i = start_idx; i < end_idx && i < constraints_.size(); i++)
      CombineConstraints(constraints, &constraints_[i]);
  }
}
//更新障碍物约束值
void NaviSpeedTsGraph::UpdateObstacleConstraints(double distance,
                                                 double safe_distance,
                                                 double following_accel_ratio,
                                                 double v,
                                                 double cruise_speed) {
  CHECK_GE(distance, 0.0);
  CHECK_GE(safe_distance, 0.0);
  CHECK_GT(following_accel_ratio, 0.0);

  // 平滑障碍跟踪
  if (distance > safe_distance &&
      distance - safe_distance < kSafeDistanceSmooth &&
      std::abs(v - start_v_) < kFollowSpeedSmooth) {
    distance = safe_distance;
    v = start_v_;
  }

  v = std::max(v, 0.0);
```

```
// 更新时间最小值 t_min
double s = 0.0;
for (auto& pc : constraints_) {
  auto t = (s - distance) / v;
  if (t >= 0.0) {
    NaviSpeedTsConstraints constraints;
    constraints.t_min = t;
    CombineConstraints(constraints, &pc);
  }
  s += s_step_;
}

// 更新速度的最优值
auto od = distance - safe_distance;
auto v0 = start_v_;
auto v1 = v;
auto r0 = (distance - safe_distance) / (distance + safe_distance);
auto vm = std::max(cruise_speed * r0, 0.0) +
          (1.0 + following_accel_ratio * r0) * v1;
auto ra = (v1 - vm) * (v1 - vm) * (v1 + v0 - 2.0 * vm) / (od * od);
auto rb = (v1 - vm) * (v1 + 2.0 * v0 - 3.0 * vm) / od;
auto rc = v0;
auto t1 = -1.0 * od / (v1 - vm);
auto s1 = ra * t1 * t1 * t1 + rb * t1 * t1 + rc * t1;
//更新障碍物约束值
double t;
double prev_v;
bool first = true;
for (auto& pc : constraints_) {
  NaviSpeedTsConstraints constraints;

  if (first) {
    first = false;
    auto cur_v = rc;
    t = 0.0;
    s = 0.0;
    prev_v = cur_v;
    constraints.v_preffered = cur_v;
  } else if (s <= s1 && t <= t1) {
    auto a = 6.0 * ra * t + 2.0 * rb;
    t += (std::sqrt(prev_v * prev_v + 2.0 * a * s_step_) - prev_v) / a;
    auto cur_v = 3.0 * ra * t * t + 2.0 * rb * t + rc;
```

```
      s += s_step_;
      prev_v = cur_v;
      constraints.v_preffered = std::max(cur_v, 0.0);
    } else {
      auto cur_v = v;
      t += 2.0 * s_step_ / (prev_v + cur_v);
      s += s_step_;
      prev_v = cur_v;
      constraints.v_preffered = cur_v;
    }

    CombineConstraints(constraints, &pc);
  }
}
//进行速度规划
Status NaviSpeedTsGraph::Solve(std::vector<NaviSpeedTsPoint>* output) {
  CHECK_NOTNULL(output);

  // 对第一个点进行约束
  auto& constraints = constraints_[0];
  constraints.v_max = start_v_;
  constraints.v_preffered = start_v_;
  constraints.a_max = start_a_;
  constraints.a_preffered = start_a_;
  constraints.da_max = start_da_;
  constraints.da_preffered = start_da_;

  // 基于 b_max 预处理 v_max
  for (ssize_t i = constraints_.size() - 2; i >= 0; i--) {
    const auto& next = constraints_[i + 1];
    auto& cur = constraints_[i];
    cur.v_max =
        std::min(std::sqrt(next.v_max * next.v_max + 2 * next.b_max * s_step_),
                 cur.v_max);
    cur.v_preffered = std::min(cur.v_max, cur.v_preffered);
  }

  // 基于 a_max 预处理 v_max
  for (size_t i = 1; i < constraints_.size(); i++) {
    const auto& prev = constraints_[i - 1];
    auto& cur = constraints_[i];
    cur.v_max =
        std::min(std::sqrt(prev.v_max * prev.v_max + 2 * cur.a_max * s_step_),
```

```
            cur.v_max);
  cur.v_preffered = std::min(cur.v_max, cur.v_preffered);
}

//基于 b_preffered 预处理 v_preffered

for (ssize_t i = constraints_.size() - 2; i >= 0; i--) {
  const auto& next = constraints_[i + 1];
  auto& cur = constraints_[i];
  cur.v_preffered = std::min(std::sqrt(next.v_preffered * next.v_preffered +
                                       2 * next.b_preffered * s_step_),
                             cur.v_preffered);
}

// 基于 a_preffered 预处理 v_preffered
for (size_t i = 1; i < constraints_.size(); i++) {
  const auto& prev = constraints_[i - 1];
  auto& cur = constraints_[i];
  cur.v_preffered = std::min(std::sqrt(prev.v_preffered * prev.v_preffered +
                                       2 * cur.a_preffered * s_step_),
                             cur.v_preffered);
}

auto& points = *output;
points.resize(constraints_.size());

// 计算第一个点的速度、位置、加速度
auto& point = points[0];
point.s = 0.0;
point.t = 0.0;
point.v = start_v_;
point.a = start_a_;
point.da = start_da_;

//计算接下来的各个点的速度、位置、加速度
for (size_t i = 1; i < points.size(); i++) {
  const auto& prev = points[i - 1];
  const auto& constraints = constraints_[i];
  auto& cur = points[i];

  // 基于 v_max 计算 t_min
  auto t_min = std::max(prev.t, constraints.t_min);
  auto v_max = constraints.v_max;
```

```
    t_min = std::max(prev.t + 2.0 * s_step_ / (prev.v + v_max), t_min);

    // 基于b_max 计算 t_max
    auto t_max = std::numeric_limits<double>::infinity();
    auto b_max = constraints.b_max;
    auto r0 = prev.v * prev.v - 2 * b_max * s_step_;
    if (r0 > 0.0) t_max = prev.t + (prev.v - std::sqrt(r0)) / b_max;

    // 如果 t_max 小于 t_min，就不能完成速度规划
    if (t_max < t_min) {
      AERROR << "failure to satisfy the constraints.";
      return Status(ErrorCode::PLANNING_ERROR,
                    "failure to satisfy the constraints.");
    }

    // 基于 v_preffered 计算 t_preffered
auto v_preffered = constraints.v_preffered;
    auto t_preffered = prev.t + 2 * s_step_ / (prev.v + v_preffered);

    cur.s = prev.s + s_step_;
    cur.t = Clamp(t_preffered, t_min, t_max);
    auto dt = cur.t - prev.t;
    cur.v = std::max(2.0 * s_step_ / dt - prev.v, 0.0);

    // 如果 t 是无穷大
    if (std::isinf(cur.t)) {
      points.resize(i + 1);
      break;
    }
  }

  for (size_t i = 1; i < points.size() - 1; i++) {
    const auto& prev = points[i - 1];
    const auto& next = points[i + 1];
    auto& cur = points[i];
    auto ds = next.s - prev.s;
    auto dt = next.t - prev.t;
    cur.v = ds / dt;
  }
 //更新 NaviSpeedTsPoint 集合数据，该集合为输出值
  auto& first = points[0];
  const auto& second = points[1];
  first.a = (second.v - first.v) / (2.0 * (second.t - first.t));
```

```
  for (size_t i = 1; i < points.size() - 1; i++) {
    const auto& prev = points[i - 1];
    const auto& next = points[i + 1];
    auto& cur = points[i];
    auto dv = next.v - prev.v;
    auto dt = next.t - prev.t;
    cur.a = dv / dt;
  }

  for (size_t i = 1; i < points.size() - 1; i++) {
    const auto& prev = points[i - 1];
    const auto& next = points[i + 1];
    auto& cur = points[i];
    auto da = next.a - prev.a;
    auto dt = next.t - prev.t;
    cur.da = da / dt;
  }

  for (size_t i = 0; i < points.size(); i++) {
    auto& point = points[i];
    point.s = std::min(kInfinityValue, point.s);
    point.t = std::min(kInfinityValue, point.t);
    point.v = std::min(kInfinityValue, point.v);
    point.a = std::min(kInfinityValue, point.a);
    point.da = std::min(kInfinityValue, point.da);

    if (std::abs(point.t - kInfinityValue) < kDoubleEpsilon)
      points.resize(i + 1);
  }

  return Status::OK();
}

}  // namespace planning
}  // namespace apollo
```

参考文献

[1] Chen Z, Huang X. End-to-end learning for lane keeping of self-driving cars[C]// 2017 IEEE Intelligent Vehicles Symposium (IV). IEEE, 2017.

[2] Arkin R C, Murphy R R. Autonomous navigation in a manufacturing environment[J]. Robotics & Automation IEEE Transactions on, 1990, 6(4):445-454.

[3] Hummel B, Kammel S, Dang T, et al. Vision-based path-planning in unstructured environments[C]// Intelligent Vehicles Symposium, 2006 IEEE. IEEE, 2006.

[4] Dolgov D, Thrun S. Autonomous driving in semi-structured environments: Mapping and planning[C]// IEEE International Conference on Robotics & Automation. IEEE, 2009.

[5] Dolgov D, Thrun S, Montemerlo M, et al. Path Planning for Autonomous Vehicles in Unknown Semi-structured Environments[J]. The International Journal of Robotics Research, 2010, 29(5):485-501.

[6] Werling M, Ziegler J, Kammel S, et al. Optimal Trajectory Generation for Dynamic Street Scenarios in a Frenet Frame[C]// Robotics and Automation (ICRA), 2010 IEEE International Conference on. IEEE, 2010.

第 15 章

无人驾驶控制策略

无人驾驶车辆控制是无人驾驶的“临门一脚”，其中的策略设计直接影响无人驾驶产品的实际体验。一般来说，无人驾驶车辆控制过程中需要考虑车辆运动学的运动规律和车辆动力学特性。PID 控制算法和 MPC 控制算法是常用的无人驾驶控制策略算法。

本章主要讲解无人驾驶的车辆模型、PID 控制算法和 MPC 控制算法。本章内容分为三部分，第一部分着重介绍无人驾驶的车辆模型的核心概念，着重分析车辆运动学模型和车辆动力学模型；第二部分着重分析 PID 控制算法的数学模型，在此基础上分析百度 Apollo 系统的无人驾驶控制模块的 PID 控制算法的实现；第三部分着重分析 MPC 控制算法的数学模型，在此基础上分析百度 Apollo 系统的无人驾驶控制模块的 MPC 控制算法实现。

15.1 车辆模型

无人驾驶车辆控制过程中需要考虑车辆运动学的运动规律和车辆动力学的特性。如果控制阶段的车辆模型的构造能够考虑车辆运动学和动力学约束，那么车辆的控制效果会更加优秀。在建立车辆模型控制算法时，必须根据无人驾驶车辆的具体行驶道路工况，通过选取相关的控制变量建立运动学和动力学两种模型。车辆的实际运动状态是十分复杂的，精确进行描述需要庞大维度的变量组合，但是控制算法的实时性是需要考虑的实车使用的关键要素，因此我们会对两种模型进行约束简化和数值近似。下面分别介绍无人驾驶车辆控制中一般使用的运动学模型和动力学模型。

15.1.1 运动学模型

运动学是从几何学的角度研究物体的运动规律，包括物体在空间的位置、速度等随时间而产生的变化。车辆转向运动模型如图 15.1 所示，在惯性坐标系 OXY 下，（X_r,Y_r）和（X_f,Y_f）分别为车辆后轴和前轴轴心的坐标，φ 为车体的航向角，δ_f 为前轮偏角，v_r 为车辆后轴的中心速度，v_f 为车辆前轴的中心速度，l 为轴距。车辆转向过程示意图如图 15.2 所示，R 为后轮转向半径，P 为车辆的瞬时转动中心，M 为车辆后轴轴心，N 为前轴轴心。此处假设转向过程中车辆质心侧偏角保持不变，即车辆瞬时转向半径和道路曲率半径相同。

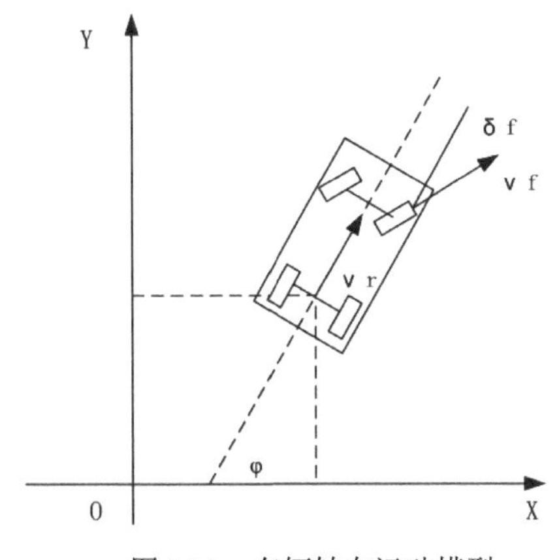

图 15.1　车辆转向运动模型

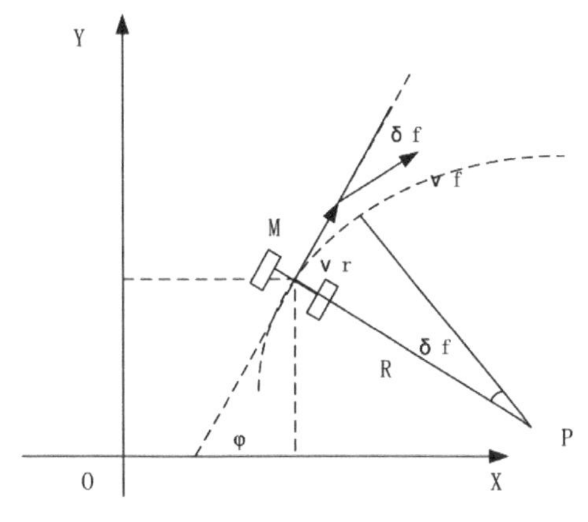

图 15.2　车辆转向过程示意图

在后轴行驶轴心（X_r,Y_r）处，速度为：

$$v_t = \dot{X}_r \cos\varphi + \dot{Y}_r \sin\varphi$$

前、后轴的运动学约束为：

$$\dot{X}_f \sin(\varphi + \delta_f) - \dot{Y}_f \cos(\varphi + \delta_f) = 0$$

$$\dot{X}_r \sin\varphi - \dot{Y}_r \cos\varphi = 0$$

根据前后轮的几何关系可得：

$$X_f = X_r + l\cos\varphi$$

$$Y_f = Y_r + l\sin\varphi$$

可以计算出车辆的运动学模型为：

$$\begin{bmatrix} \dot{X}_r \\ \dot{Y}_r \\ \dot{\varphi} \end{bmatrix} = \begin{bmatrix} \cos\varphi \\ \sin\varphi \\ 0 \end{bmatrix} v_r + \begin{bmatrix} 0 \\ 0 \\ 1 \end{bmatrix} \omega$$

其中，ω 为车辆航向角速度，根据上面得到的公式可以清楚地看到，在运动学模型中会使用 v_r 和 ω 作为车辆控制的输入值。

15.1.2 动力学模型

自动驾驶汽车的车辆动力学模型一般研究车辆在纵向和侧向的动力学特征，参考模型需要减少控制算法的计算量，从而能够作为车辆控制的实时性输入。因此，需要在进行车辆动力学建模时进行以下理想化的假设：

（1）假设无人驾驶车辆在平坦路面上行驶，忽略车辆垂向运动。

（2）悬架系统及车辆是刚性的，忽略悬架运动及其对耦合关系的影响。

（3）只考虑纯侧偏轮胎特性，忽略轮胎力的纵横向耦合关系。

（4）用单轨模型来描述车辆运动，不考虑载荷的左右转移。

（5）假设车辆行驶速度变化缓慢，忽略前后轴的载荷转移。

（6）忽略纵向和横向空气动力学。

基于以上 6 点理想假设，平面运动车辆只具有 3 个方向的运动，即纵向、横向和横摆运动。设定车辆为前轮驱动，满足以上设定的平面运动车辆模型如图 15.3 所示。其中，坐标 $OXYZ$ 为固定车身的车辆坐标系。XOZ 处于车辆左右对称的平面内。车辆质心所在点为坐标原点 O，X 轴沿车辆纵轴，Y 轴与车辆纵轴方向垂直，而 Z 轴满足右手法则，垂直于 XOY 且向上。坐标系 OXY 为基于地面的惯性坐标系，也满足右手法则。轮胎受力包含如下四个方面：

- 前、后轮胎受到的纵向力。
- 前、后轮胎受到的侧向力。
- 前、后轮胎受到的 X 方向的力。
- 前、后轮胎受到的 Y 方向的力。

根据牛顿第二定律，轮胎分别沿 X 轴、Y 轴和绕 Z 轴达到受力平衡。

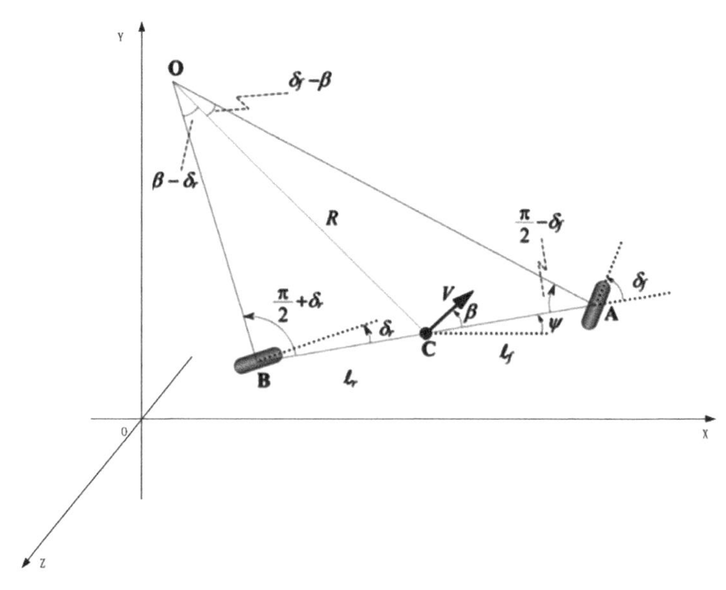

图 15.3 车辆单轨模型

15.2　PID 控制算法

PID（Proportion Integration Differentiation）其实就是指比例、积分、微分控制。总的来说，当得到系统的输出后，将输出经过比例、积分、微分 3 种运算方式叠加到输入中，从而控制系统的行为。下面用一个简单的实例来说明。

15.2.1　比例控制算法

我们先用一个经典的例子来介绍 PID 中简单的比例控制，抛开其他两个不谈。假设有一个水缸，最终的控制目的是保证水缸里的水位永远维持在 1 米的高度。假设初始时刻，水缸里的水位是 0.2 米，当前时刻的水位和目标水位之间存在一个偏差（error），且 error 为 0.8 米。这个时候，假设旁边站着一个人，这个人通过往缸里加水的方式来控制水位。如果单纯地用比例控制算法，就是指加入的水量 u 和偏差 error 是成正比的。所以，比例控制算法如下：

$$u=k_p*error$$

假设 k_p 取 0.5，那么 t=1 时（表示第 1 次加水，也就是第一次对系统施加控制），u=0.5×0.8=0.4 米，所以这一次加入的水量会使水位在 0.2 米的基础上上升 0.4 米，达到 0.6 米。

接着，t=2 时（第 2 次施加控制），当前水位是 0.6 米，所以 error 是 0.4 米。u=0.5×0.4=0.2 米，会使水位再次上升 0.2 米，达到 0.8 米。

如此循环下去，就是比例控制算法的运行方法。从计算过程不难推断出最终水位会达到我们需要的 1 米。

但是，单单的比例控制存在着一些不足，最重要的缺点就是稳态误差会很大。

像上述的例子，根据 k_p 取值不同，系统最后都会达到 1 米，不会有稳态误差。但是，考虑另一种情况，假设这个水缸在加水的过程中存在漏水的情况，每次加水的过程都会漏掉 0.1 米高度的水。仍然假设 k_p 取 0.5，那么存在某种情况，经过几次加水，水缸中的水位到 0.8 米时，水位将不再变化。因为若水位为 0.8 米，则误差 error=0.2 米，所以每次往水缸中加水的量为 u=0.5×0.2=0.1 米。同时，每次水缸里又会流出去 0.1 米的水，加入的水和流出的水相抵消，水位将不再变化。

也就是说，我们的目标是 1 米，但是最后水位达到 0.8 米就不再变化了，且系统已经达到稳定。由此产生的误差就是稳态误差了。

在实际情况中，这种类似水缸漏水的情况往往更加常见，比如控制汽车运动，摩擦阻力就相当于是“漏水”，控制机械臂、无人机的飞行，各类阻力和消耗都可以理解为本例中的“漏水”。所以，单独的比例控制在很多时候并不能满足要求。

15.2.2　积分控制算法

运用上面的例子，如果仅仅用比例，可以发现存在暂态误差，最后的水位就卡在 0.8 米了。于是，在控制中，我们再引入一个分量，该分量和误差的积分呈正比。所以，比例+积分控制的算法为：

$$u=k_p*error+ k_i*\int error$$

还是用上面的例子来说明，第一次的误差是 0.8，第二次的误差是 0.4，至此，误差的积分（离散情况下积分其实就是进行累加）：$\int error=0.8+0.4=1.2$。这个时候的控制量除了比例的那一部分外，还有一部分就是一个系数 k_i 乘以这个积分项。由于这个积分项会将前面若干次的误差进行累计，因此可以很好地消除稳态误差（假设在仅有比例项的情况下，系统卡在稳态误差，即上例中的 0.8 米，由于加入了积分项，会使输入增大，从而使得水缸的水位可以大于 0.8 米，渐渐到达目标的 1.0 米），这就是积分项的作用。

15.2.3 微分控制算法

考虑另一个例子：汽车刹车的情况。平稳地驾驶车辆，当发现前面有红灯时，为了使得行车平稳，基本上提前几十米就放松油门并踩刹车了。当车辆离停车线非常近的时候，使劲踩刹车，使车辆停下来。整个过程可以看作一个加入微分的控制策略。

微分，在离散情况下，就是结果误差（error，记为 err）的差值，对于 t 时刻和 t-1 时刻，可得 $u=k_d*$（error（t）-error（t-1）），其中 k_d 是一个系数项。可以看到，在刹车过程中，因为 error 是越来越小的，所以这个微分控制项一定是负数，在控制中加入一个负数项，作用就是为了防止汽车由于刹车不及时而闯过了线。从常识上可以理解，越是靠近停车线，越是应该注意踩刹车，不能让车过线，所以这个微分项的作用就可以理解为刹车，当车离停车线很近并且车速很快时，这个微分项的绝对值（实际上是一个负数）就会很大，从而表示应该用力踩刹车才能让车停下来。

切换到上面给水缸加水的例子，就是当发现水缸里的水快要接近 1 的时候，加入微分项，可以防止把水缸里的水加到超过 1 米的高度，其实就是减少控制过程中的震荡。

PID 控制对应的最终数学表达式为：

$$U(t) = k_p\big(err(t)\big) + \frac{1}{T_I}\int err(t)d_t + \frac{T_D \cdot d(err(t))}{d_t}$$

15.2.4 百度 Apollo 相关源码分析

在 Apollo 系统中，PID 车辆控制算法集中在 LatController 和 LonController 对象中，对的应源代码文件为 modules/control/controller/lat_controller.cpp 和 modules/control/controller/lon_controller.cpp。下面以注释的方式来了解代码的具体执行流程。

```
//LatController 的控制算法实现（控制车辆转弯），输出控制命令 ControlCommand 对象
Status LatController::ComputeControlCommand(
    const localization::LocalizationEstimate *localization,
    const canbus::Chassis *chassis,
    const planning::ADCTrajectory *planning_published_trajectory,
    ControlCommand *cmd) {
  auto vehicle_state = VehicleStateProvider::Instance();
```

```
  auto target_tracking_trajectory = *planning_published_trajectory;

  if (FLAGS_use_navigation_mode &&
      FLAGS_enable_navigation_mode_position_update) {
    auto time_stamp_diff =
        planning_published_trajectory->header().timestamp_sec() -
        current_trajectory_timestamp_;

    auto curr_vehicle_x = localization->pose().position().x();
    auto curr_vehicle_y = localization->pose().position().y();

    double curr_vehicle_heading = 0.0;
    const auto &orientation = localization->pose().orientation();
    if (localization->pose().has_heading()) {
      curr_vehicle_heading = localization->pose().heading();
    } else {
      curr_vehicle_heading =
          common::math::QuaternionToHeading(orientation.qw(),
orientation.qx(),
                                            orientation.qy(), orientation.qz());
    }

    // 新规划轨迹

   if (time_stamp_diff > 1.0e-6) {
      init_vehicle_x_ = curr_vehicle_x;
      init_vehicle_y_ = curr_vehicle_y;
      init_vehicle_heading_ = curr_vehicle_heading;

      current_trajectory_timestamp_ =
          planning_published_trajectory->header().timestamp_sec();
    } else {
      auto x_diff_map = curr_vehicle_x - init_vehicle_x_;
      auto y_diff_map = curr_vehicle_y - init_vehicle_y_;
      auto theta_diff = curr_vehicle_heading - init_vehicle_heading_;

      auto cos_map_veh = std::cos(init_vehicle_heading_);
      auto sin_map_veh = std::sin(init_vehicle_heading_);

      auto x_diff_veh = cos_map_veh * x_diff_map + sin_map_veh * y_diff_map;
      auto y_diff_veh = -sin_map_veh * x_diff_map + cos_map_veh * y_diff_map;
```

```
        auto cos_theta_diff = std::cos(-theta_diff);
        auto sin_theta_diff = std::sin(-theta_diff);

        auto tx = -(cos_theta_diff * x_diff_veh - sin_theta_diff * y_diff_veh);
        auto ty = -(sin_theta_diff * x_diff_veh + cos_theta_diff * y_diff_veh);

        auto ptr_trajectory_points =
            target_tracking_trajectory.mutable_trajectory_point();
        std::for_each(
            ptr_trajectory_points->begin(), ptr_trajectory_points->end(),
            [&cos_theta_diff, &sin_theta_diff, &tx, &ty,
             &theta_diff](common::TrajectoryPoint &p) {
              auto x = p.path_point().x();
              auto y = p.path_point().y();
              auto theta = p.path_point().theta();

              auto x_new = cos_theta_diff * x - sin_theta_diff * y + tx;
              auto y_new = sin_theta_diff * x + cos_theta_diff * y + ty;
              auto theta_new = common::math::WrapAngle(theta - theta_diff);

              p.mutable_path_point()->set_x(x_new);
              p.mutable_path_point()->set_y(y_new);
              p.mutable_path_point()->set_theta(theta_new);
            });
      }
    }

    trajectory_analyzer_ =
        std::move(TrajectoryAnalyzer(&target_tracking_trajectory));

    UpdateDrivingOrientation();

    SimpleLateralDebug *debug =
cmd->mutable_debug()->mutable_simple_lat_debug();
    debug->Clear();

    // 更新状态信息格式
    // Update state = [Lateral Error, Lateral Error Rate, Heading Error, Heading
    // Error Rate, preview lateral error1 , preview lateral error2, ...]
    UpdateState(debug);

    UpdateMatrix();
```

```
//具有道路预览模型的复合离散矩阵
UpdateMatrixCompound();

// 调整矩阵在倒车时更新
int q_param_size = control_conf_->lat_controller_conf().matrix_q_size();
int reverse_q_param_size =
    control_conf_->lat_controller_conf().reverse_matrix_q_size();
if (VehicleStateProvider::Instance()->gear() ==
    canbus::Chassis::GEAR_REVERSE) {
  for (int i = 0; i < reverse_q_param_size; ++i) {
    matrix_q_(i, i) =
        control_conf_->lat_controller_conf().reverse_matrix_q(i);
  }
} else {
  for (int i = 0; i < q_param_size; ++i) {
    matrix_q_(i, i) = control_conf_->lat_controller_conf().matrix_q(i);
  }
}

//添加用于高速转向的增益调度程序
if (FLAGS_enable_gain_scheduler) {
  matrix_q_updated_(0, 0) =
      matrix_q_(0, 0) *
lat_err_interpolation_->Interpolate(vehicle_state->linear_velocity());
  matrix_q_updated_(2, 2) =
      matrix_q_(2, 2) * heading_err_interpolation_->Interpolate(
                            vehicle_state->linear_velocity());
  common::math::SolveLQRProblem(matrix_adc_, matrix_bdc_,
matrix_q_updated_,
                                matrix_r_, lqr_eps_, lqr_max_iteration_,
                                &matrix_k_);
} else {
  common::math::SolveLQRProblem(matrix_adc_, matrix_bdc_, matrix_q_,
                                matrix_r_, lqr_eps_, lqr_max_iteration_,
                                &matrix_k_);
}

// feedback = - K * state
// 将车辆转向角从 rad 转换为 degree，然后转换为 steer degree，然后转换为 100%的比率
const double steer_angle_feedback=-(matrix_k_ * matrix_state_)(0,0) * 180 /
                                   M_PI * steer_ratio_ /
                                   steer_single_direction_max_degree_ * 100;
```

```
    const double steer_angle_feedforward =
ComputeFeedForward(debug->curvature());

    // 夹住转向角从-100.0 至 100.0
    double steer_angle = 0.0;
    bool enable_leadlag = control_conf_->lat_controller_conf()
                              .enable_reverse_leadlag_compensation();
    if (enable_leadlag) {
      steer_angle = common::math::Clamp(
          leadlag_controller_.Control(steer_angle_feedback, ts_) +
              steer_angle_feedforward,
          -100.0, 100.0);
    } else {
      steer_angle = common::math::Clamp(
          steer_angle_feedback + steer_angle_feedforward, -100.0, 100.0);
    }

    if (FLAGS_set_steer_limit) {
      const double steer_limit = std::atan(max_lat_acc_ * min_turn_radius_ /
                                           (vehicle_state->linear_velocity() *
                                            vehicle_state->linear_velocity())) *
                                 steer_ratio_ * 180 / M_PI /
                                 steer_single_direction_max_degree_ * 100;

      // 夹住转向角
      double steer_angle_limited =
          common::math::Clamp(steer_angle, -steer_limit, steer_limit);
      steer_angle_limited = digital_filter_.Filter(steer_angle_limited);
      steer_angle = steer_angle_limited;
      debug->set_steer_angle_limited(steer_angle_limited);
    } else {
      steer_angle = digital_filter_.Filter(steer_angle);
    }

    if (std::abs(vehicle_state->linear_velocity()) < FLAGS_lock_steer_speed &&
        (vehicle_state->gear() == canbus::Chassis::GEAR_DRIVE ||
         vehicle_state->gear() == canbus::Chassis::GEAR_REVERSE) &&
        chassis->driving_mode() == canbus::Chassis::COMPLETE_AUTO_DRIVE) {
      steer_angle = pre_steer_angle_;
    }
    pre_steer_angle_ = steer_angle;
    cmd->set_steering_target(steer_angle);
```

```
  cmd->set_steering_rate(FLAGS_steer_angle_rate);
  // 计算日志记录和调试的额外信息
  const double steer_angle_lateral_contribution =
     -matrix_k_(0, 0) * matrix_state_(0, 0) * 180 / M_PI * steer_ratio_ /
     steer_single_direction_max_degree_ * 100;

  const double steer_angle_lateral_rate_contribution =
     -matrix_k_(0, 1) * matrix_state_(1, 0) * 180 / M_PI * steer_ratio_ /
     steer_single_direction_max_degree_ * 100;

  const double steer_angle_heading_contribution =
     -matrix_k_(0, 2) * matrix_state_(2, 0) * 180 / M_PI * steer_ratio_ /
     steer_single_direction_max_degree_ * 100;

  const double steer_angle_heading_rate_contribution =
     -matrix_k_(0, 3) * matrix_state_(3, 0) * 180 / M_PI * steer_ratio_ /
     steer_single_direction_max_degree_ * 100;

  debug->set_heading(driving_orientation_);
  debug->set_steer_angle(steer_angle);
  debug->set_steer_angle_feedforward(steer_angle_feedforward);

debug->set_steer_angle_lateral_contribution(steer_angle_lateral_contribution);
  debug->set_steer_angle_lateral_rate_contribution(
     steer_angle_lateral_rate_contribution);

debug->set_steer_angle_heading_contribution(steer_angle_heading_contribution);
  debug->set_steer_angle_heading_rate_contribution(
     steer_angle_heading_rate_contribution);
  debug->set_steer_angle_feedback(steer_angle_feedback);
  debug->set_steering_position(chassis->steering_percentage());
  debug->set_ref_speed(vehicle_state->linear_velocity());

  ProcessLogs(debug, chassis);
  return Status::OK();
 }
 //LonController 的控制算法实现（控制车辆加减速度），输出控制命令 ControlCommand 对象

 Status LonController::ComputeControlCommand(
    const localization::LocalizationEstimate *localization,
    const canbus::Chassis *chassis,
    const planning::ADCTrajectory *planning_published_trajectory,
```

```
    control::ControlCommand *cmd) {
  localization_ = localization;
  chassis_ = chassis;

  trajectory_message_ = planning_published_trajectory;
//函数脏检测
  if (!control_interpolation_) {
    AERROR << "Fail to initialize calibration table.";
    return Status(ErrorCode::CONTROL_COMPUTE_ERROR,
                  "Fail to initialize calibration table.");
  }

  if (trajectory_analyzer_ == nullptr ||
      trajectory_analyzer_->seq_num() !=
          trajectory_message_->header().sequence_num()) {
    trajectory_analyzer_.reset(new TrajectoryAnalyzer(trajectory_message_));
  }
  const LonControllerConf &lon_controller_conf =
      control_conf_->lon_controller_conf();

  auto debug = cmd->mutable_debug()->mutable_simple_lon_debug();
  debug->Clear();
//加减速控制变量初始化
  double brake_cmd = 0.0;
  double throttle_cmd = 0.0;
  double ts = lon_controller_conf.ts();
  double preview_time = lon_controller_conf.preview_window() * ts;
  bool enable_leadlag =
      lon_controller_conf.enable_reverse_leadlag_compensation();

  if (preview_time < 0.0) {
    const auto error_msg = common::util::StrCat(
        "Preview time set as: ", preview_time, " less than 0");
    AERROR << error_msg;
    return Status(ErrorCode::CONTROL_COMPUTE_ERROR, error_msg);
  }
  ComputeLongitudinalErrors(trajectory_analyzer_.get(), preview_time, ts,
                            debug);

  double station_error_limit = lon_controller_conf.station_error_limit();
  double station_error_limited = 0.0;
  if (FLAGS_enable_speed_station_preview) {
    station_error_limited =
```

```
        common::math::Clamp(debug->preview_station_error(),
                            -station_error_limit, station_error_limit);
    } else {
      station_error_limited = common::math::Clamp(
          debug->station_error(), -station_error_limit, station_error_limit);
    }
  //PID 控制
    if (trajectory_message_->gear() == canbus::Chassis::GEAR_REVERSE) {
      station_pid_controller_.SetPID(
          lon_controller_conf.reverse_station_pid_conf());

speed_pid_controller_.SetPID(lon_controller_conf.reverse_speed_pid_conf());
      if (enable_leadlag) {
        station_leadlag_controller_.SetLeadlag(
            lon_controller_conf.reverse_station_leadlag_conf());
        speed_leadlag_controller_.SetLeadlag(
            lon_controller_conf.reverse_speed_leadlag_conf());
      }
    } else if (VehicleStateProvider::Instance()->linear_velocity() <=
               lon_controller_conf.switch_speed()) {
      speed_pid_controller_.SetPID(lon_controller_conf.low_speed_pid_conf());
    } else {
      speed_pid_controller_.SetPID(lon_controller_conf.high_speed_pid_conf());
    }

    double speed_offset =
        station_pid_controller_.Control(station_error_limited, ts);
    if (enable_leadlag) {
      speed_offset = station_leadlag_controller_.Control(speed_offset, ts);
    }

    double speed_controller_input = 0.0;
    double speed_controller_input_limit =
        lon_controller_conf.speed_controller_input_limit();
    double speed_controller_input_limited = 0.0;
    if (FLAGS_enable_speed_station_preview) {
      speed_controller_input = speed_offset + debug->preview_speed_error();
    } else {
      speed_controller_input = speed_offset + debug->speed_error();
    }
    speed_controller_input_limited =
        common::math::Clamp(speed_controller_input,
-speed_controller_input_limit,
```

```
                         speed_controller_input_limit);

  double acceleration_cmd_closeloop = 0.0;
//加速控制过程
  acceleration_cmd_closeloop =
      speed_pid_controller_.Control(speed_controller_input_limited, ts);
  debug->set_pid_saturation_status(
      speed_pid_controller_.IntegratorSaturationStatus());
  if (enable_leadlag) {
    acceleration_cmd_closeloop =
        speed_leadlag_controller_.Control(acceleration_cmd_closeloop, ts);
    debug->set_leadlag_saturation_status(
        speed_leadlag_controller_.InnerstateSaturationStatus());
  }

  double slope_offset_compenstaion = digital_filter_pitch_angle_.Filter(
      GRA_ACC * std::sin(VehicleStateProvider::Instance()->pitch()));

  if (std::isnan(slope_offset_compenstaion)) {
    slope_offset_compenstaion = 0;
  }

  debug->set_slope_offset_compensation(slope_offset_compenstaion);

  double acceleration_cmd =
      acceleration_cmd_closeloop + debug->preview_acceleration_reference() +
      FLAGS_enable_slope_offset * debug->slope_offset_compensation();
  debug->set_is_full_stop(false);
  GetPathRemain(debug);

  if ((trajectory_message_->trajectory_type() ==
       apollo::planning::ADCTrajectory::NORMAL) &&
      ((std::fabs(debug->preview_acceleration_reference()) <=
            FLAGS_max_acceleration_when_stopped &&
        std::fabs(debug->preview_speed_reference()) <=
            vehicle_param_.max_abs_speed_when_stopped()) ||
       std::abs(debug->path_remain()) < 0.3)) {
    acceleration_cmd = lon_controller_conf.standstill_acceleration();
    AINFO << "Stop location reached";
    debug->set_is_full_stop(true);
  }

  double throttle_lowerbound =
```

```
      std::max(vehicle_param_.throttle_deadzone(),
               lon_controller_conf.throttle_minimum_action());
  double brake_lowerbound =
      std::max(vehicle_param_.brake_deadzone(),
               lon_controller_conf.brake_minimum_action());
  double calibration_value = 0.0;
  double acceleration_lookup =
      (chassis->gear_location() == canbus::Chassis::GEAR_REVERSE)
          ? -acceleration_cmd
          : acceleration_cmd;

  if (FLAGS_use_preview_speed_for_table) {
    calibration_value = control_interpolation_->Interpolate(
        std::make_pair(debug->preview_speed_reference(),
acceleration_lookup));
  } else {
    calibration_value = control_interpolation_->Interpolate(
        std::make_pair(chassis_->speed_mps(), acceleration_lookup));
  }
//确定节气门控制参数和刹车控制参数
  if (acceleration_lookup >= 0) {
    if (calibration_value >= 0) {
      throttle_cmd = std::max(calibration_value, throttle_lowerbound);
    } else {
      throttle_cmd = throttle_lowerbound;
    }
    brake_cmd = 0.0;
  } else {
    throttle_cmd = 0.0;
    if (calibration_value >= 0) {
      brake_cmd = brake_lowerbound;
    } else {
      brake_cmd = std::max(-calibration_value, brake_lowerbound);
    }
  }
//控制调试输出过程
  debug->set_station_error_limited(station_error_limited);
  debug->set_speed_offset(speed_offset);

debug->set_speed_controller_input_limited(speed_controller_input_limited);
  debug->set_acceleration_cmd(acceleration_cmd);
  debug->set_throttle_cmd(throttle_cmd);
  debug->set_brake_cmd(brake_cmd);
```

```
    debug->set_acceleration_lookup(acceleration_lookup);
    debug->set_speed_lookup(chassis_->speed_mps());
    debug->set_calibration_value(calibration_value);
    debug->set_acceleration_cmd_closeloop(acceleration_cmd_closeloop);

    if (FLAGS_enable_csv_debug && speed_log_file_ != nullptr) {
      fprintf(speed_log_file_,
              "%.6f, %.6f, %.6f, %.6f, %.6f, %.6f, %.6f, %.6f, %.6f, %.6f,"
              "%.6f, %.6f, %.6f, %.6f, %.6f, %.6f, %.6f, %d,\r\n",
              debug->station_reference(), debug->station_error(),
              station_error_limited, debug->preview_station_error(),
              debug->speed_reference(), debug->speed_error(),
              speed_controller_input_limited, debug->preview_speed_reference(),
              debug->preview_speed_error(),
              debug->preview_acceleration_reference(),
acceleration_cmd_closeloop,
              acceleration_cmd, debug->acceleration_lookup(),
              debug->speed_lookup(), calibration_value, throttle_cmd, brake_cmd,
              debug->is_full_stop());
    }

    // 需要车辆加速的情况下放开刹车和节气门
    cmd->set_throttle(throttle_cmd);
    cmd->set_brake(brake_cmd);
    cmd->set_acceleration(acceleration_cmd);

    if (std::fabs(VehicleStateProvider::Instance()->linear_velocity()) <=
            vehicle_param_.max_abs_speed_when_stopped() ||
        chassis->gear_location() == trajectory_message_->gear() ||
        chassis->gear_location() == canbus::Chassis::GEAR_NEUTRAL) {
      cmd->set_gear_location(trajectory_message_->gear());
    } else {
      cmd->set_gear_location(chassis->gear_location());
    }

    return Status::OK();
  }
```

15.3 MPC 控制算法

在当今过程控制领域中，PID 是用得最多的控制方法，但 MPC 也超过了 10%的占有率。MPC 是一个总称，有着各种各样的算法，其动态矩阵控制（DMC）是代表作。DMC 采用的是系统的阶跃响应曲线，其突出的特点是解决了约束控制问题。那么 DMC 是怎么解决约束的呢？在这里只给出宏观的解释，而不进行详细的说明。DMC 把线性规划和控制问题结合起来，用线性规划解决输出约束的问题，同时解决了静态最优的问题，一箭双雕，因此在工业界取得了极大的成功。

15.3.1 MPC 的控制原理

MPC 作用机理描述为在每一个采样时刻，根据获得的当前的测量信息在线求解一个有限时间开环优化问题，并将得到的控制序列的第一个元素作用于被控对象。在下一个采样时刻，重复上述过程，即用新的测量值作为此时预测系统未来动态的初始条件，刷新优化问题并重新求解。

MPC 算法包括 3 个步骤：

（1）预测系统的未来动态。

（2）（数值）求解开环优化问题。

（3）将优化解的第一个元素（或者说第一部分）作用于系统。

这 3 步是在每个采样时刻重复进行的，且无论采用什么样的模型，每个采样时刻得到的测量值都作为当前时刻预测系统未来动态的初始条件。

在线求解开环优化问题获得开环优化序列是 MPC 和传统控制方法的主要区别，因为后者通常是离线求解一个反馈控制律，并将得到的反馈控制律一直作用于系统。

15.3.2 百度 Apollo 相关源码分析

在 Apollo 系统中，MPC 车辆控制算法集中在 MPCController 对象中，对应源代码文件为 modules/control/controller/mpc_controller.cpp。下面以注释的方式来了解一下代码的具体执行流程。

```
//MPCController 的控制算法实现（控制整车的所有驾驶参数），输出控制命令 ControlCommand 对象

Status MPCController::ComputeControlCommand(
    const localization::LocalizationEstimate *localization,
    const canbus::Chassis *chassis,
    const planning::ADCTrajectory *planning_published_trajectory,
    ControlCommand *cmd) {
  trajectory_analyzer_ =
      std::move(TrajectoryAnalyzer(planning_published_trajectory));

  SimpleMPCDebug *debug = cmd->mutable_debug()->mutable_simple_mpc_debug();
  debug->Clear();

  ComputeLongitudinalErrors(&trajectory_analyzer_, debug);

  //更新状态
```

```
  UpdateState(debug);

  UpdateMatrix(debug);

  FeedforwardUpdate(debug);

  // 添加用于高速转向的增益调度程序
  if (FLAGS_enable_gain_scheduler) {
    matrix_q_updated_(0, 0) =
        matrix_q_(0, 0) *
        lat_err_interpolation_->Interpolate(
            VehicleStateProvider::Instance()->linear_velocity());
    matrix_q_updated_(2, 2) =
        matrix_q_(2, 2) *
        heading_err_interpolation_->Interpolate(
            VehicleStateProvider::Instance()->linear_velocity());
    steer_angle_feedforwardterm_updated_ =
        steer_angle_feedforwardterm_ *
        feedforwardterm_interpolation_->Interpolate(
            VehicleStateProvider::Instance()->linear_velocity());
    matrix_r_updated_(0, 0) =
        matrix_r_(0, 0) *
        steer_weight_interpolation_->Interpolate(
            VehicleStateProvider::Instance()->linear_velocity());
  } else {
    matrix_q_updated_ = matrix_q_;
    matrix_r_updated_ = matrix_r_;
    steer_angle_feedforwardterm_updated_ = steer_angle_feedforwardterm_;
  }

  debug->add_matrix_q_updated(matrix_q_updated_(0, 0));
  debug->add_matrix_q_updated(matrix_q_updated_(1, 1));
  debug->add_matrix_q_updated(matrix_q_updated_(2, 2));
  debug->add_matrix_q_updated(matrix_q_updated_(3, 3));

  debug->add_matrix_r_updated(matrix_r_updated_(0, 0));
  debug->add_matrix_r_updated(matrix_r_updated_(1, 1));
//控制模型矩阵构造
  Matrix control_matrix = Matrix::Zero(controls_, 1);
  std::vector<Matrix> control(horizon_, control_matrix);

  Matrix control_gain_matrix = Matrix::Zero(controls_, basic_state_size_);
  std::vector<Matrix> control_gain(horizon_, control_gain_matrix);
```

```
  Matrix addition_gain_matrix = Matrix::Zero(controls_, 1);
  std::vector<Matrix> addition_gain(horizon_, addition_gain_matrix);

  Matrix reference_state = Matrix::Zero(basic_state_size_, 1);
  std::vector<Matrix> reference(horizon_, reference_state);

  Matrix lower_bound(controls_, 1);
  lower_bound << -wheel_single_direction_max_degree_, max_deceleration_;

  Matrix upper_bound(controls_, 1);
  upper_bound << wheel_single_direction_max_degree_, max_acceleration_;

  double mpc_start_timestamp = Clock::NowInSeconds();
  double steer_angle_feedback = 0.0;
  double acc_feedback = 0.0;
  double steer_angle_ff_compensation = 0.0;
  double unconstraint_control_diff = 0.0;
  double control_gain_truncation_ratio = 0.0;
  double unconstraint_control = 0.0;
  const double v = VehicleStateProvider::Instance()->linear_velocity();
 // 使用线性模型进行控制命令求解
  if (!common::math::SolveLinearMPC(matrix_ad_, matrix_bd_, matrix_cd_,
                                    matrix_q_updated_, matrix_r_updated_,
                                    lower_bound, upper_bound, matrix_state_,
                                    reference, mpc_eps_, mpc_max_iteration_,
                                    &control, &control_gain, &addition_gain)) {
    AERROR << "MPC solver failed";
  } else {
 ADEBUG << "MPC problem solved! ";
 //MPC 计算成功，从计算结果中转化出控制相关参数
    steer_angle_feedback = Wheel2SteerPct(control[0](0, 0));
    acc_feedback = control[0](1, 0);
    for (int i = 0; i < basic_state_size_; ++i) {
      unconstraint_control += control_gain[0](0, i) * matrix_state_(i, 0);
    }
    unconstraint_control += addition_gain[0](0, 0) * v * debug->curvature();
    if (enable_mpc_feedforward_compensation_) {
      unconstraint_control_diff =
          Wheel2SteerPct(control[0](0, 0) - unconstraint_control);
      if (fabs(unconstraint_control_diff) <= unconstraint_control_diff_limit_)
{
        steer_angle_ff_compensation =
```

```
          Wheel2SteerPct(debug->curvature() *
                        (control_gain[0](0, 2) *
                              (lr_ - lf_ / cr_ * mass_ * v * v / wheelbase_) -
                         addition_gain[0](0, 0) * v));
      } else {
        control_gain_truncation_ratio = control[0](0, 0) /
unconstraint_control;
        steer_angle_ff_compensation =
            Wheel2SteerPct(debug->curvature() *
                        (control_gain[0](0, 2) *
                              (lr_ - lf_ / cr_ * mass_ * v * v / wheelbase_) -
                         addition_gain[0](0, 0) * v) *
                        control_gain_truncation_ratio);
      }
    } else {
      steer_angle_ff_compensation = 0.0;
    }
  }

  double mpc_end_timestamp = Clock::NowInSeconds();

  ADEBUG << "MPC core algorithm: calculation time is: "
         << (mpc_end_timestamp - mpc_start_timestamp) * 1000 << " ms.";

  // 评估结果后是否需要添加样条线平滑
  double steer_angle = steer_angle_feedback +
                       steer_angle_feedforwardterm_updated_ +
                       steer_angle_ff_compensation;

  // 夹住转向角从-100.0至100.0
  steer_angle = common::math::Clamp(steer_angle, -100.0, 100.0);

  if (FLAGS_set_steer_limit) {
    const double steer_limit =
        std::atan(max_lat_acc_ * wheelbase_ /
                  (VehicleStateProvider::Instance()->linear_velocity() *
                   VehicleStateProvider::Instance()->linear_velocity())) *
        steer_ratio_ * 180 / M_PI / steer_single_direction_max_degree_ * 100;

  // 夹住转向角
    double steer_angle_limited =
        common::math::Clamp(steer_angle, -steer_limit, steer_limit);
    steer_angle_limited = digital_filter_.Filter(steer_angle_limited);
```

```
      cmd->set_steering_target(steer_angle_limited);
      debug->set_steer_angle_limited(steer_angle_limited);
    } else {
      steer_angle = digital_filter_.Filter(steer_angle);
      cmd->set_steering_target(steer_angle);
    }

    debug->set_acceleration_cmd_closeloop(acc_feedback);

    double acceleration_cmd = acc_feedback + debug->acceleration_reference();

    debug->set_is_full_stop(false);
    if (std::fabs(debug->acceleration_reference()) <=
            FLAGS_max_acceleration_when_stopped &&
        std::fabs(debug->speed_reference()) <=
            vehicle_param_.max_abs_speed_when_stopped()) {
      acceleration_cmd = standstill_acceleration_;
      AINFO << "Stop location reached";
      debug->set_is_full_stop(true);
    }

    debug->set_acceleration_cmd(acceleration_cmd);

    double calibration_value = 0.0;
    if (FLAGS_use_preview_speed_for_table) {
      calibration_value = control_interpolation_->Interpolate(
          std::make_pair(debug->speed_reference(), acceleration_cmd));
    } else {
      calibration_value = control_interpolation_->Interpolate(std::make_pair(
          VehicleStateProvider::Instance()->linear_velocity(),
acceleration_cmd));
    }

    debug->set_calibration_value(calibration_value);

  //计算最终的节气门控制命令和刹车控制命令
   double throttle_cmd = 0.0;
    double brake_cmd = 0.0;
    if (calibration_value >= 0) {
      throttle_cmd = std::max(calibration_value, throttle_lowerbound_);
      brake_cmd = 0.0;
    } else {
      throttle_cmd = 0.0;
```

```
      brake_cmd = std::max(-calibration_value, brake_lowerbound_);
    }

    cmd->set_steering_rate(FLAGS_steer_angle_rate);
    // 如果车辆处于加速状态，那么提供命令是节气门和刹车都放开
    cmd->set_throttle(throttle_cmd);
    cmd->set_brake(brake_cmd);
    cmd->set_acceleration(acceleration_cmd);

    debug->set_heading(VehicleStateProvider::Instance()->heading());
    debug->set_steering_position(chassis->steering_percentage());
    debug->set_steer_angle(steer_angle);
    debug->set_steer_angle_feedforward(steer_angle_feedforwardterm_updated_);

debug->set_steer_angle_feedforward_compensation(steer_angle_ff_compensation);
    debug->set_steer_unconstraint_control_diff(unconstraint_control_diff);
    debug->set_steer_angle_feedback(steer_angle_feedback);
    debug->set_steering_position(chassis->steering_percentage());

    if (std::fabs(VehicleStateProvider::Instance()->linear_velocity()) <=
            vehicle_param_.max_abs_speed_when_stopped() ||
        chassis->gear_location() == planning_published_trajectory->gear() ||
        chassis->gear_location() == canbus::Chassis::GEAR_NEUTRAL) {
      cmd->set_gear_location(planning_published_trajectory->gear());
    } else {
      cmd->set_gear_location(chassis->gear_location());
    }

    ProcessLogs(debug, chassis);
    return Status::OK();
  }
```

参考文献

[1] O'Dwyer. Handbook of PI and PID controller tuning rules[J]. Automatica, 2006, 41(2):355-356.

[2] Skogestad S. Simple analytic rules for model reduction and PID controller tuning[J]. Modeling, Identification and Control, 2003, 13(4):291-309.

[3] Chen G. Conventional and fuzzy PID controllers: An overview[C]// Int J Intell Control Syst. 1996.

[4] Borrelli F, Falcone P, T. Keviczky†. MPC-Based Approach to Active Steering for Autonomous Vehicle Systems[J]. International Journal of Vehicle Autonomous Systems, 2005, 3(2/3/4):265.

[5] Nayl T, Nikolakopoulos G, Gustafsson T . Effect of kinematic parameters on MPC based on-line motion planning for an articulated vehicle[J]. Robotics and Autonomous Systems, 2015, 70:16-24.

[6] Kong J, Pfeiffer M, Schildbach G, et al. [IEEE 2015 IEEE Intelligent Vehicles Symposium (IV) - Seoul, South Korea (2015.6.28-2015.7.1)] 2015 IEEE Intelligent Vehicles Symposium (IV) - Kinematic and dynamic vehicle models for autonomous driving control design[C]// 2015 IEEE Intelligent Vehicles Symposium (IV). IEEE, 2015:1094-1099.

[7] Rajamani R. Vehicle Dynamics and Control[M]// VEHICLE DYNAMICS AND CONTROL. 2006.

[8] Falcone P, Eric Tseng H, Borrelli F, et al. MPC-based yaw and lateral stabilisation via active front steering and braking[J]. Vehicle System Dynamics, 2008, 46(sup1):611-628.

附录 A

强化学习：贪吃蛇 AI 完整游戏逻辑代码

```
FPS = 15
##WINDOWWIDTH = 640
#WINDOWHEIGHT = 480
WINDOWWIDTH = 640
WINDOWHEIGHT = 480
CELLSIZE = 40
assert WINDOWWIDTH % CELLSIZE == 0, "Window width must be a multiple of cell size."
assert WINDOWHEIGHT % CELLSIZE == 0, "Window height must be a multiple of cell size."
CELLWIDTH = int(WINDOWWIDTH / CELLSIZE)
CELLHEIGHT = int(WINDOWHEIGHT / CELLSIZE)

#            R    G    B
WHITE     = (255, 255, 255)
BLACK     = (  0,   0,   0)
RED       = (255,   0,   0)
GREEN     = (  0, 255,   0)
DARKGREEN = (  0, 155,   0)
DARKGRAY  = ( 40,  40,  40)
BGCOLOR = BLACK

UP = 'up'
DOWN = 'down'
LEFT = 'left'
RIGHT = 'right'

# 神经网络的输出
```

```
MOVE_UP = [1, 0, 0, 0]
MOVE_DOWN = [0, 1, 0, 0]
MOVE_LEFT = [0, 0, 1, 0]
MOVE_RIGHT = [0, 0, 0, 1]

def getRandomLocation(worm):
    temp = {'x': random.randint(0, CELLWIDTH - 1), 'y': random.randint(0,
CELLHEIGHT - 1)}
    while test_not_ok(temp, worm):
        temp = {'x': random.randint(0, CELLWIDTH - 1), 'y': random.randint(0,
CELLHEIGHT - 1)}
    return temp

def test_not_ok(temp, worm):
    for body in worm:
        if temp['x'] == body['x'] and temp['y'] == body['y']:
            return True
    return False

HEAD = 0 # syntactic sugar: index of the worm's head
 # Set a random start point.
startx = random.randint(5, CELLWIDTH - 6)
starty = random.randint(5, CELLHEIGHT - 6)
wormCoords = [{'x': startx,    'y': starty},
              {'x': startx - 1, 'y': starty},
              {'x': startx - 2, 'y': starty}]
direction = RIGHT

# Start the apple in a random place.
apple = getRandomLocation(wormCoords)

def game():
    global FPSCLOCK, DISPLAYSURF, BASICFONT

    pygame.init()
    FPSCLOCK = pygame.time.Clock()
    DISPLAYSURF = pygame.display.set_mode((WINDOWWIDTH, WINDOWHEIGHT))
    BASICFONT = pygame.font.Font('freesansbold.ttf', 18)
    pygame.display.set_caption('Snaky')

    showStartScreen()
    # while True:
    #    runGame()
```

```
    #     # showGameOverScreen()

def runGame(action = MOVE_UP):
    global direction, wormCoords, apple
    # while True: # main game loop
    pre_direction = direction
    if action == MOVE_LEFT and direction != RIGHT:
        direction = LEFT
    elif action == MOVE_RIGHT and direction != LEFT:
        direction = RIGHT
    elif action == MOVE_UP and direction != DOWN:
        direction = UP
    elif action == MOVE_DOWN and direction != UP:
        direction = DOWN
    reward = 0
     # check if the worm has hit itself or the edge
    if wormCoords[HEAD]['x'] == -1 or wormCoords[HEAD]['x'] == CELLWIDTH or
wormCoords[HEAD]['y'] == -1 or wormCoords[HEAD]['y'] == CELLHEIGHT:
        reward = -1
        startx = random.randint(5, CELLWIDTH - 6)
        starty = random.randint(5, CELLHEIGHT - 6)
        wormCoords = [{'x': startx,     'y': starty},
                      {'x': startx - 1, 'y': starty},
                      {'x': startx - 2, 'y': starty}]
        direction = RIGHT

        # Start the apple in a random place.
        apple = getRandomLocation(wormCoords)
        screen_image = pygame.surfarray.array3d(pygame.display.get_surface())
        pygame.display.update()
        FPSCLOCK.tick(FPS)
        return reward, screen_image
     for wormBody in wormCoords[1:]:
         if wormBody['x'] == wormCoords[HEAD]['x'] and wormBody['y'] ==
wormCoords[HEAD]['y']:
            reward = -1
            startx = random.randint(5, CELLWIDTH - 6)
            starty = random.randint(5, CELLHEIGHT - 6)
            wormCoords = [{'x': startx,     'y': starty},
                          {'x': startx - 1, 'y': starty},
                          {'x': startx - 2, 'y': starty}]
            direction = RIGHT
```

```
            # Start the apple in a random place.
            apple = getRandomLocation(wormCoords)
            apple = getRandomLocation(wormCoords)
            screen_image =
pygame.surfarray.array3d(pygame.display.get_surface())
            pygame.display.update()
            FPSCLOCK.tick(FPS)
            return reward, screen_image

        # check if worm has eaten an apply
        if wormCoords[HEAD]['x'] == apple['x'] and wormCoords[HEAD]['y'] ==
apple['y']:
            # don't remove worm's tail segment
            apple = getRandomLocation(wormCoords) # set a new apple somewhere
            reward = 1    # 击中奖励
        else:
            del wormCoords[-1] # remove worm's tail segment

         # move the worm by adding a segment in the direction it is moving
        if not examine_direction(direction, pre_direction):
            direction = pre_direction
        if direction == UP:
            newHead = {'x': wormCoords[HEAD]['x'], 'y': wormCoords[HEAD]['y'] - 1}
        elif direction == DOWN:
            newHead = {'x': wormCoords[HEAD]['x'], 'y': wormCoords[HEAD]['y'] + 1}
        elif direction == LEFT:
            newHead = {'x': wormCoords[HEAD]['x'] - 1, 'y': wormCoords[HEAD]['y']}
        elif direction == RIGHT:
            newHead = {'x': wormCoords[HEAD]['x'] + 1, 'y': wormCoords[HEAD]['y']}
        wormCoords.insert(0, newHead)
        DISPLAYSURF.fill(BGCOLOR)
        drawGrid()
        drawWorm(wormCoords)
        drawApple(apple)
        drawScore(len(wormCoords) - 3)
         # 获得游戏界面像素
        screen_image = pygame.surfarray.array3d(pygame.display.get_surface())
        pygame.display.update()
        FPSCLOCK.tick(FPS)

         # 返回游戏界面像素和对应的奖励
        return reward, screen_image
```

```
def examine_direction(temp , direction):
    if direction == UP:
        if temp == DOWN:
            return False
    elif direction == RIGHT:
        if temp == LEFT:
            return False
    elif direction == LEFT:
        if temp == RIGHT:
            return False
    elif direction == DOWN:
        if temp == UP:
            return False
    return True

def drawPressKeyMsg():
    pressKeySurf = BASICFONT.render('Press a key to play.', True, DARKGRAY)
    pressKeyRect = pressKeySurf.get_rect()
    pressKeyRect.topleft = (WINDOWWIDTH - 200, WINDOWHEIGHT - 30)
    DISPLAYSURF.blit(pressKeySurf, pressKeyRect)

def checkForKeyPress():
    if len(pygame.event.get(QUIT)) > 0:
        terminate()

    keyUpEvents = pygame.event.get(KEYUP)
    if len(keyUpEvents) == 0:
        return None
    if keyUpEvents[0].key == K_ESCAPE:
        terminate()
    return keyUpEvents[0].key

def showStartScreen():
    titleFont = pygame.font.Font('freesansbold.ttf', 100)
    titleSurf1 = titleFont.render('Snaky!', True, WHITE, DARKGREEN)
    titleSurf2 = titleFont.render('Snaky!', True, GREEN)

    degrees1 = 0
    degrees2 = 0
    while True:
        DISPLAYSURF.fill(BGCOLOR)
```

```
        rotatedSurf1 = pygame.transform.rotate(titleSurf1, degrees1)
        rotatedRect1 = rotatedSurf1.get_rect()
        rotatedRect1.center = (WINDOWWIDTH / 2, WINDOWHEIGHT / 2)
        DISPLAYSURF.blit(rotatedSurf1, rotatedRect1)

        rotatedSurf2 = pygame.transform.rotate(titleSurf2, degrees2)
        rotatedRect2 = rotatedSurf2.get_rect()
        rotatedRect2.center = (WINDOWWIDTH / 2, WINDOWHEIGHT / 2)
        DISPLAYSURF.blit(rotatedSurf2, rotatedRect2)

        drawPressKeyMsg()

        if checkForKeyPress():
            pygame.event.get() # clear event queue
            return
        pygame.display.update()
        FPSCLOCK.tick(FPS)
        degrees1 += 3 # rotate by 3 degrees each frame
        degrees2 += 7 # rotate by 7 degrees each frame

def terminate():
    pygame.quit()
    sys.exit()


# def getRandomLocation(worm):
#     temp = {'x': random.randint(0, CELLWIDTH - 1), 'y': random.randint(0,
CELLHEIGHT - 1)}
#     while test_not_ok(temp, worm):
#         temp = {'x': random.randint(0, CELLWIDTH - 1), 'y': random.randint(0,
CELLHEIGHT - 1)}
#     return temp

# def test_not_ok(temp, worm):
#     for body in worm:
#         if temp['x'] == body['x'] and temp['y'] == body['y']:
#             return True
#     return False

def showGameOverScreen():
    gameOverFont = pygame.font.Font('freesansbold.ttf', 150)
    gameSurf = gameOverFont.render('Game', True, WHITE)
    overSurf = gameOverFont.render('Over', True, WHITE)
```

```
    gameRect = gameSurf.get_rect()
    overRect = overSurf.get_rect()
    gameRect.midtop = (WINDOWWIDTH / 2, 10)
    overRect.midtop = (WINDOWWIDTH / 2, gameRect.height + 10 + 25)

    DISPLAYSURF.blit(gameSurf, gameRect)
    DISPLAYSURF.blit(overSurf, overRect)
    drawPressKeyMsg()
    pygame.display.update()
    pygame.time.wait(500)
    checkForKeyPress() # clear out any key presses in the event queue

    while True:
        if checkForKeyPress():
            pygame.event.get() # clear event queue
            return

def drawScore(score):
    scoreSurf = BASICFONT.render('Score: %s' % (score), True, WHITE)
    scoreRect = scoreSurf.get_rect()
    scoreRect.topleft = (WINDOWWIDTH - 120, 10)
    DISPLAYSURF.blit(scoreSurf, scoreRect)

def drawWorm(wormCoords):
    for coord in wormCoords:
        x = coord['x'] * CELLSIZE
        y = coord['y'] * CELLSIZE
        wormSegmentRect = pygame.Rect(x, y, CELLSIZE, CELLSIZE)
        pygame.draw.rect(DISPLAYSURF, DARKGREEN, wormSegmentRect)
        wormInnerSegmentRect = pygame.Rect(x+4,y+4,CELLSIZE-8,CELLSIZE - 8)
        pygame.draw.rect(DISPLAYSURF, GREEN, wormInnerSegmentRect)

def drawApple(coord):
    x = coord['x'] * CELLSIZE
    y = coord['y'] * CELLSIZE
    appleRect = pygame.Rect(x, y, CELLSIZE, CELLSIZE)
    pygame.draw.rect(DISPLAYSURF, RED, appleRect)

def drawGrid():
    for x in range(0, WINDOWWIDTH, CELLSIZE): # draw vertical lines
        pygame.draw.line(DISPLAYSURF, DARKGRAY, (x, 0), (x, WINDOWHEIGHT))
    for y in range(0, WINDOWHEIGHT, CELLSIZE): # draw horizontal lines
        pygame.draw.line(DISPLAYSURF, DARKGRAY, (0, y), (WINDOWWIDTH, y))
```

附录 B

CyberRT 系统核心 API 字典

```
Node API
For additional information and examples, refer to Node
API List
//create writer with user-define attr and message type
  auto CreateWriter(const proto::RoleAttributes& role_attr)
      -> std::shared_ptr<transport::Writer<MessageT>>;
  //create reader with user-define attr, callback and message type
  auto CreateReader(const proto::RoleAttributes& role_attr,
      const croutine::CRoutineFunc<MessageT>& reader_func)
      -> std::shared_ptr<transport::Reader<MessageT>>;
  //create writer with specific channel name and message type
  auto CreateWriter(const std::string& channel_name)
      -> std::shared_ptr<transport::Writer<MessageT>>;
  //create reader with specific channel name, callback and message type
  auto CreateReader(const std::string& channel_name,
      const croutine::CRoutineFunc<MessageT>& reader_func)
      -> std::shared_ptr<transport::Reader<MessageT>>;
  //create reader with user-define config, callback and message type
  auto CreateReader(const ReaderConfig& config,
                    const CallbackFunc<MessageT>& reader_func)
      -> std::shared_ptr<cybertron::Reader<MessageT>>;
  //create service with name and specific callback
  auto CreateService(const std::string& service_name,
      const typename service::Service<Request, Response>::ServiceCallback&
service_calllback)
      -> std::shared_ptr<service::Service<Request, Response>>;
  //create client with name to send request to server
  auto CreateClient(const std::string& service_name)
      -> std::shared_ptr<service::Client<Request, Response>>;
```

Writer API

```
For additional information and examples, refer to Writer
```

```
API List
bool Write(const std::shared_ptr<MessageT>& message);
```

Client API

```
For additional information and examples, refer to Client
API List
SharedResponse SendRequest(SharedRequest request,
const std::chrono::seconds& timeout_s =
std::chrono::seconds(5));SharedResponse SendRequest(const Request& request,
const std::chrono::seconds& timeout_s = std::chrono::seconds(5));
```

Parameter API

```
The interface that the user uses to perform parameter related operations:
Set the parameter related API.
Read the parameter related API.
Create a ParameterService to provide parameter service related APIs for other
nodes.
Create a ParameterClient that uses the parameters provided by other nodes to
service related APIs.
For additional information and examples, refer to Parameter
API List - Setting parameters
 Parameter(); // Name is empty, type is NOT_SET
  explicit Parameter(const Parameter& parameter);
  explicit Parameter(const std::string& name); // Type is NOT_SET
  Parameter(const std::string& name, const bool bool_value);
  Parameter(const std::string& name, const int int_value);
  Parameter(const std::string& name, const int64_t int_value);
  Parameter(const std::string& name, const float double_value);
  Parameter(const std::string& name, const double double_value);
  Parameter(const std::string& name, const std::string& string_value);
  Parameter(const std::string& name, const char* string_value);
  Parameter(const std::string& name, const std::string& msg_str,
          const std::string& full_name, const std::string& proto_desc);
  Parameter(const std::string& name, const google::protobuf::Message& msg);
API List - Reading parameters
  inline ParamType type() const;
  inline std::string TypeName() const;
  inline std::string Descriptor() const;
  inline const std::string Name() const;
  inline bool AsBool() const;
  inline int64_t AsInt64() const;
  inline double AsDouble() const;
  inline const std::string AsString() const;
```

```
    std::string DebugString() const;
    template <typename Type>
    typename std::enable_if<std::is_base_of<google::protobuf::Message,
Type>::value, Type>::type
    value() const;
    template <typename Type>
    typename std::enable_if<std::is_integral<Type>::value && !std::is_same<Type,
bool>::value, Type>::type
    value() const;
    template <typename Type>
    typename std::enable_if<std::is_floating_point<Type>::value, Type>::type
    value() const;
    template <typename Type>
    typename std::enable_if<std::is_convertible<Type, std::string>::value,
const std::string&>::type
    value() const;
    template <typename Type>
    typename std::enable_if<std::is_same<Type, bool>::value, bool>::type
    value() const;
  API List - Creating parameter service
    explicit ParameterService(const std::shared_ptr<Node>& node);
    void SetParameter(const Parameter& parameter);
    bool GetParameter(const std::string& param_name, Parameter* parameter);
    bool ListParameters(std::vector<Parameter>* parameters);
  API List - Creating parameter client
   ParameterClient(const std::shared_ptr<Node>& node, const std::string&
service_node_name);
    bool SetParameter(const Parameter& parameter);
    bool GetParameter(const std::string& param_name, Parameter* parameter);
    bool ListParameters(std::vector<Parameter>* parameters);
```

Timer API

```
  You can set the parameters of the Timer and call the start and stop interfaces
to start the timer and stop the timer. For additional information and examples,
refer to Timer
  API List
   Timer(uint32_t period, std::function<void()> callback, bool oneshot);
    Timer(TimerOption opt);
    void SetTimerOption(TimerOption opt);
    void Start();
    void Stop();
```

Time API

```
For additional information and examples, refer to Time
API List
  static const Time MAX;
  static const Time MIN;
  Time() {}
  explicit Time(uint64_t nanoseconds);
  explicit Time(int nanoseconds);
  explicit Time(double seconds);
  Time(uint32_t seconds, uint32_t nanoseconds);
  Time(const Time& other);
  static Time Now();
  static Time MonoTime();
  static void SleepUntil(const Time& time);
  double ToSecond() const;
  uint64_t ToNanosecond() const;
  std::string ToString() const;
  bool IsZero() const;
```

Duration API

```
    Interval-related interface, used to indicate the time interval, can be
initialized according to the specified nanosecond or second.
    API List
      Duration() {}
      Duration(int64_t nanoseconds);
      Duration(int nanoseconds);
      Duration(double seconds);
      Duration(uint32_t seconds, uint32_t nanoseconds);
      Duration(const Rate& rate);
      Duration(const Duration& other);
      double ToSecond() const;
      int64_t ToNanosecond() const;
      bool IsZero() const;
      void Sleep() const;
```

Rate API

```
    The frequency interface is generally used to initialize the time of the sleep
frequency after the object is initialized according to the specified frequency.
    API List
      Rate(double frequency);
      Rate(uint64_t nanoseconds);
      Rate(const Duration&);
      void Sleep();
```

```
    void Reset();
    Duration CycleTime() const;
    Duration ExpectedCycleTime() const { return expected_cycle_time_; }
  RecordReader API
  The interface for reading the record file is used to read the message and channel
information in the record file.
  API List
    RecordReader();
    bool Open(const std::string& filename, uint64_t begin_time = 0,
              uint64_t end_time = UINT64_MAX);
    void Close();
    bool ReadMessage();
    bool EndOfFile();
    const std::string& CurrentMessageChannelName();
    std::shared_ptr<RawMessage> CurrentRawMessage();
    uint64_t CurrentMessageTime();
```

RecordWriter API

```
  The interface for writing the record file, used to record the message and channel
information into the record file.
  API List
   RecordWriter();
    bool Open(const std::string& file);
    void Close();
    bool WriteChannel(const std::string& name, const std::string& type,
                      const std::string& proto_desc);
    template <typename MessageT>
    bool WriteMessage(const std::string& channel_name, const MessageT& message,
                      const uint64_t time_nanosec,
                      const std::string& proto_desc = "");
    bool SetSizeOfFileSegmentation(uint64_t size_kilobytes);
    bool SetIntervalOfFileSegmentation(uint64_t time_sec);
```